GRAPHIC COMMUNICATION
for
TECHNICAL DESIGN

GRAPHIC COMMUNICATION for TECHNICAL DESIGN

K. Simms, B.Sc., C.Eng. M.I.Prod.E.

M

First published 1981 by
THE MACMILLAN PRESS LTD
London and Basingstoke
Associated companies in Delhi Dublin
Hong Kong Johannesburg Lagos Melbourne
New York Singapore and Tokyo

ISBN 0 333 29181 6

Printed in Hong Kong

Typeset by *Illustrated Arts*

CONTENTS

PREFACE

In order to communicate ideas on design the engineer makes use of drawings, in either pictorial or orthographic projection, which is really the common language of the technologist, draughtsman or design engineer. Hence, in the title 'Graphic Communication for Technical Design' there is the connotation of a language — a common language — which may be understood by anyone who takes the trouble to learn the simple principles of its translation.

The order of treatment of the various topics is logical and progressive, ranging from simple basic geometrical constructions to preparation of the various types of pictorial projections and engineering drawing assemblies in orthographic projection. This self tutoring style work course has been class room tested for some years and in the opinion of the author forms an ideal basic course for use with pupils of mixed ability who may be studying for different examinations. The latest drawing office standards are used and emphasised throughout the work.

Both first and third angle systems of orthographic projection have been freely used in view of the fact that there is still no firm ruling from standards authorities for the adoption of one universal system. From the educational point of view the author believes that the student should be encouraged to become competent in either system of projection. The word 'view' has been employed in place of the word 'elevation', in keeping with the latest trend, except in the chapter on Auxiliary Projection where the use of the word 'view' could lead to some confusion.

The basic graphic communication topics have been treated adequately for the examinations already mentioned. In fact some topics have been taken a stage further, to add interest and to cater for the more able student who may have in mind more advanced work in technical design.

I wish to thank the engineering firms who kindly supplied photographs and permission to use diagrams, for which acknowledgement is given in the text. I also wish to thank the NI Schools Examinations Council for permission to use examination questions as indicated.

K. Simms

1 BASIC INFORMATION AND CONSTRUCTIONS

Certain basic information and constructions, which are essential knowledge to later graphic communication work, will be given in these opening pages.

PAPER SIZES AND DRAWING BOARDS

The most usual size of drawing board for use by the beginner is that which will accommodate A2 size drawing sheets. The 'A' sizes of paper now in general use are based on the rectangle having sides in the ratio $1 : \sqrt{2}$. This rectangle proportion has the property that if sheets are halved they still retain the original proportion. The paper sizes in the 'A' system are as follows:

A0 841 mm x 1189 mm
A1 594 mm x 841 mm
A2 420 mm x 594 mm
A3 297 mm x 420 mm
A4 210 mm x 297 mm

A0 size paper is one square metre in area. The A2 size is quite commonly used in schools and colleges whilst the A3 size is probably most widely used because of its convenient size for handling.

The drawing board and Tee square are fast disappearing in favour of a board having some form of parallel-motion straight-edge fitted. Most professional design drawing offices are equipped with expensive and highly accurate draughting machines, but this standard of equipment is rarely found in schools for student use. A very common drawing unit suitable for elementary work in schools and colleges consists of a plastic faced drawing board fixed to an adjustable stand, or desk, with a parallel-motion straight-edge of perspex controlled by wire running on pulley wheels fixed to the four corners of the board. This type of equipment is sufficiently accurate and is more convenient to use than the board with separate Tee square. It is preferable to use spring steel clips to fasten drawing paper to the board, rather than drawing pins which damage the surface.

DRAWING INSTRUMENTS

The following are considered essential for elementary drawing work:

(i) An accurate ruler having graduations of millimetres along one edge and inches along the other. It is not necessary to purchase an expensive engine-divided ruler since most cheap transparent plastic rulers are sufficiently accurate.
Also there is no parallax error in reading a transparent ruler having graduation marks on the under side. At a later stage the more advanced student may find it necessary to purchase rules incorporating scales.

(ii) A good quality pair of general purpose compasses which are rigid and have a smooth action. The test is will the compasses draw just one circular arc at any particular setting? Usually compasses are made to accept a short piece of pencil lead and have a fine, shouldered steel point.
As an aid to drawing small circles, spring bow compasses operated by a screw are sometimes employed. These are not considered essential for the beginner. Larger type spring bows are available but the author has found that these tend to slow down drawing work and are not recommended at the early stages. The student, or the benevolent parents, should avoid the temptation to buy a large, plush-lined box of so-called drawing instruments. These beautiful boxes are usually filled with poorly engineered goods most of which are non-essential even assuming that they did perform the job for which they were designed!

(iii) Two set-squares, one 45° of about 200 mm size and one 60° of about 250 mm size are suitable for use on the A2 size board. Perspex set-squares are preferable but are more expensive than the cheap injection moulded type. However, the cheap set-square is usually very accurately manufactured and if carefully handled should give reasonable service.

(iv) Two good quality 'lead' pencils, one of 2H grade and the other of H grade. The 'lead' is really a compressed mixture of graphite and special clay with other additions so that it is important to have pencils of good quality with 'lead' free from any hard grittiness.

Pencils should be sharpened using a good quality bench fixed sharpener or an efficient pocket type. The point may be finished if necessary using a penknife. Professional people sometimes prefer to use a chisel-edged point on the pencil, which enables many more lines of uniform thickness to be drawn before re-sharpening is necessary. This practice is not really helpful to the beginner at the early stages of his training.

As an alternative to the cedar wood pencil, clutch type lead holders are available, complete with lead pointer, at very reasonable prices. There are also automatic press-button pencils designed to use short, fine leads fed from the magazine inside the pencil body. These are fairly expensive to buy, very expensive to keep up on leads and not as reliable as the clutch type or the wood body pencil.

The degree of hardness of a pencil lead depends upon the number of H's in the grading and the degree of softness on the number of B's. Thus a 2H grade pencil has a harder lead than a grade H. The HB grade is the pencil in everyday use but is just a little too soft for technical drawing work. Grades B, 2B, 3B, etc. are used by artists for shading drawings. Of the two pencils recommended for technical work the 2H is used mainly for fine preliminary construction lines and the H to give a good black finishing line or for lettering.

(v) A good quality eraser made of vinyl plastic. This type of eraser does not spoil the paper surface and is very efficient at cleaning up as well as removing even heavy pencil lines.

Erasure of india ink lines from tracing paper or cloth is accomplished by first of all gently scraping the ink error with the corner of a razor blade and then smoothing the erased surface with a hard rubber. Drawing offices are usually equipped with electric erasers to speed up this process.

(vi) A good quality semi-circular protractor for the setting off and measurement of angles.

ADDITIONAL DRAWING AIDS

A special set-square having a number of uses is illustrated in figure 1.1. This is a perspex 45° set-square with a centre line and 5 mm spaced parallel lines marked on the underside. This instrument serves as a useful parallel ruler and usually a protractor is incorporated.

An adjustable set-square is shown in figure 1.2. This is particularly useful for setting off angles and for drawing parallel lines quickly. Although not absolutely necessary for the beginner this instrument will be found much more convenient to use than a pair of set-squares for drawing parallel lines.

For all stages of engineering drawing a radius aid is indispensable. Its use is illustrated by figures 1.3 to 1.5. The curve is drawn, figure 1.4, and the finished straight lines are then drawn to blend in with the curve thus avoiding any unsightly 'joints' such as may be obtained using compasses. Another important use of the radius aid is in drawing nuts and bolts in engineering drawing assemblies. However, for the necessarily more accurate geometrical construction work it is recommended that small spring bow compasses be used for small circles rather than the radius aid.

French curves, made from transparent plastic, are used for drawing curved lines which are not composed of circular arcs. A typical example of a french curve is shown in figure 1.6. One or more of these curves, perhaps even a dozen of different shapes, comprise a set designed for specific jobs such as architecture, naval architecture, aircraft design work, etc. French curves are a slow means of curve drawing and to save time it is preferable to become practised at the freehand drawing of curved lines. In any case before french curves can be employed to finish a curved line it is necessary for the draughtsman to sketch an approximate curve through the plotted points. A 'flexi-curve', that is a pliable plastic ruler, may be used in place of french curves.

Templates are sometimes employed for drawing small circles, ellipses, hexagonal nuts and bolt heads, etc.

5mm SPACING PARALLEL LINES

PROTRACTOR GRADUATIONS

SYMMETRICAL mm SCALE WITH ZERO ON CENTRE

FIG 1.1

A SPECIAL SET-SQUARE FOR PERPENDICULAR LINES, PARALLEL LINES AND ANGLES

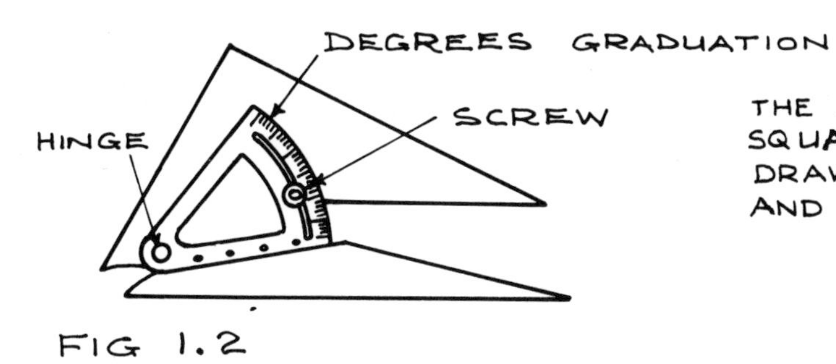

FIG 1.2

THE ADJUSTABLE SET-SQUARE FOR RAPID DRAWING OF PARALLELS AND FOR ANGLES

THE RADIUS AID

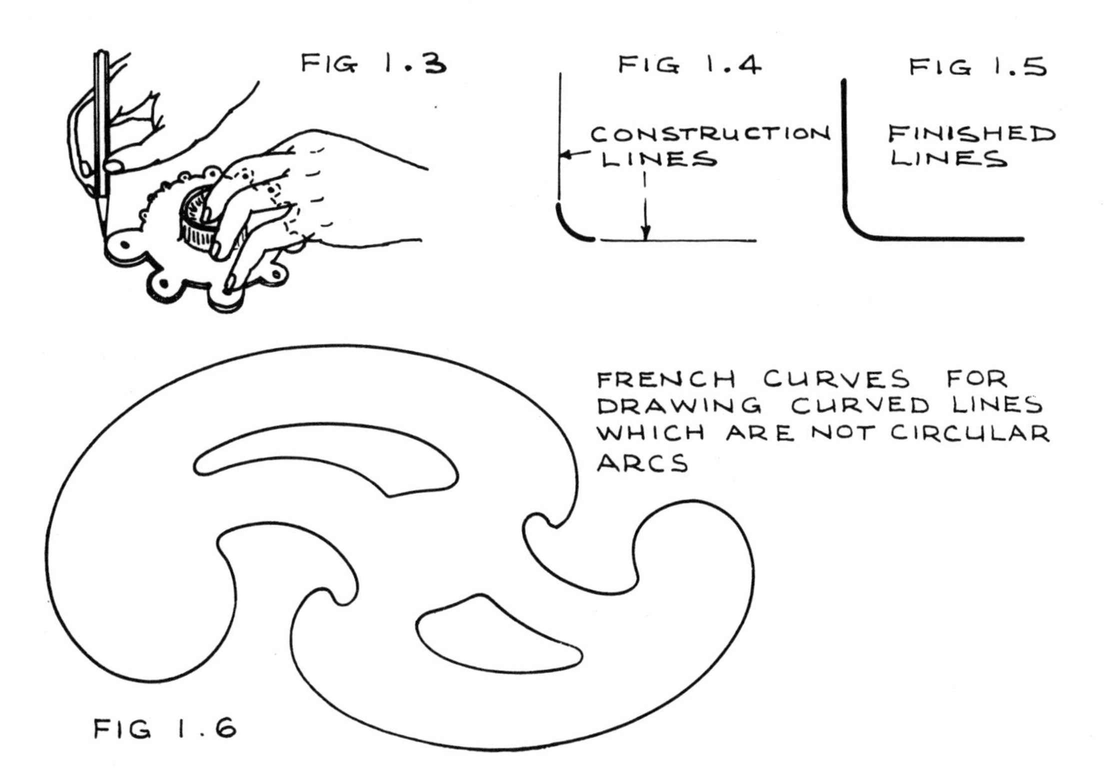

FRENCH CURVES FOR DRAWING CURVED LINES WHICH ARE NOT CIRCULAR ARCS

FIG 1.6

DRAWING PAPERS

White cartridge paper, made in three thicknesses, is generally used for pencil drawing work such as is carried out in schools and colleges. A weight of around 125 grammes per square metre is suitable since both sides may be used for economy and storage convenience. It is false economy to use light weight cartridge paper except perhaps for preliminary layouts and freehand sketches.

There is a right and wrong side to cartridge paper. If a drawing is to be finished in ink it would be best done on the right side which is the side not having a very slightly, regularly pitted surface.

Tracing paper is semi-transparent and thus drawings placed underneath may be clearly visible for the purpose of copying by tracing. Tracings finished in black drawing ink, called india ink, can be used as negatives for production of further copies by photo-printing. Quite acceptable photocopies may be obtained from a tracing finished in dense black pencil lines.

Tracing cloth, which is specially treated to be semi-transparent, is used for important master negatives which may require to be stored for long periods and be subject to fairly frequent handling. Ink tracings made on either paper or cloth are more easily completed if the surface is first rubbed with french chalk applied sparingly on a cloth duster.

Tracing film, having stretch-proof and moisture-proof properties, is increasingly being used in place of tracing cloth. The most common plastic draughting film is the chemically coated polyester type which may have a matt surface on one or both sides.

DRAWING IN INK

Ink drawings are seldom done in normal school course work, except perhaps for exhibition purposes. Should the student require to complete a drawing in india ink then a reservoir type pen fitted with a tubular nib unit should be used. These nib units are interchangeable and may be obtained in sizes ranging from about 0.2 mm to more than 1 mm line thickness. Tubular nib units may also be fitted to special compass attachments. The old-fashioned 'tweezer' type nibs found on pens and compasses of instrument sets already referred to require a great deal of skill to use successfully. In fact it is a very lengthy and frustrating experience to attempt an ink drawing using poor quality 'tweezer' type pens. In addition to the reservoir type pen with one or two tubular nib units the student will find an ordinary fine-nibbed fountain pen invaluable for lettering purposes. However, it should be remembered that normal black drawing ink, or india ink, will quickly ruin the ink feed mechanism. For this reason the fountain pen should be filled with non-clog drawing ink specially manufactured for the fountain pen.

PRINTS FROM DRAWINGS

A common means of obtaining copies of a tracing is by the *Diazo* process. Printing paper coated with diazonium salts reacts to ultraviolet light when the tracing is superimposed in a printing machine. The black lines of the tracing, shield the printing paper coating from the ultraviolet light which bleaches all the remainder of the coating. A chemical called a *coupler* is then applied to transform the remaining shielding coating into permanent *Azo* dye lines of a darker colour.

MICROFILMING

This is a means of storing original drawings and technical information on film, reduced in size to perhaps one-twenty-fifth of the original size. The films may be mounted on cards and conveniently filed. Prints may be obtained at any time from the film card without recourse to handling the large and expensive originals. There are savings of space, time, cost of working drawings, and greater security for original drawings and technical information. By use of a micro film reader an enlarged image of a film card or strip may be studied without the expense of making a print.

LINES AND LETTERING

All finished linework should be uniformly black. Particular care should be taken with circular arcs to ensure that the proper grade of pencil be used to achieve line work quality for good print production or for microfilming. The same applies to drawings made in ink.

Two line thicknesses, thick lines and thin lines, are recommended by the British Standards Institution in their publication BS 308: 1972 'Engineering drawing practice'. The student is referred to the extremely helpful condensed version of this standard published by B.S.I. Education Section, PD 7308: 1978.

Thick lines are used for all visible outlines and edges.

Thin lines are used for projection lines, dimension lines, leader lines, hatching, etc. A fuller explanation of these applications will be given later in the work.

Hidden edges are indicated by a thin line in the form of short dashes.

Centre lines are indicated by a thin line in the form of a chain line.

The relationships between thicknesses of the two recommended lines is that the thick line should be about two-and-a-half times the thickness of the thin line.

Note. The foregoing recommendations regarding lines and their thickness refer to British standards for drawing office practice in engineering drawing. In the case of geometrical drawing constructions, to be dealt with in this opening chapter, a bold outline answer is desirable but extremely fine lines are used for construction work. Thick lines are not generally used in geometrical drawing work because extreme accuracy is required.

LETTERING

A simple style of capital lettering, which is recommended for general use in technical drawings is shown in figure 1.7. All lettering on a drawing should be clearly spaced, in good style and of such a height as to be suitable for microfilming if required. Lower case lettering is shown in figure 1.8. The recommended style for numerals, which should be done with special clarity, is also shown in figure 1.9.

The most common mistake of the beginner is that of not allowing proper spacing between lines of lettering. It is only necessary to notice that the print lines of a book or newspaper are spaced by an amount approximately equal to the height of the print.

Another tendency for the beginner is to make the lettering, particularly titles, too large. In order to emphasise words it is not necessary to increase the height of the lettering. Any necessary emphasis can be obtained by increasing width as shown in figure 1.10.

The recommended minimum character height for dimensions and notes on a drawing is 3 mm. Titles and drawing numbers may be made about 5 mm tall.

The British Standards Institution recommends that, in general, capital letters be used on drawings. In some cases lower case lettering may be used for the longer notes. For obvious reasons no handwriting, except the draughtsman's signature, should appear on a drawing.

The general appearance of a finished drawing may be enhanced or marred by the lettering. The student should endeavour to cultivate a neat style of freehand lettering. If his style is naturally forward sloping, this is acceptable provided that this style is consistently used throughout the drawing. Stencils are available as aids to lettering but these should not be resorted to at the elementary stage. Indeed, a good freehand style is more pleasing and quicker than use of a stencil. Adhesive lettering sets are available but these also are not recommended for students' use.

Underlining of notes on a drawing is not generally recommended.

LETTERING AND NUMERALS

ABCDEFGHIJKLMNOPQRSTUVWXYZ

FIG 1.7

aabcdeefghijklmnopqrstuvwxyz

FIG 1.8

1234567890 (!:;?&%+-=)

FIG 1.9

ALPHABET ALPHABET ALPHABET

FIG 1.10

Note: All lettering, numerals and symbols shown on this page, with the exception of Figure 1.10, were produced with the aid of a lettering stencil and a 0.6 mm tubular nib, reservoir type pen containing black india ink. This is the type of equipment sometimes employed by the professional draughtsman, but the beginner is recommended to practise a freehand style based on these standardized shapes.

BASIC GEOMETRIC CONSTRUCTIONS

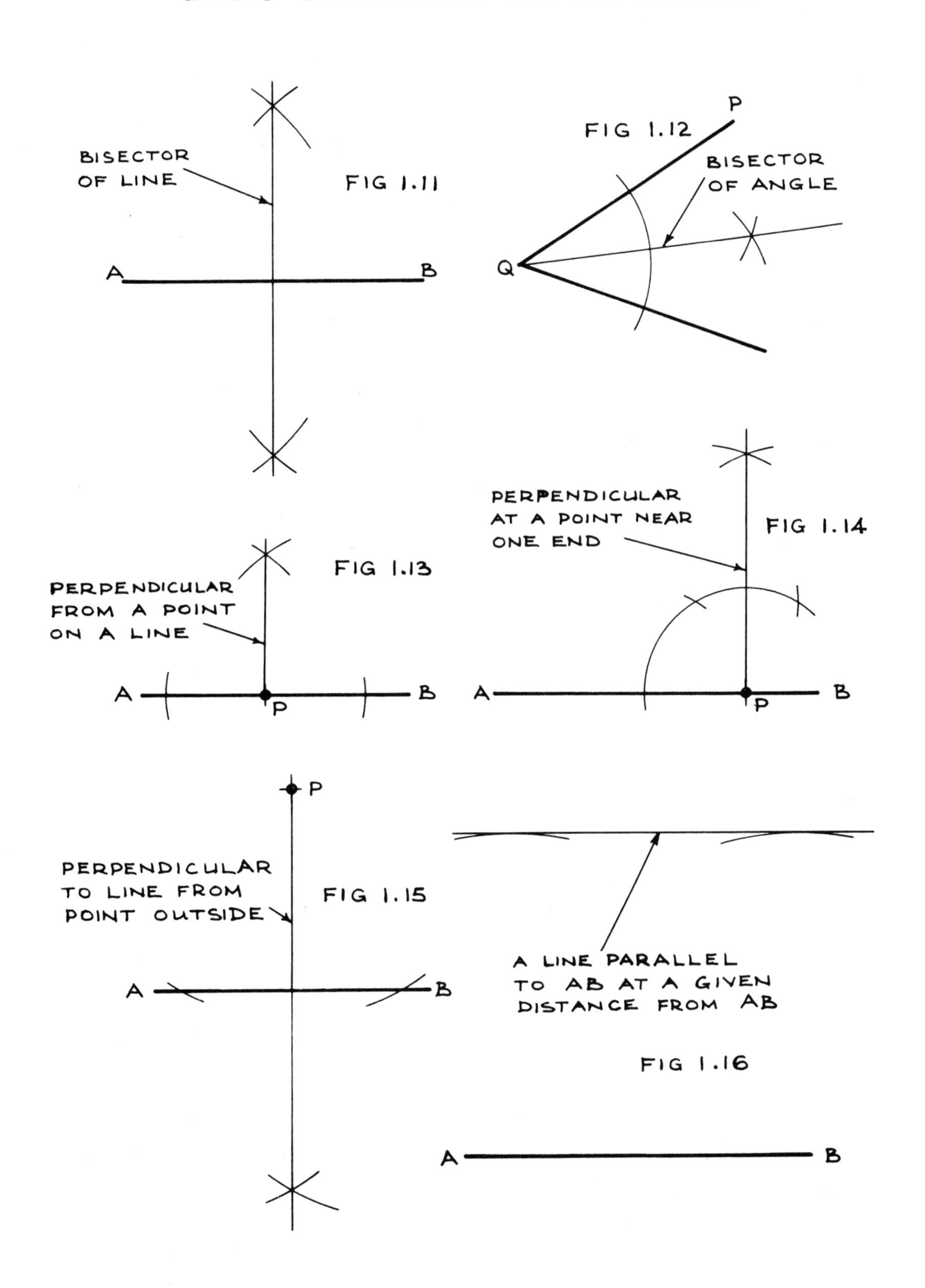

BASIC GEOMETRIC CONSTRUCTIONS

DIVISION OF A LINE INTO
ANY NUMBER OF PARTS

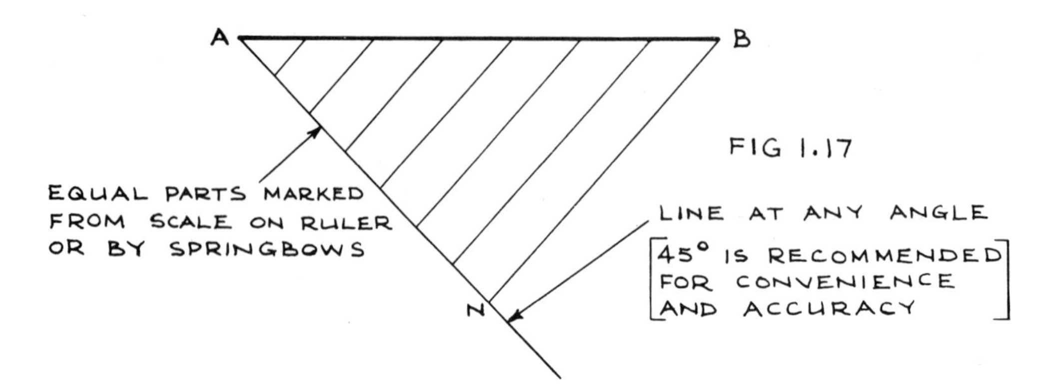

FIG 1.17

PROCEDURE:
Draw a line at any angle to AB and mark the required number of equal parts along this line.
Join point N to B and draw parallels to NB to complete the division of AB.

DIVISION OF SPACE BETWEEN
PARALLEL LINES AB AND CD

FIG 1.18

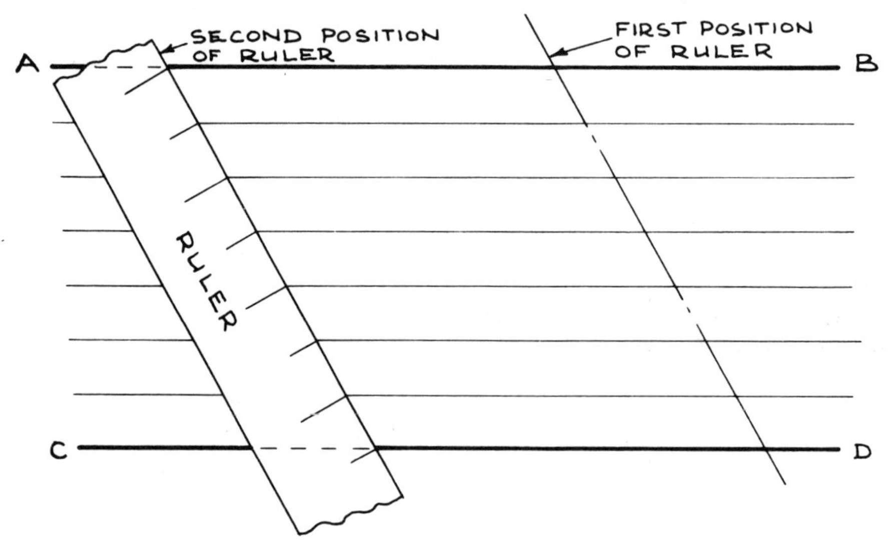

PROCEDURE:
Place ruler in the first position so that the correct number of scale divisions lies between AB and CD.
Mark the divisions using a sharp pencil point and repeat for the second position of the ruler.
Draw the required parallel dividing lines by joining appropriate points.

BASIC GEOMETRIC CONSTRUCTIONS

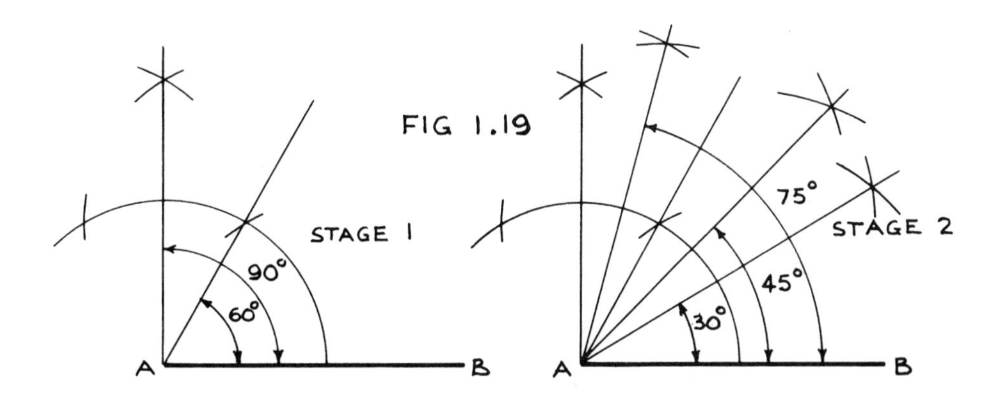

CONSTRUCTION OF ANGLES USING COMPASSES

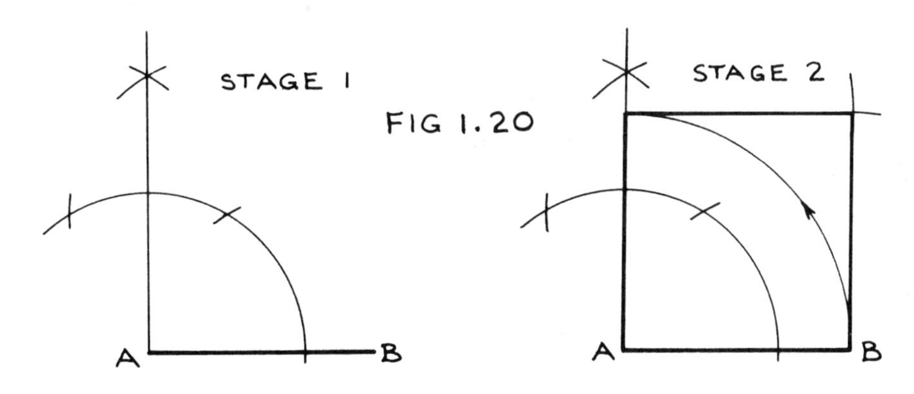

CONSTRUCTION OF A SQUARE USING COMPASSES

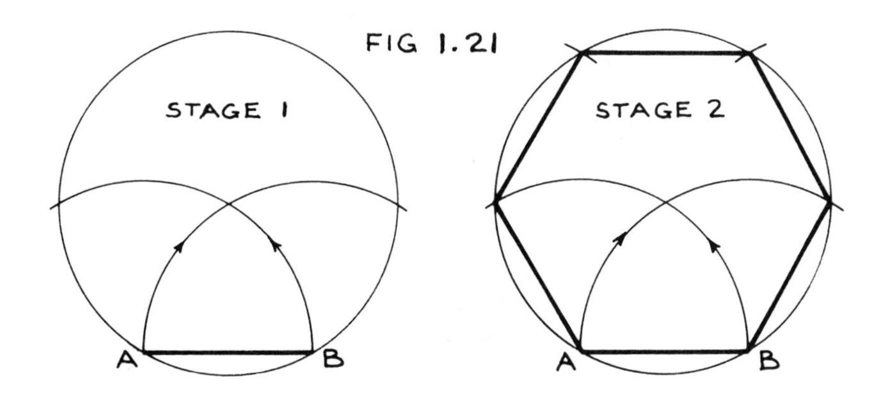

COMPASS CONSTRUCTION FOR A REGULAR HEXAGON

NOTE: Throughout the construction keep compasses set at given regular hexagon base length AB.

BASIC GEOMETRIC CONSTRUCTIONS

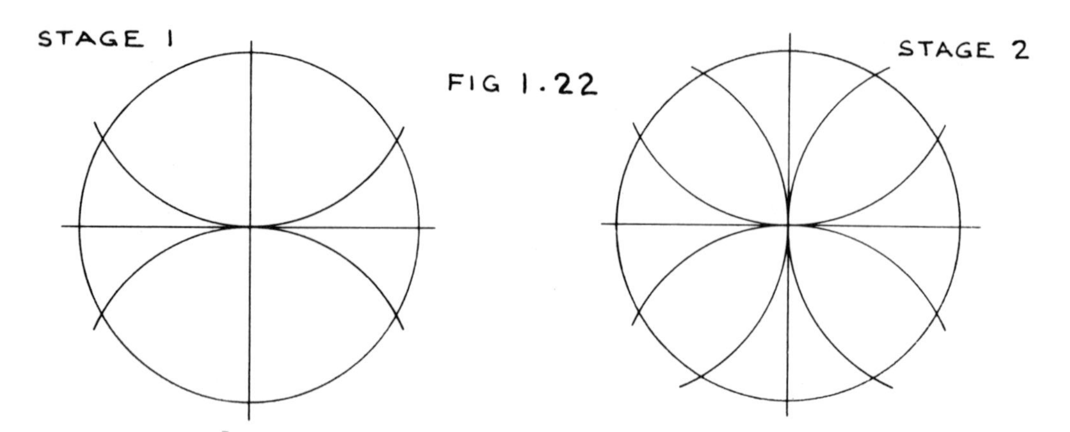

PROCEDURE: Use the same radius as the circle for the construction arcs. Complete arcs as shown are not necessary but are shown to demonstrate the method.

COMPASS DIVISION OF CIRCLE INTO TWELVE PARTS

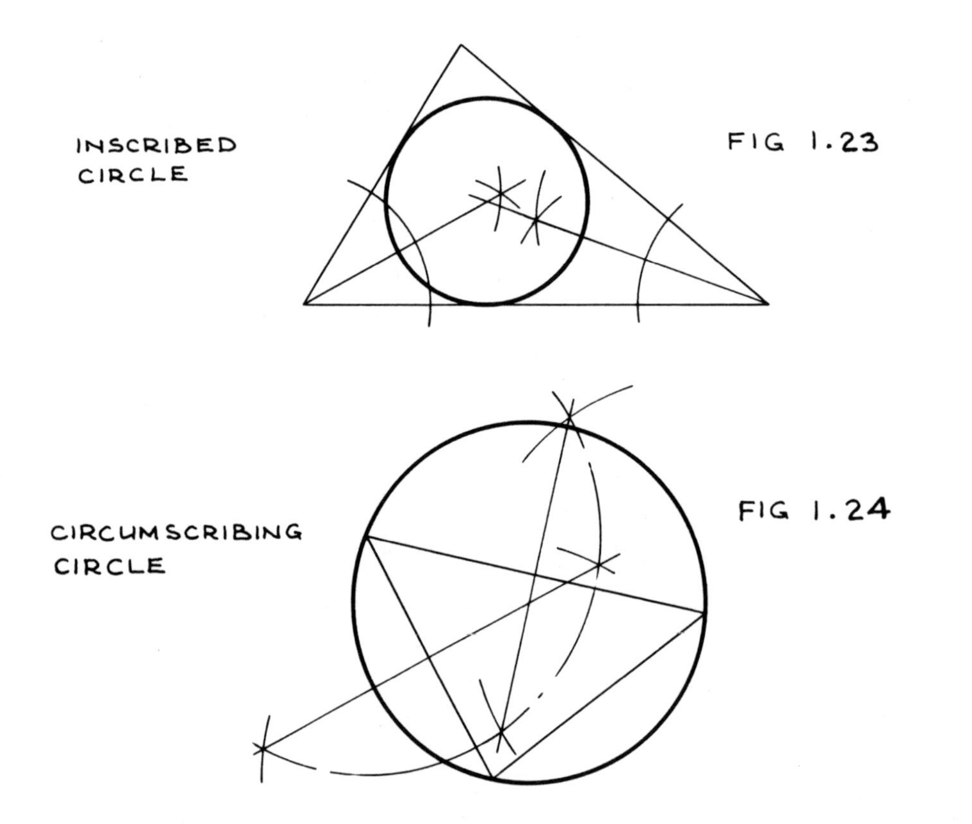

INSCRIBED CIRCLE AND CIRCUMSCRIBING CIRCLE TO A TRIANGLE

NOTE: The constructions shown in Figs. 1.23 and 1.24 serve as excellent elementary work accuracy tests.

SCALES EXERCISES

Exercise 1. Construct a scale in which 25 mm represents 1 m, reading up to 5 m and showing 200 mm sub-divisions.

Exercise 2. If 50 mm is to represent 1 m in a drawing of an object having measurements varying between 50 mm and 3 m, draw a suitable scale. What is the representative fraction of this scale?

Exercise 3. A front view of a garden shed is shown in figure 1.27, with the dimensions given in metres. Construct a plain scale of 40 mm representing 1 m and use this scale to make a front view scale drawing of the shed.

Exercise 4. Construct a diagonal scale to read up to 120 mm and show single millimetres.

Exercise 5. Draw a diagonal scale of 50 mm to represent 1 m, showing tenths and hundredths of a metre and to read up to 3 m. Mark on your scale 1.67 m and 2.14 m.

Exercise 6. Construct a diagonal scale of 75 mm representing 1 m to read up to 2 m and showing 0.01 m. Mark on the scale 1.04 m and 1.36 m.

Exercise 7. Using a proportional scale enlarge the drawing shown in figure 1.35 so that each line of the new drawing becomes 15/8 times the original size.

FURTHER EXAMPLES

Example 1. Draw a scale of 30 mm representing 1 m to read up to 4 m, the first unit being sub-divided to show 0.1 m. State the representative fraction of this scale.

Example 2. The line AB, 0.1 m long on a scale drawing, represents an actual length of 30 m. Draw AB on your paper and produce it to represent an actual length of 57 m.

Example 3. Draw a diagonal scale, full size, to read up to 150 mm, showing single millimetres.

Example 4. If 40 mm represents 1 m construct a diagonal scale to read to 4 m showing tenths and hundredths of a metre. State the representative fraction of this scale.

Example 5. A diagram of a roof truss is shown. The triangle ABC is equilateral and the other triangles are right-angled. Prepare a scale of 15 mm to represent 1 m, showing tenths of a metre, and draw this roof truss. State the representative fraction of your scale. What would be the actual lengths of members AC and CD scaled from your drawing?

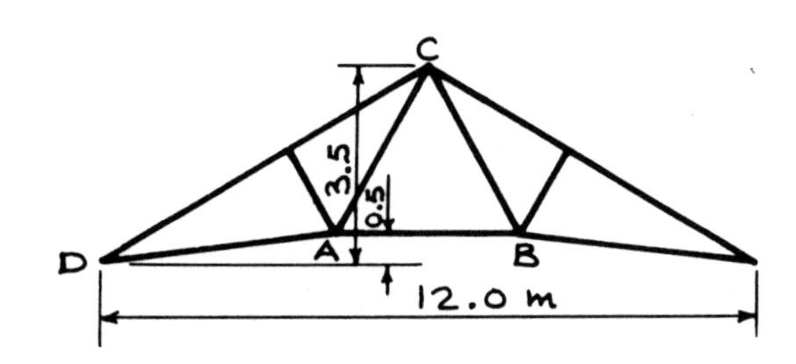

PLAIN SCALES

Scales are used in drawing to enable a large object, such as a building or a ship, to be represented on a sheet of paper.

Sometimes a scale may be used to produce an enlarged drawing of a very small object such as a sewing machine part.

EXERCISE 1 SOLUTION

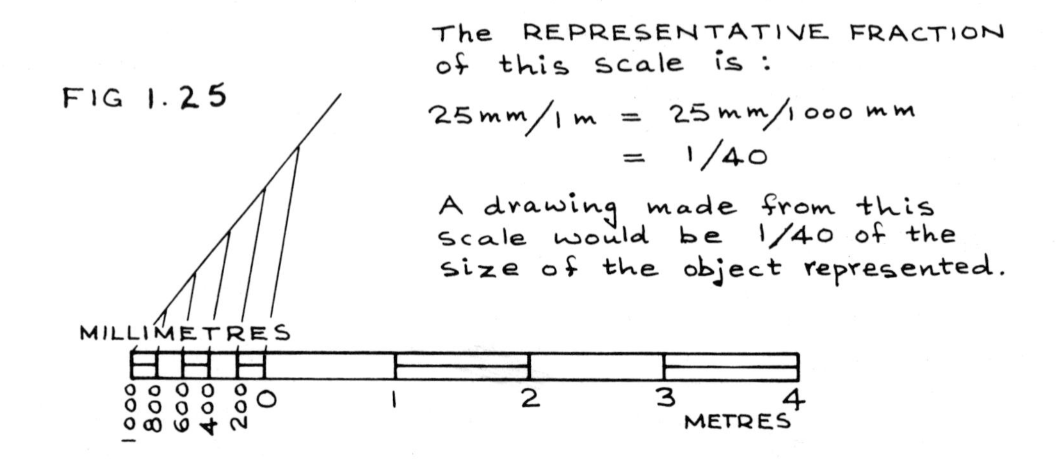

EXERCISE 2 SOLUTION

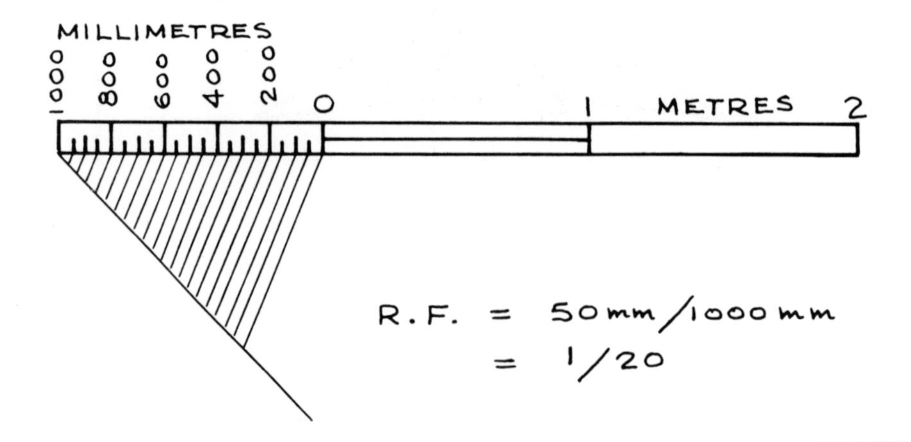

FIG 1.26

EXERCISE 3 SOLUTION

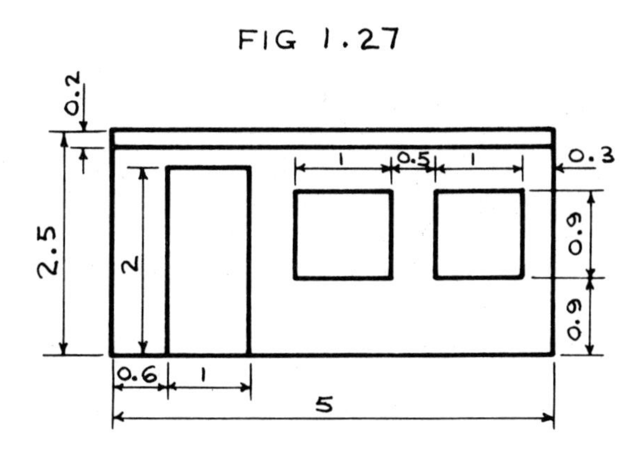

FIG 1.27

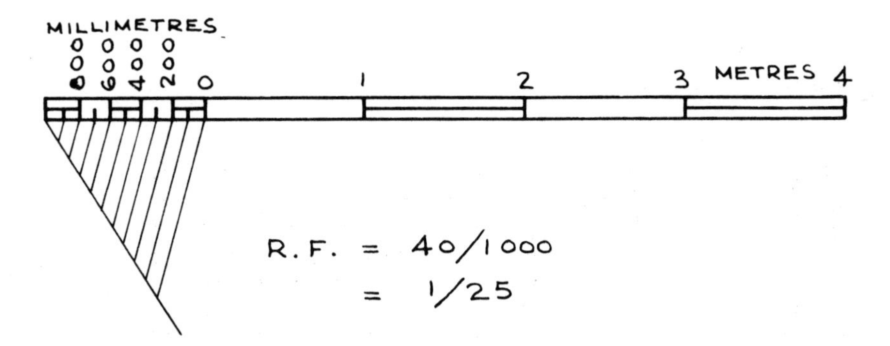

FIG 1.28

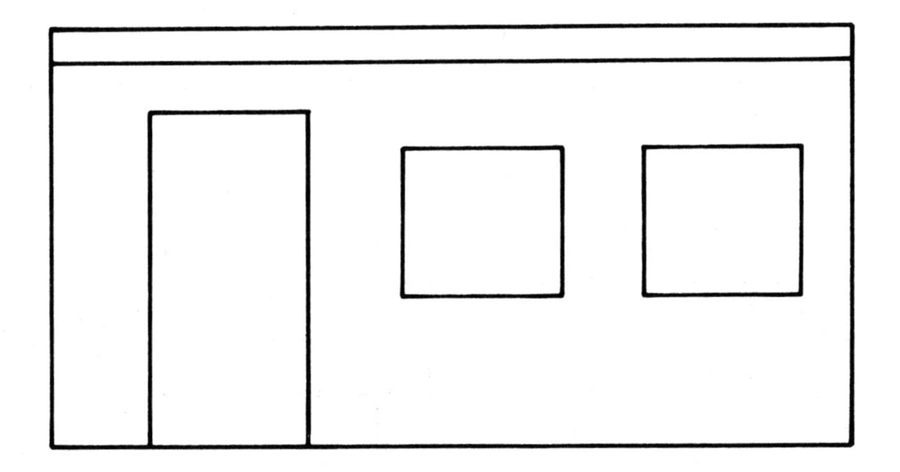

FIG 1.29

DIAGONAL SCALES

An accurate diagonal scale is convenient to construct and easy to read. The principle of the diagonal scale is based on the property of similar triangles.

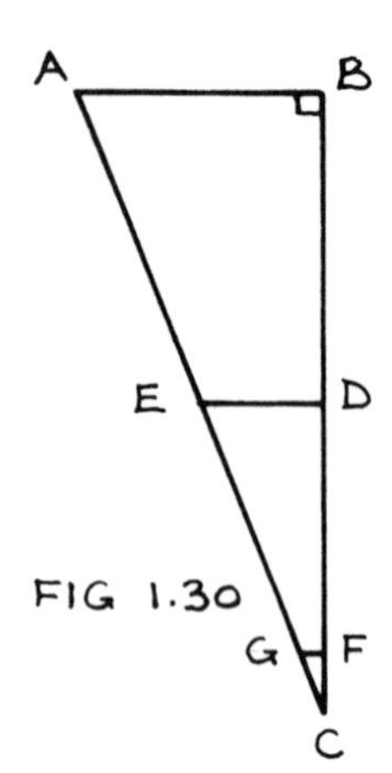

FIG 1.30

In the triangle ABC suppose AB = 10 mm and BC is any length.
If, through D the mid-point of BC, a line is drawn parallel to AB, then the length DE = 5 mm.
Again, if CF = BC/10, then FG = AB/10, that is, FG = 1 mm.
Now suppose that AB is much smaller than 10 mm, say 1 mm, then FG would be 1/10 mm.

EXERCISE 4 SOLUTION

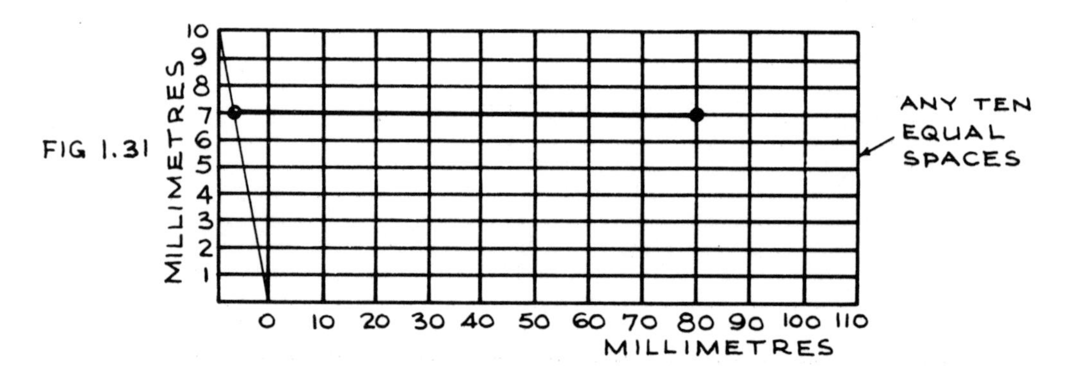

FIG 1.31

NOTE: Figures 1.31 and 1.32 give alternative solutions to this exercise.

The distance between the small circles is 87 mm

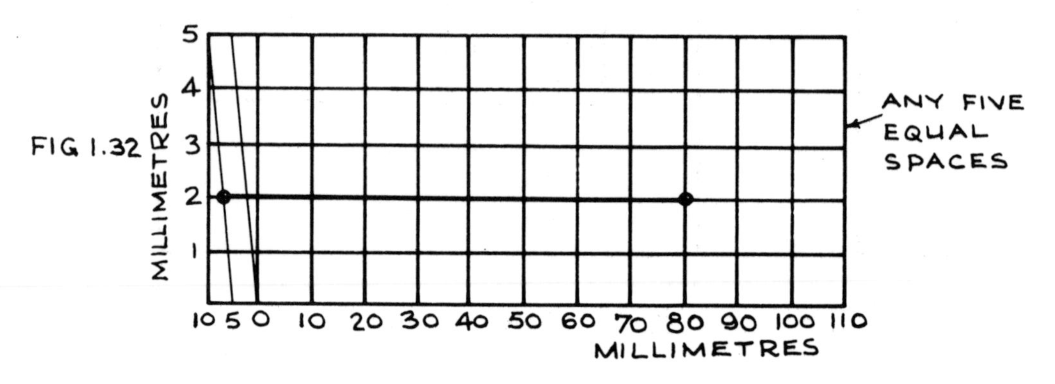

FIG 1.32

EXERCISE 5 SOLUTION

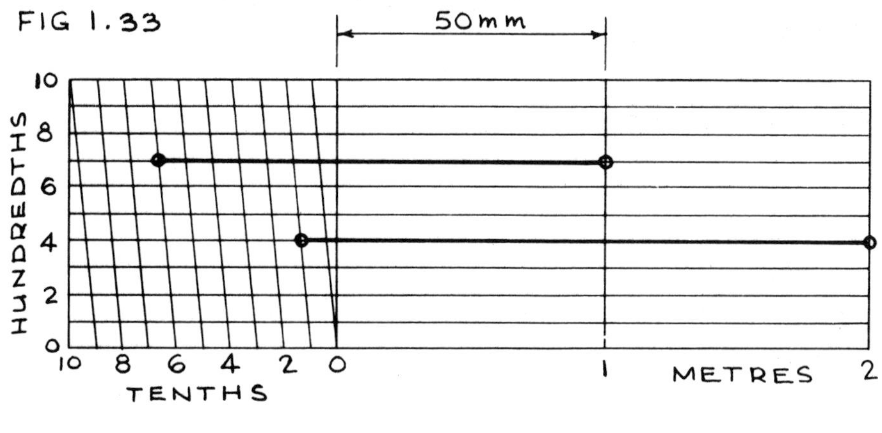

R.F. = 50/1000
= 1/20

EXERCISE 6 SOLUTION

R.F. = 75/1000
= 3/40

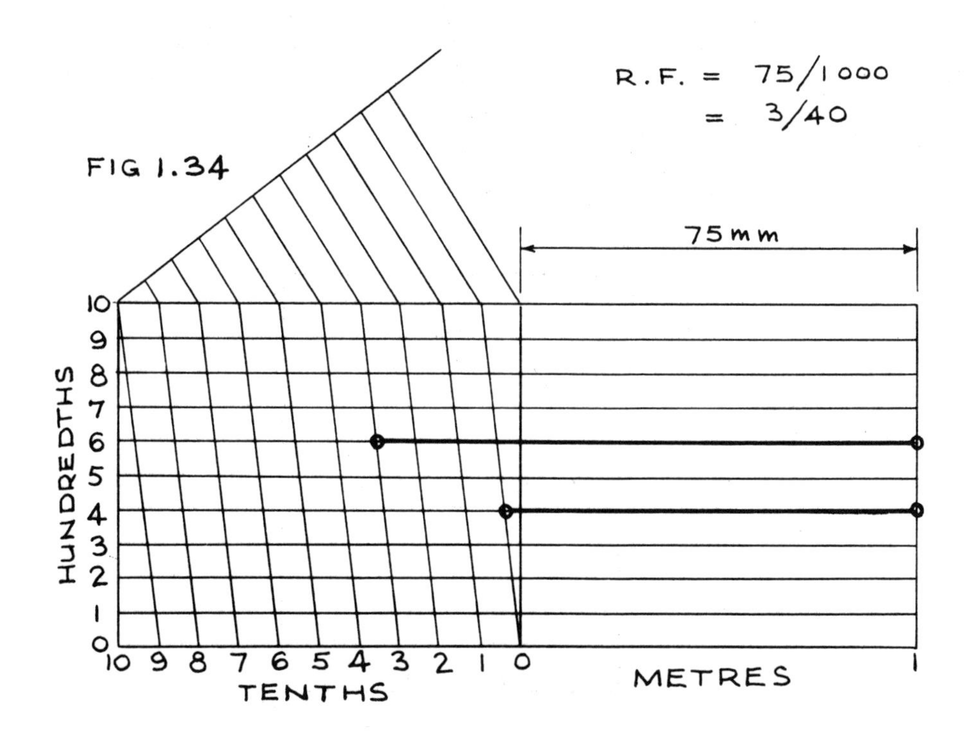

PROPORTIONAL SCALES

A proportional scale, consisting essentially of parallel lines drawn between two lines at any angle, is an extremely simple means of enlarging or reducing the size of a given drawing. Exercise 7 gives a simple illustration of the application of a proportional scale.

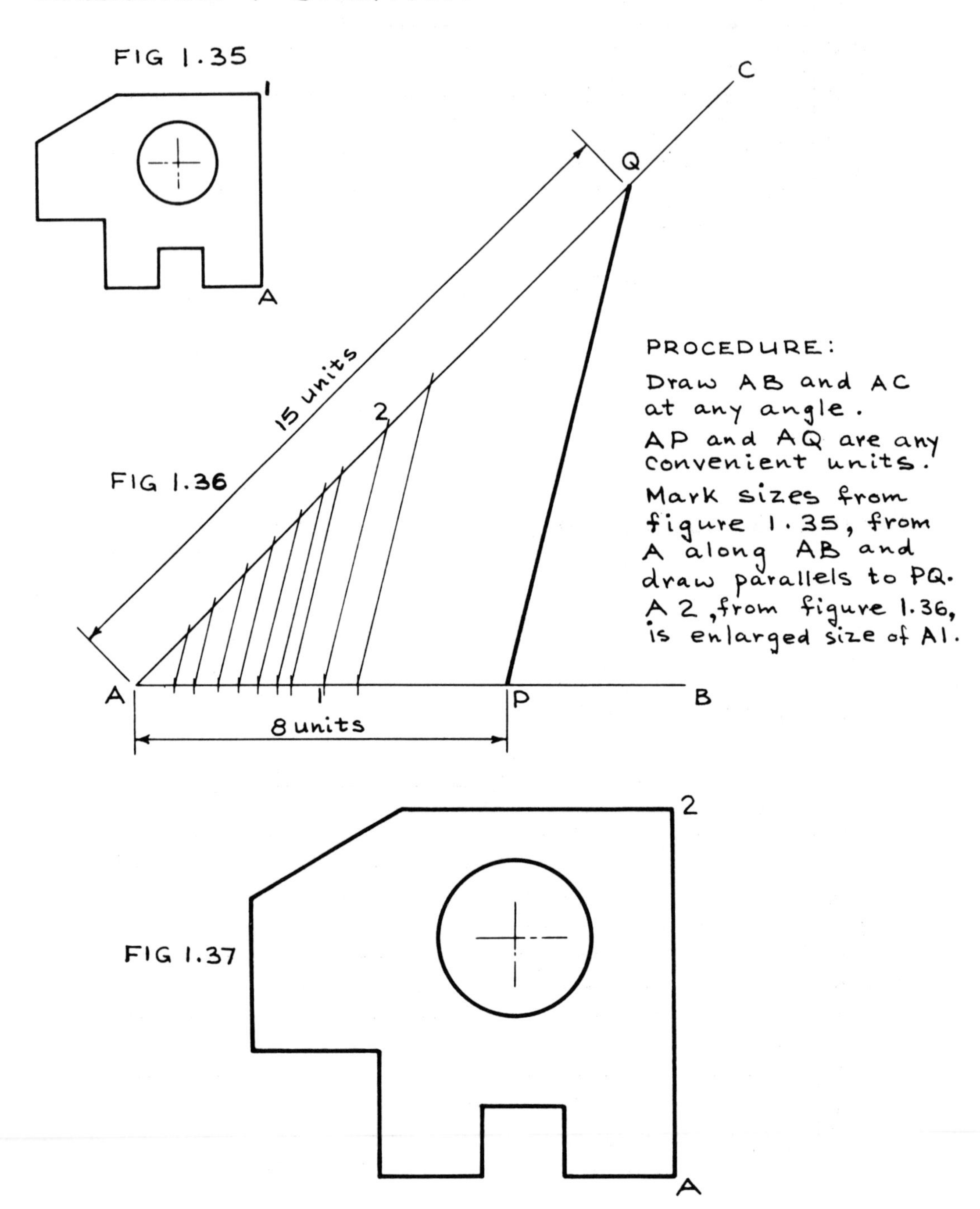

TRIANGLE AND POLYGON EXERCISES

Construct triangles according to the information given in the following Exercises 8 to 13.

Exercise 8. Perimeter 150 mm and the ratio of the sides 5:7:9.

Exercise 9. Base 87 mm, base angle 53° and the sum of the other two sides 150 mm.

Exercise 10. Base 115 mm, base angle 50° and the difference of the other two sides 28 mm.

Exercise 11. Perimeter 180 mm, altitude 50 mm and base angle 60°.

Exercise 12. Base 85 mm, altitude 56 mm and vertical angle 68°.

Exercise 13. Perimeter 150 mm and the base angles 47° and 75°.

Exercise 14. Construct the polygon ABCDE from the following data: AB = 70; BC = 59; AC = 106; AE = 70; BE = 106; DE = 80 mm and angle EDC = 90°.

Exercise 15. To construct any regular polygon on a given base. (Draw a regular heptagon on a 50 mm base.)

Exercise 16. To inscribe a regular octagon in a given square.

Exercise 17. To inscribe any regular polygon in a given circle.

FURTHER EXAMPLES

Example 1. Construct a triangle with AB = 85 mm, BC = 106 mm and CA = 73 mm.

Example 2. Construct a right triangle given the hypotenuse 125 mm and one side 80 mm. Use compasses and rule only.

Example 3. Draw an equilateral triangle of 100 mm side and inscribe three equal circles each touching two sides and two other circles.

Example 4. A triangle has a perimeter of 150 mm, one base angle of 50° and one side of 56 mm. Find the lengths of the two remaining sides and the sizes of the two remaining angles.

Example 5. Construct a regular pentagon on a 60 mm base. Inscribe in the pentagon five equal circles each touching one side and two other circles. Measure and state the diameter of one of the equal circles.

Example 6. Construct regular hexagons (i) given that the distance between two parallel sides is 55 mm, and (ii) given that the distance between opposite corners is 70 mm. Measure and state the length of side in each case.

Example 7. Construct the pentagon ABCDE from the following data: AB = 77; BC = 74; AC = 124; AE = 78; BE = 116; ED = 91 mm and angle EDC = 90°.

EXERCISE 8 SOLUTION

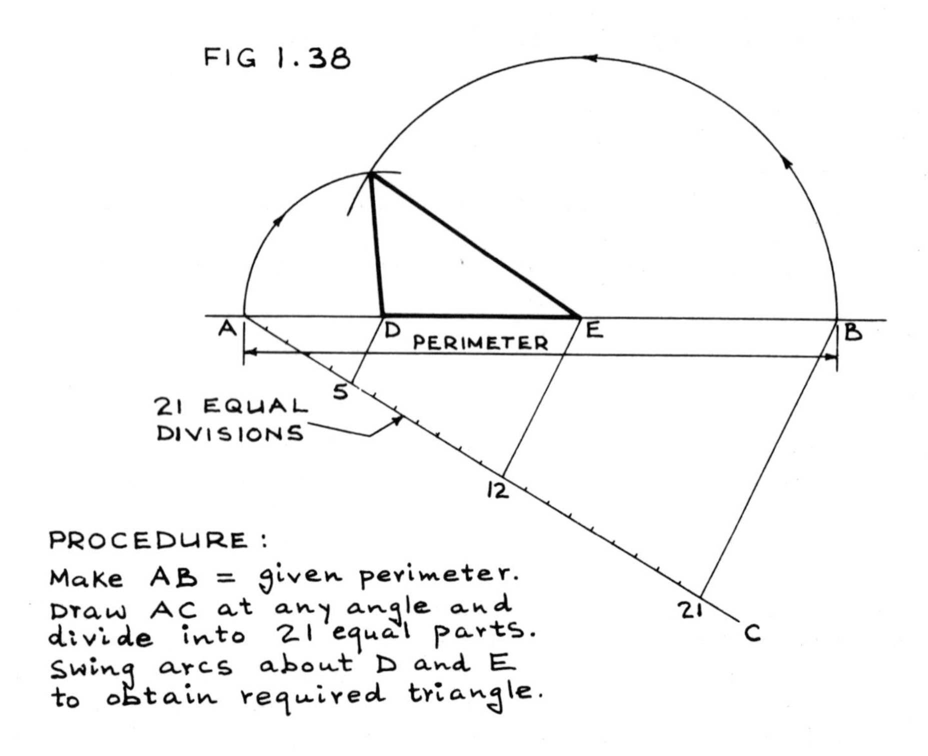

FIG 1.38

PROCEDURE:

Make AB = given perimeter.
Draw AC at any angle and divide into 21 equal parts.
Swing arcs about D and E to obtain required triangle.

EXERCISE 9 SOLUTION

PROCEDURE:

Set off AB, angle BAC and AC as indicated in figure 1.39.
Join CB and bisect.
ABD is the required triangle.

NOTE: This construction is based on the properties of the isosceles triangle.

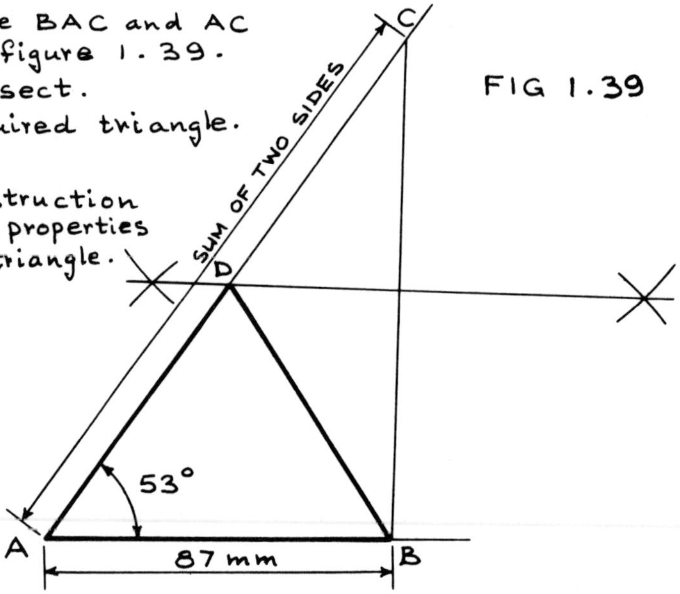

FIG 1.39

SUM OF TWO SIDES

C

D

53°

A

87 mm

B

EXERCISE 10 SOLUTION

PROCEDURE:
Set off the given base, base angle and the difference between the two sides as indicated in figure 1.40.
Bisect CB to find D.
ADB is required triangle.

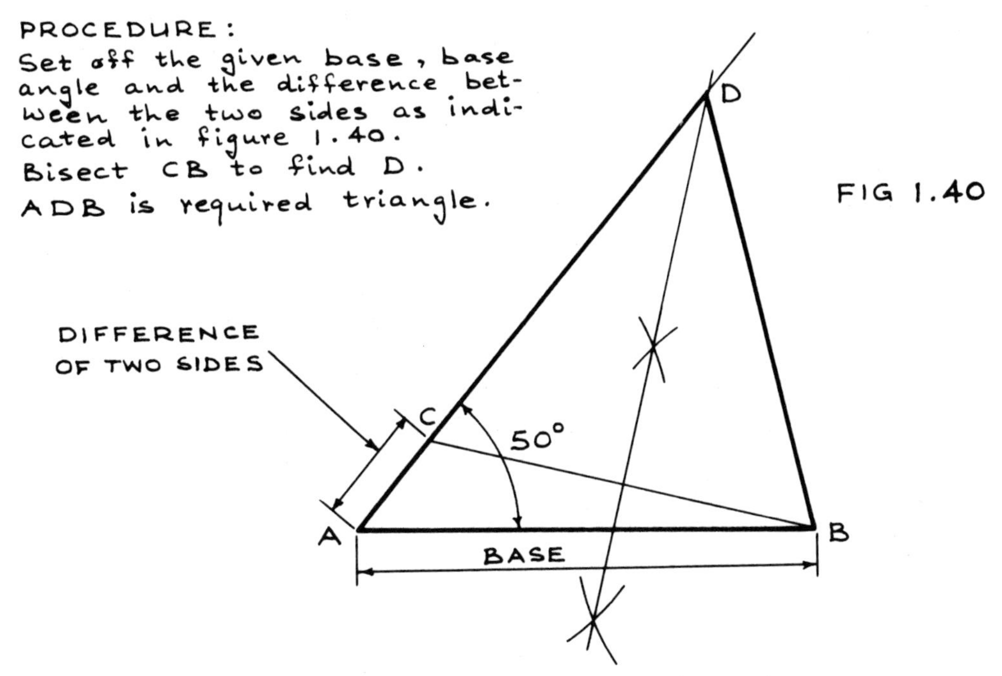

NOTE: Exercises 10 and 11 are also based on the properties of the isosceles triangle.

EXERCISE 11 SOLUTION

PROCEDURE:
Set off the given sizes as shown in figure 1.41.
Mark the altitude line at 50 mm from AB.
Make BD = AC and bisect CD to find E.
ACE is required triangle.

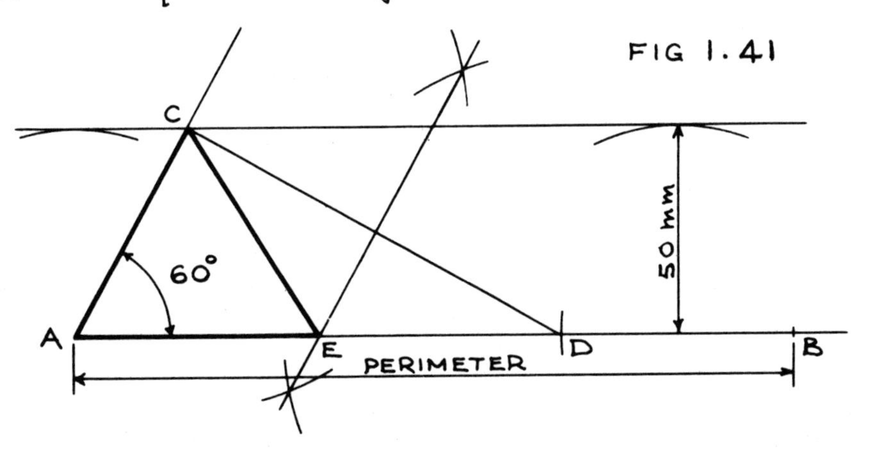

EXERCISE 12 SOLUTION

PROCEDURE:
Make angle BAC = 68° and angle CAD = 90°.
Bisect AB to obtain arc centre O and draw arc, radius OA.
Draw altitude line parallel to AB and at distance 56 mm.
ABE is one of two possible triangle answers.

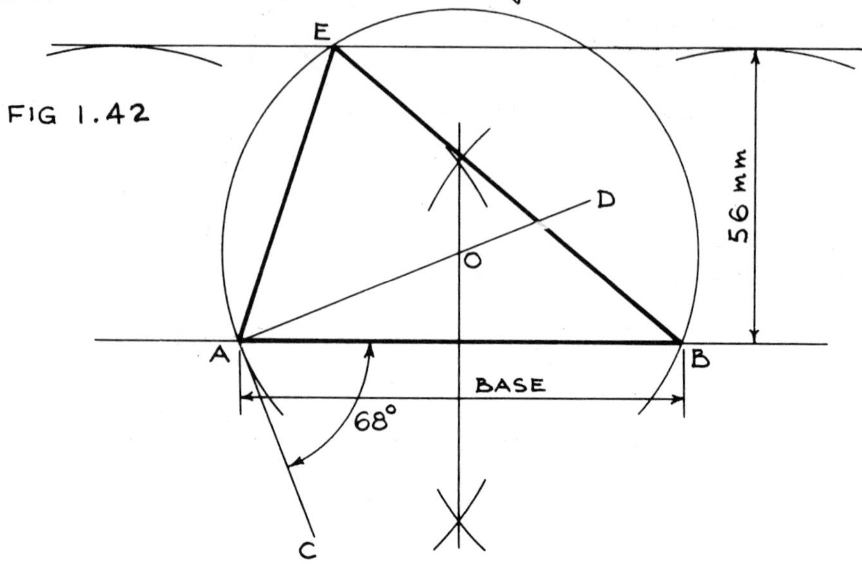

NOTE: This construction is based on the fact that the angle between a chord and a tangent is equal to the angle in the alternate segment. [Alternate segment theorem]

EXERCISE 13 SOLUTION

PROCEDURE:
Set off given perimeter AB and make angles at A and B equal to the given base angles.
Bisect these angles.
Through E draw EF and EG parallel to CA and DB respectively.

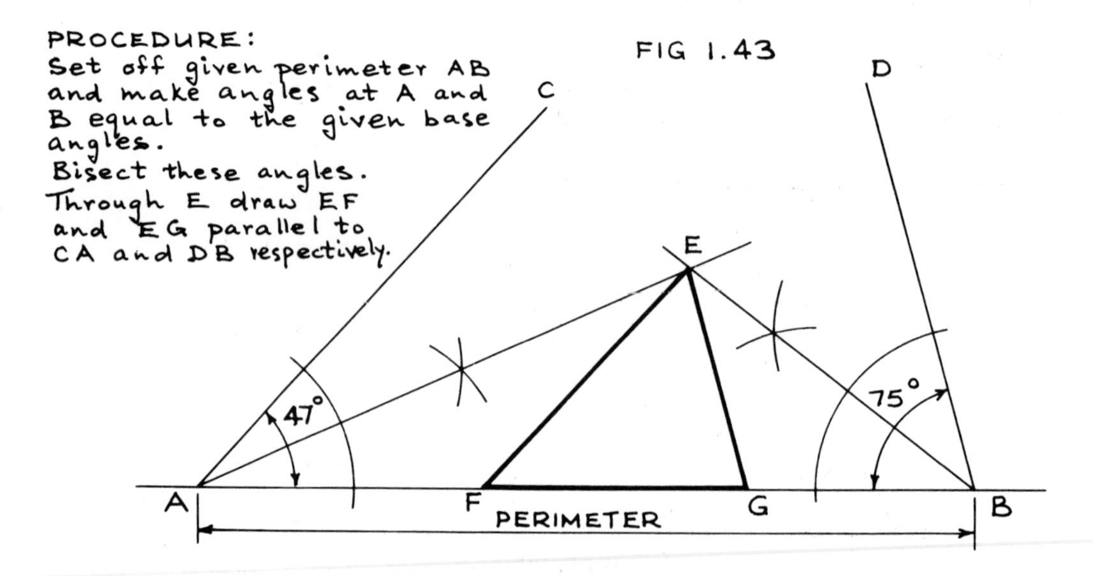

NOTE: This construction is based on the properties of the rhombus, that is, a parallelogram with equal sides.

EXERCISE 14 SOLUTION

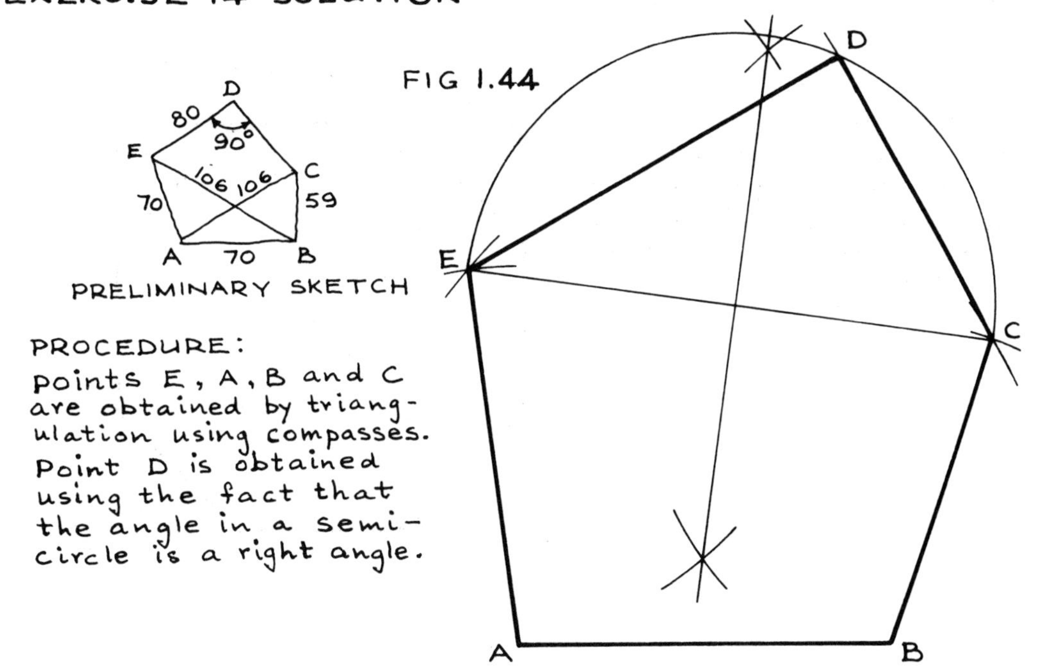

EXERCISE 15 SOLUTION

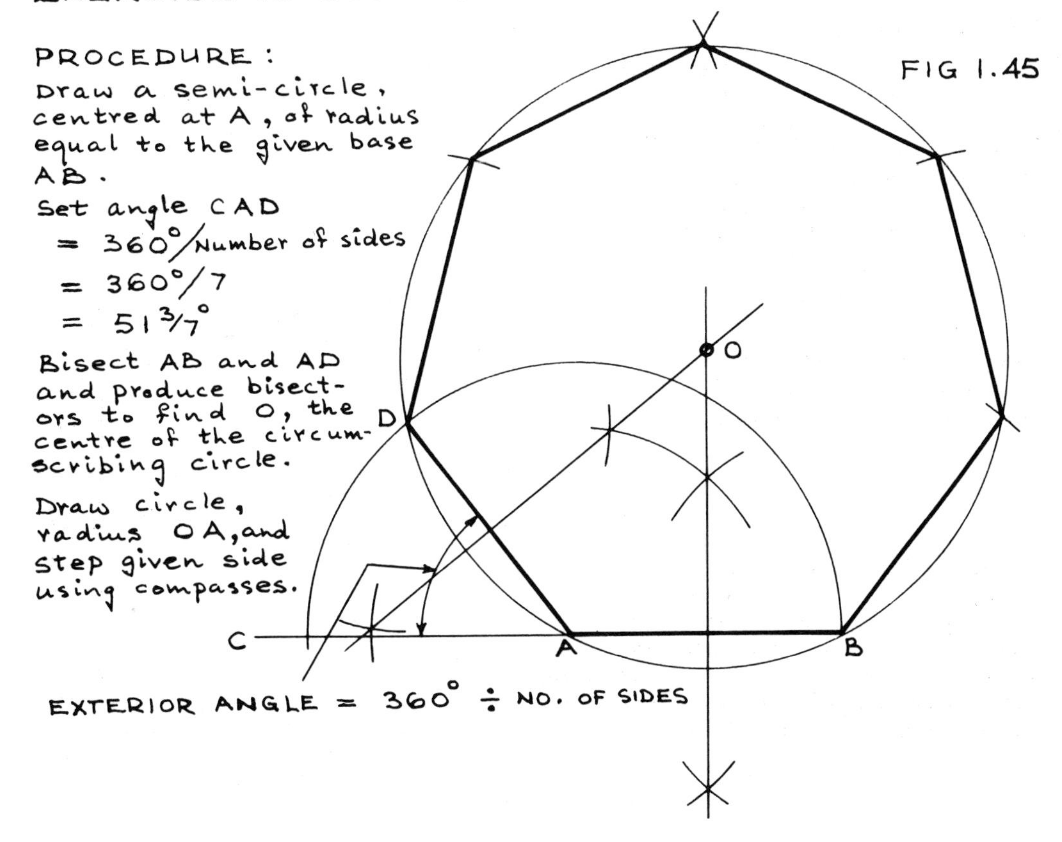

EXERCISE 16 SOLUTION

PROCEDURE:

First draw the diagonals of the given square.

Draw arcs of radius equal to half length of diagonals, centred at A, B, C and D.

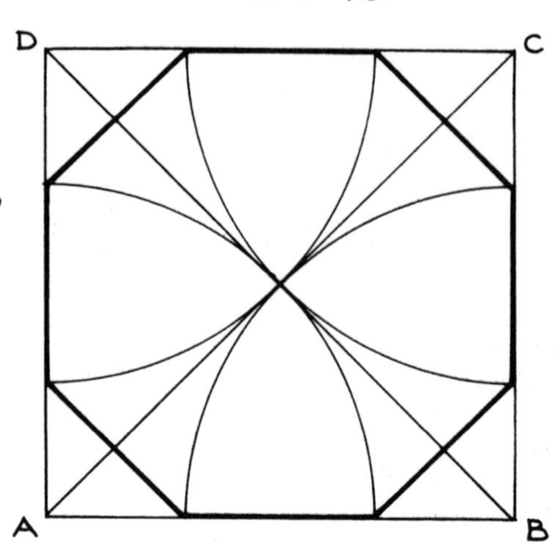

FIG 1.46

EXERCISE 17 SOLUTION

PROCEDURE:

Divide the diameter AB of the given circle into the same number of equal parts as sides required.

Obtain point C by intersection of arcs centred at A and B respectively.

Join C to point 2 and produce to cut circle in D.

AD gives the length of polygon side to step round the given circle.

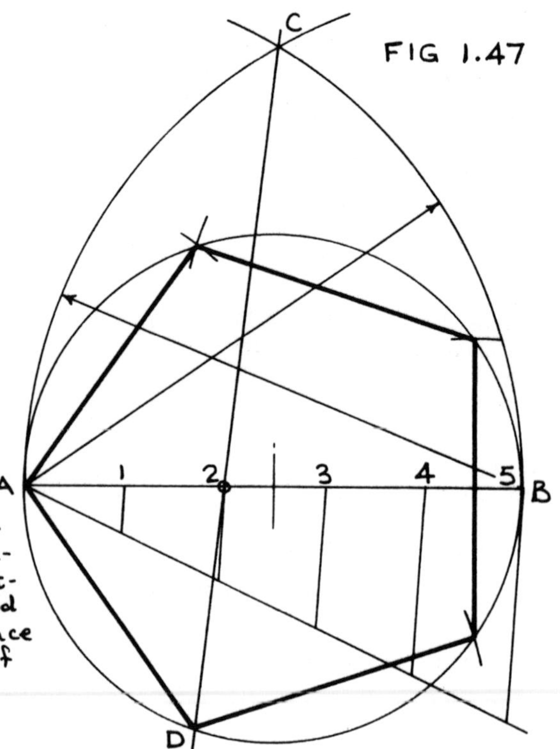

FIG 1.47

NOTE: This method is rather complicated and not particularly accurate. An alternative and much simpler method is to divide the circumference into the required number of parts using a protractor.

RADIAL PROJECTION AND AREAS EXERCISES

Exercise 18. By radial projection, to reduce the given hexagon ABCDEF to a similar figure having a base length AG.
Exercise 19. To enlarge the given figure ABCDEFGHIJK, by radial projection, so that length AE becomes AL.
Exercise 20. To reduce a given quadrilateral ABCD to a triangle having the same area.
Exercise 21. To reduce a given pentagon ABCDE to a triangle having the same area.
Exercise 22. To construct a rectangle having the same area as a given triangle ABC.
Exercise 23. To construct a square having the same area as a given rectangle ABCD.
Exercise 24. To construct a square having an area equal to the sum of the areas of two given squares.
Exercise 25. To divide a given triangle ABC into two equal areas by a straight line passing through a given point D in AC.
Exercise 26. To divide a quadrilateral ABCD into two equal areas by a straight line from corner D.
Exercise 27. To enlarge, or reduce, a given polygon ABCDEF so that the areas are in a given ratio.

FURTHER EXAMPLES

Example 1. Pin-point the given polygon ABCDEF, figure 1.60, on to your sheet and, by radial projection, enlarge it so that the new length of AB is 85 mm.
Example 2. Pin-point the figure 1.61 on to your sheet and reduce it, by radial projection, so that length AB becomes AB′.
Example 3. A field in the shape of a quadrilateral ABCD has the following measurements: AB = 114 m; AC = 142 m; BC = 43 m; AD = 105 m; DC = 83 m. Draw the field to a suitable scale and divide it into two equal areas by means of a straight line through the corner D.
Example 4. Pin-point the pentagon, figure 1.62, on to your paper and reduce it first of all to a triangle of equal area, then to a rectangle and finally to a square. Measure and state the length of the side of the square.
Example 5. Pin-point figure 1.62 on to your paper and reduce it to a similar polygon having half the area of ABCDE. (Ratio of areas 1:2 in this case.).

EXERCISE 18 SOLUTION

PROCEDURE:
Draw radial lines from A to each corner of the given figure.
Through G draw GH parallel to BC, H I parallel to CD, etc.

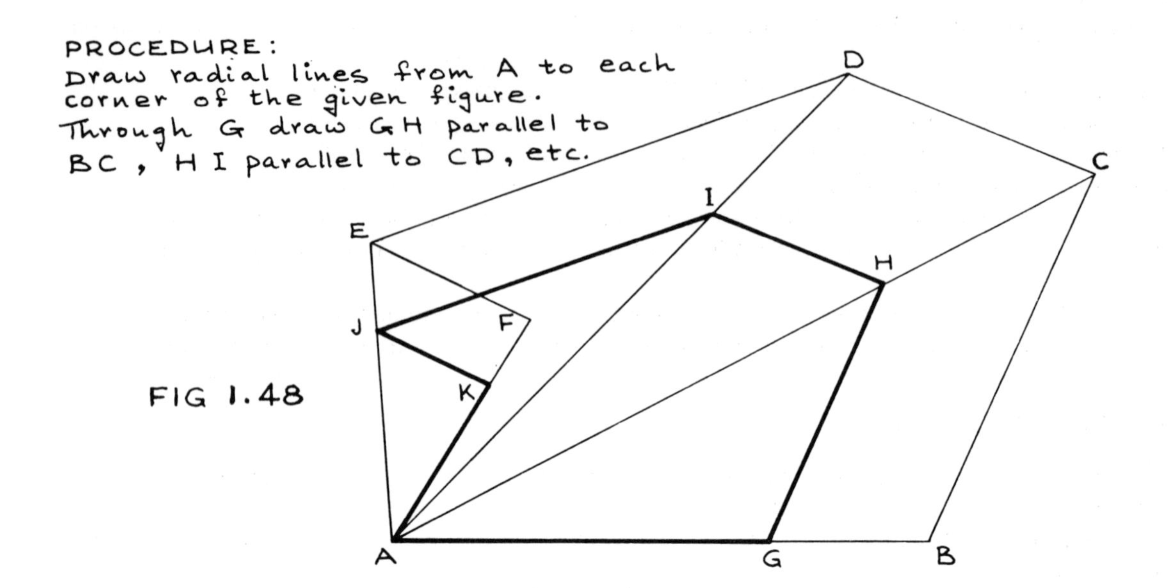

FIG 1.48

EXERCISE 19 SOLUTION

PROCEDURE:
Draw radial lines from corner A to each corner of the given figure and produce each line.
Commence at given point L and draw parallels to meet the appropriate radial line, as in the previous exercise.
Notice the extra piece of construction necessary to locate the new position of vertex C.

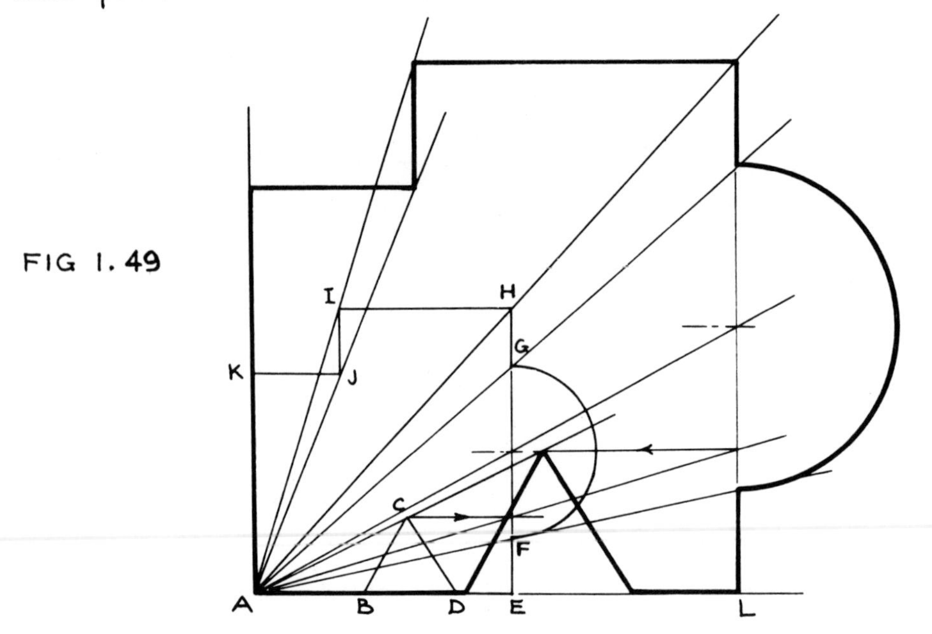

FIG 1.49

EXERCISE 20 SOLUTION

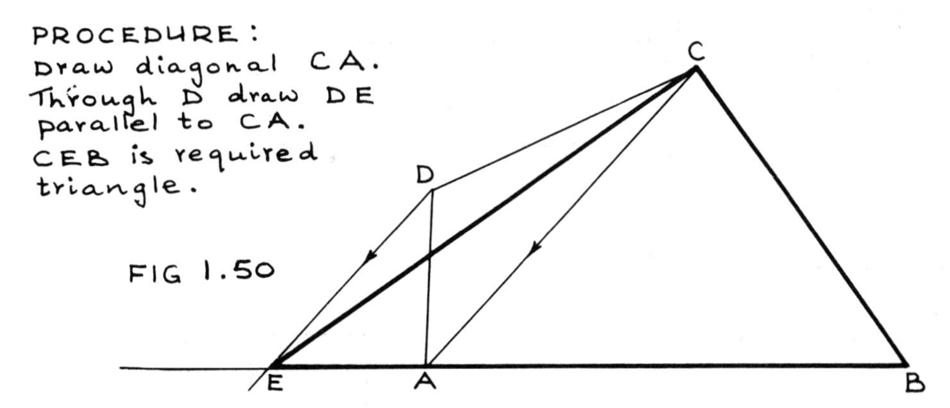

NOTE: Area ABC is common to both figures.
Triangles AEC and ADC are equal in area.

EXERCISE 21 SOLUTION

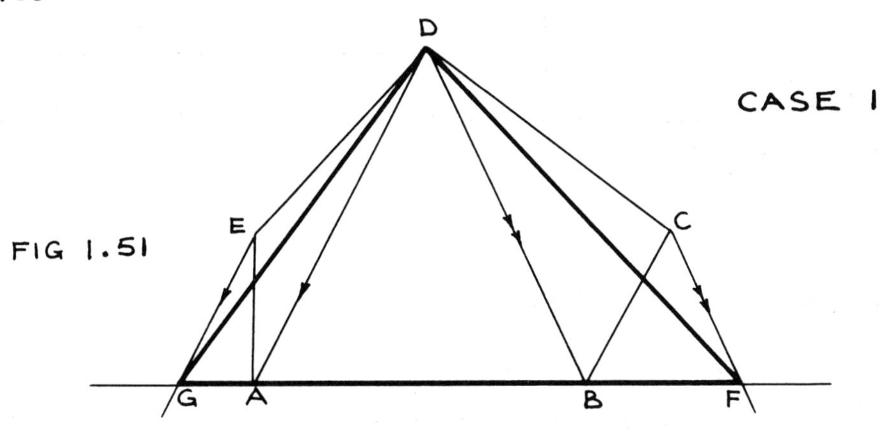

PROCEDURE:
Draw diagonal DB and through C draw CF parallel to DB.
Draw diagonal DA and through E draw EG parallel to DA.
DGF is the required triangle.

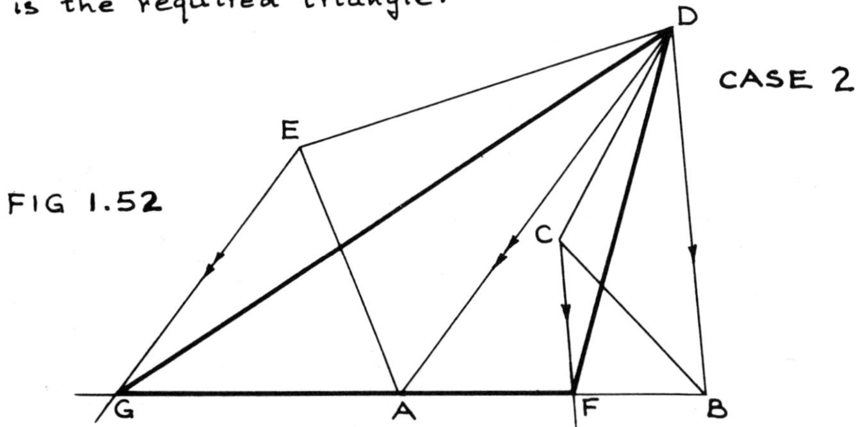

EXERCISE 22 SOLUTION

PROCEDURE:
Bisect BC at D.
Through D draw EF parallel to BA.
BE and AF are at right angles to AB.

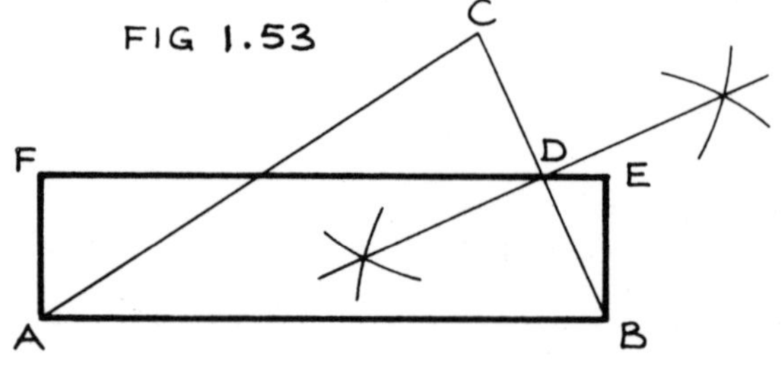

FIG 1.53

EXERCISE 23 SOLUTION

PROCEDURE:
Produce AB and mark off BE = BC.
Bisect AE and draw a semi-circle.
Produce BC to find F.
BGHF is the required square.

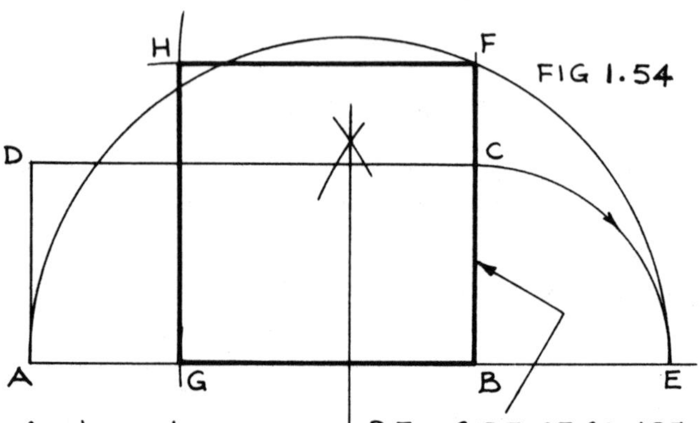

FIG 1.54

NOTE: This construction is based on the property of intersecting chords of a circle.

EXERCISE 24 SOLUTION

PROCEDURE:
Construct a right-angled triangle ABC with sides containing the right-angle equal to the sides of the given squares.
Construct square ACDE on hypotenuse of triangle ABC.

FIG 1.55

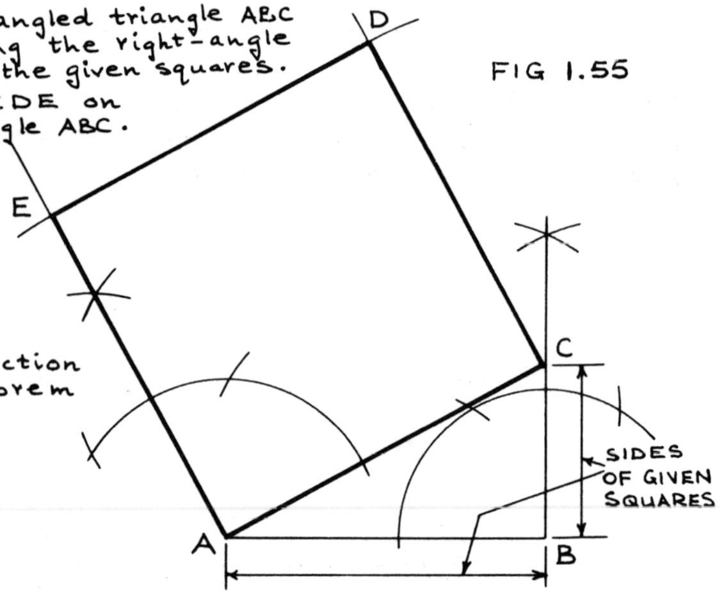

NOTE: This construction is based on the theorem of Pythagoras.

SIDES OF GIVEN SQUARES

EXERCISE 25 SOLUTION

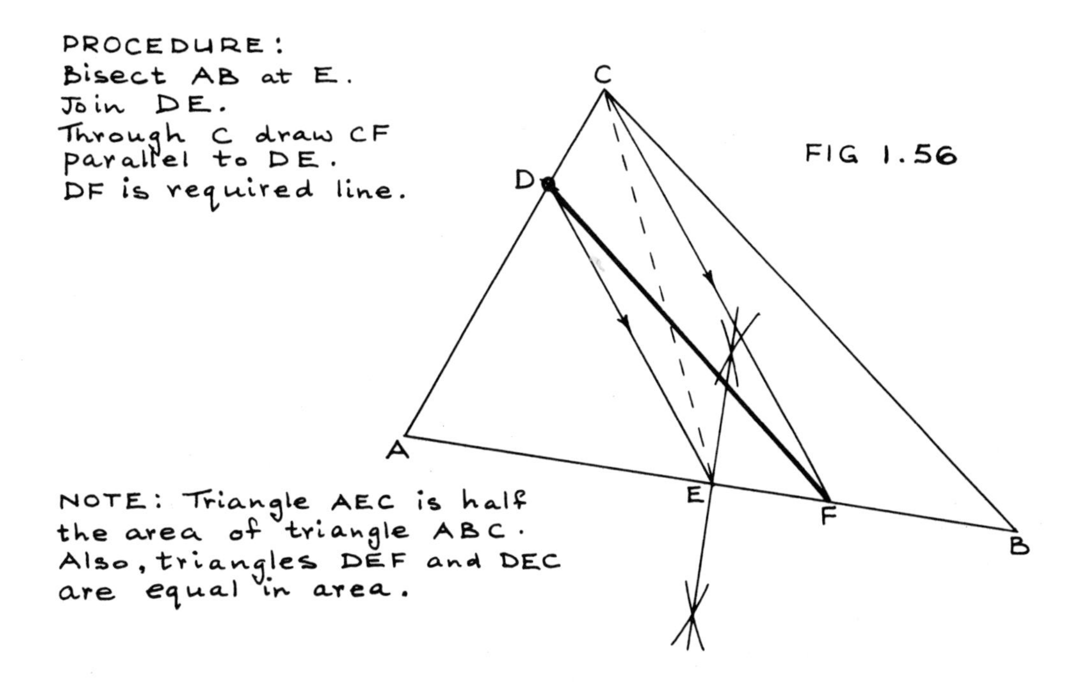

EXERCISE 26 SOLUTION

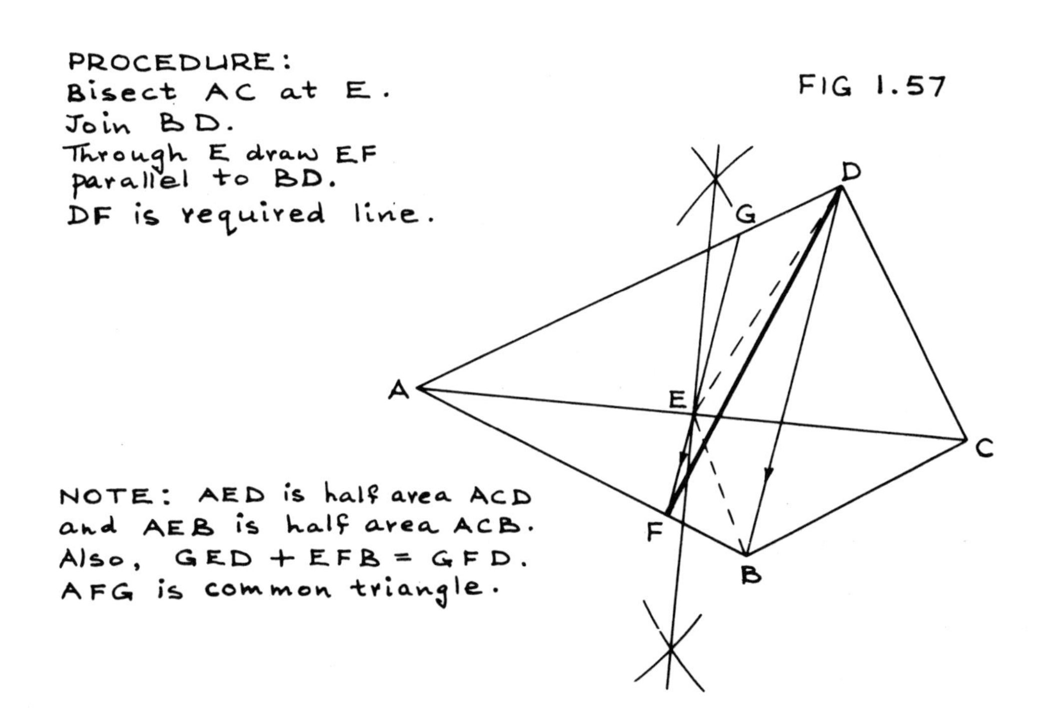

EXERCISE 27 SOLUTION

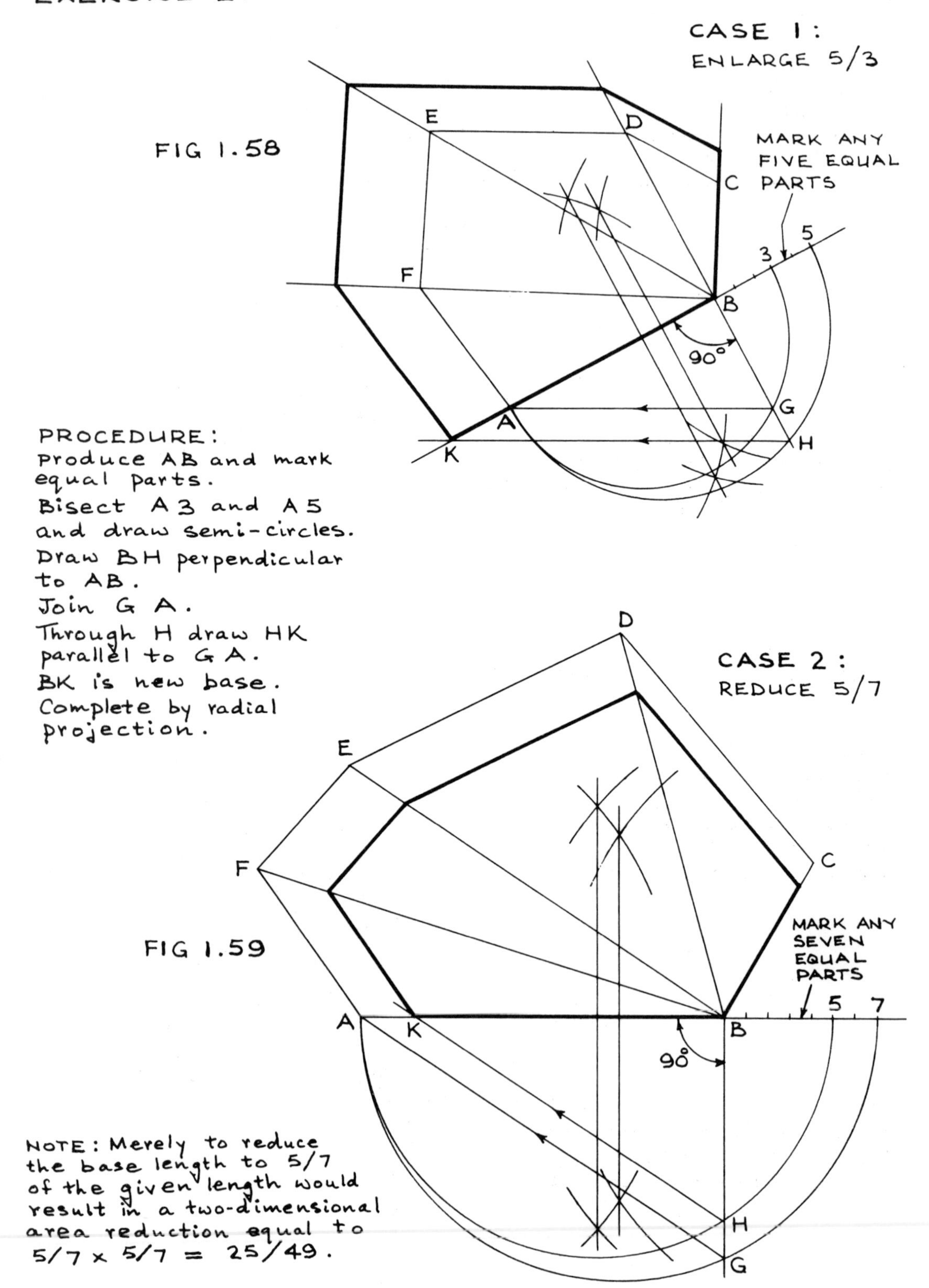

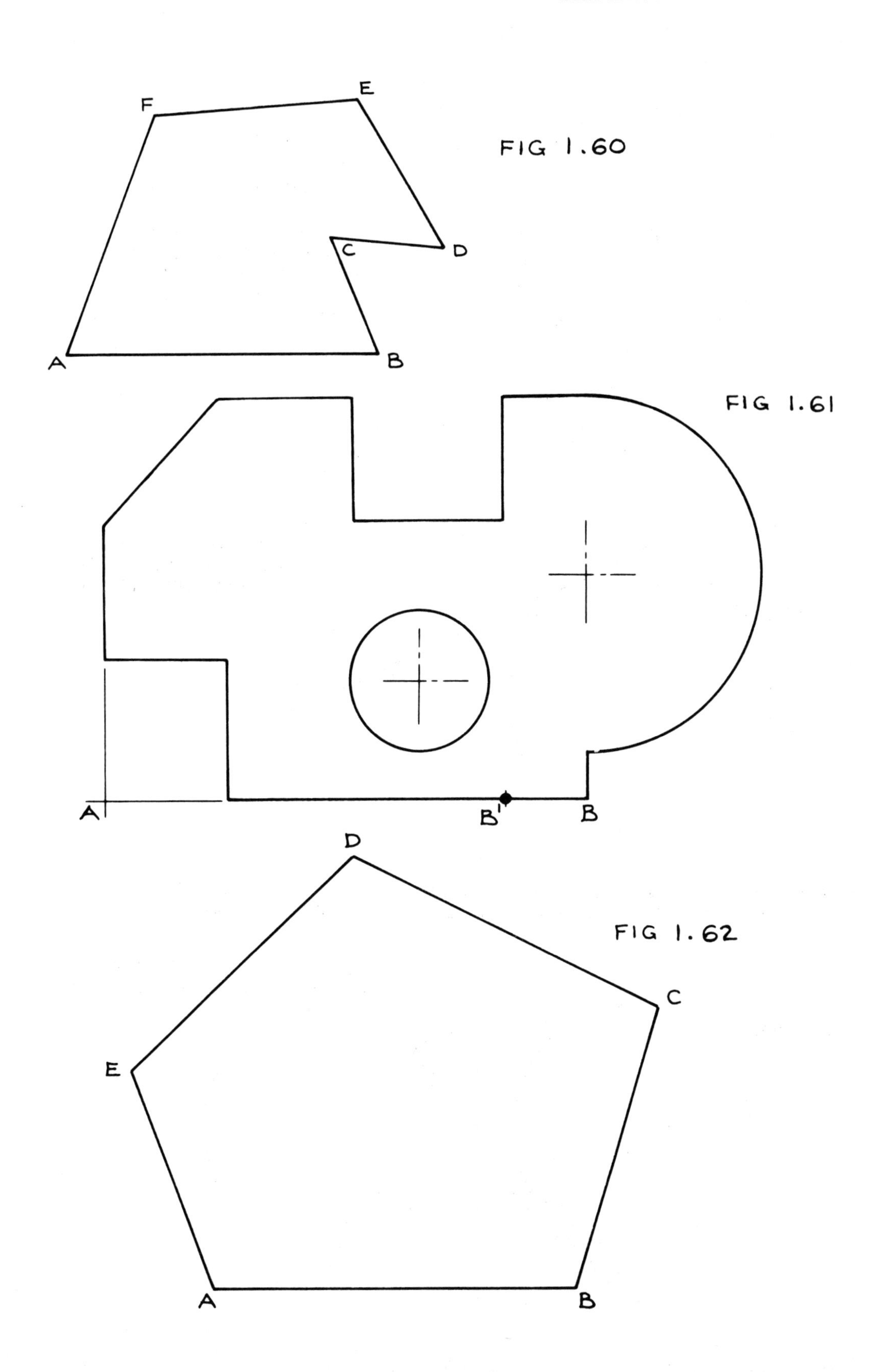
F
E
FIG 1.60
C
D
A
B
FIG 1.61
A
B'
B
D
FIG 1.62
C
E
A
B

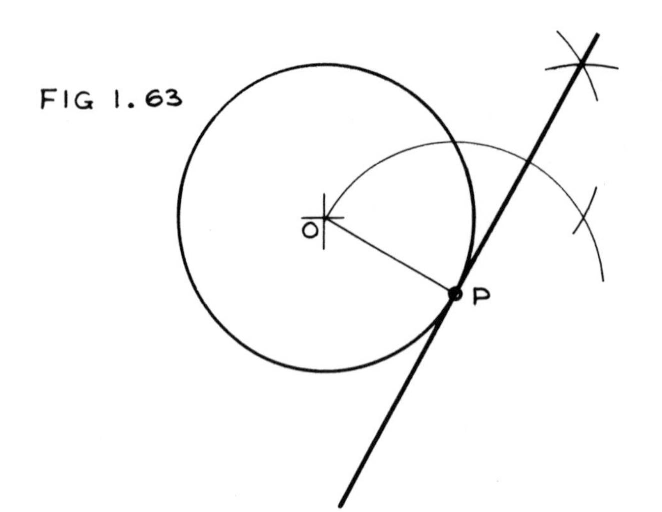

A TANGENT IS A STRAIGHT LINE TO TOUCH A CIRCLE

A TANGENT IS AT RIGHT-ANGLES TO THE RADIUS DRAWN TO THE POINT OF CONTACT

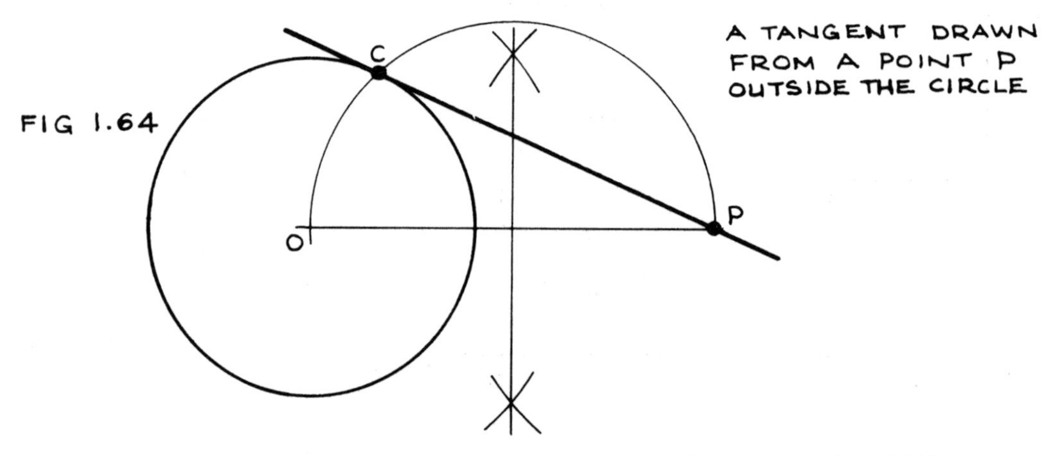

A TANGENT DRAWN FROM A POINT P OUTSIDE THE CIRCLE

NOTE: THESE CONSTRUCTIONS USE THE FACT THAT THE ANGLE IN A SEMI-CIRCLE IS A RIGHT-ANGLE

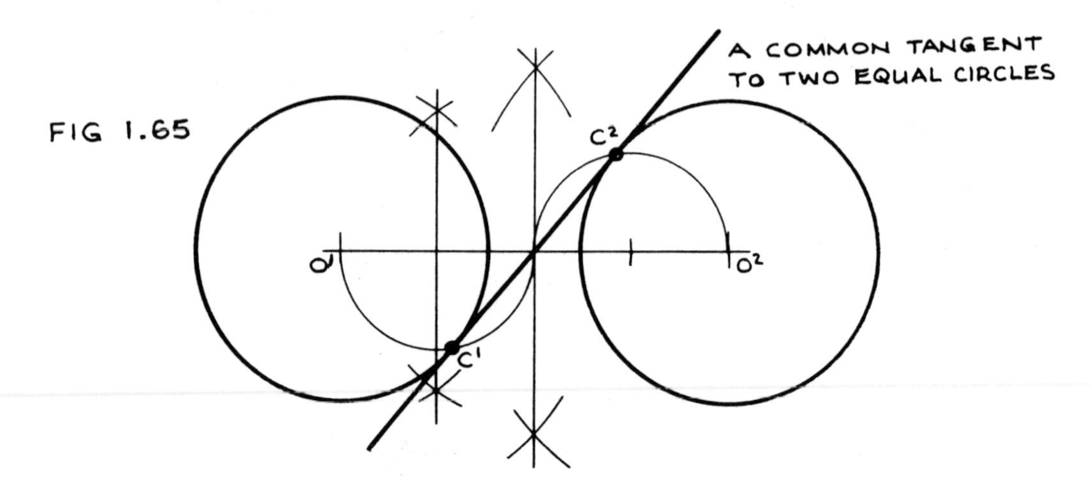

A COMMON TANGENT TO TWO EQUAL CIRCLES

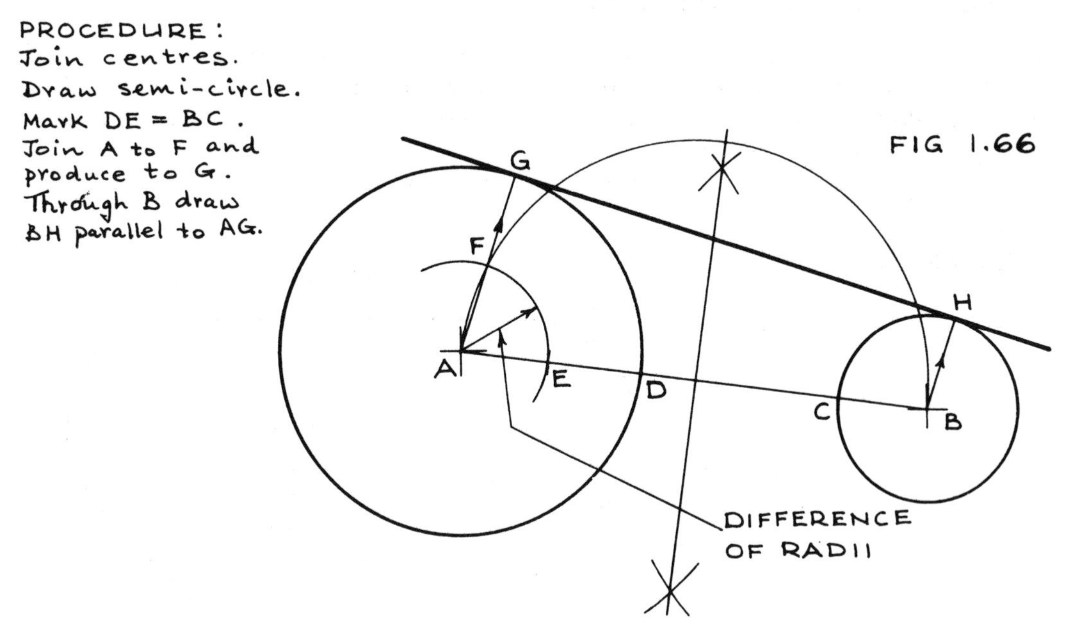

COMMON TANGENTS TO UNEQUAL CIRCLES

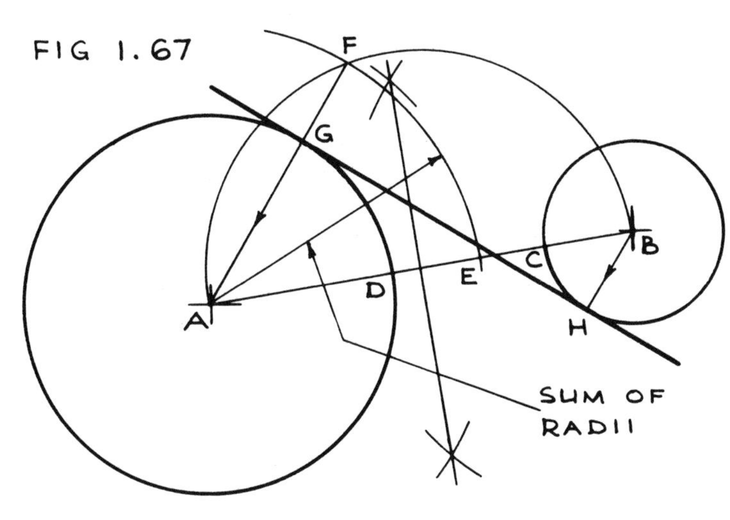

PROCEDURE:
Join centres.
Draw semi-circle.
Mark DE = BC.
Join FA.
Through B draw BH parallel to FA.

SPECIAL NOTE ON TANGENTS TO CIRCLES

The construction shown in figure 1.66 may not be particularly accurate. This is so, for example, when the circles are nearly equal, in which case AF would be a short line and its directional accuracy extremely doubtful.

This highlights the relevance of tangent construction in general. Since a circle is a curve which can be drawn with extreme accuracy it is only necessary to use a good straight-edge to obtain an accurate tangent. A draughtsman would never resort to such constructions as are shown in figures 1.63 to 1.67, but these are included in this work to satisfy syllabus demands.

CIRCLES AND TANGENCY EXERCISES

Exercise 28. To draw a succession of circles to touch two converging lines and one another.
Exercise 29. To draw a circle to touch two converging lines and to pass through a given point P.
Exercise 30. To draw a circle of given radius, 60 mm, to touch two given circles, 20 and 15 mm radius respectively, having centres 50 mm apart.
Exercise 31. To draw a circle to touch a given line AB at a given point P and also to touch a given circle.

FURTHER EXAMPLES

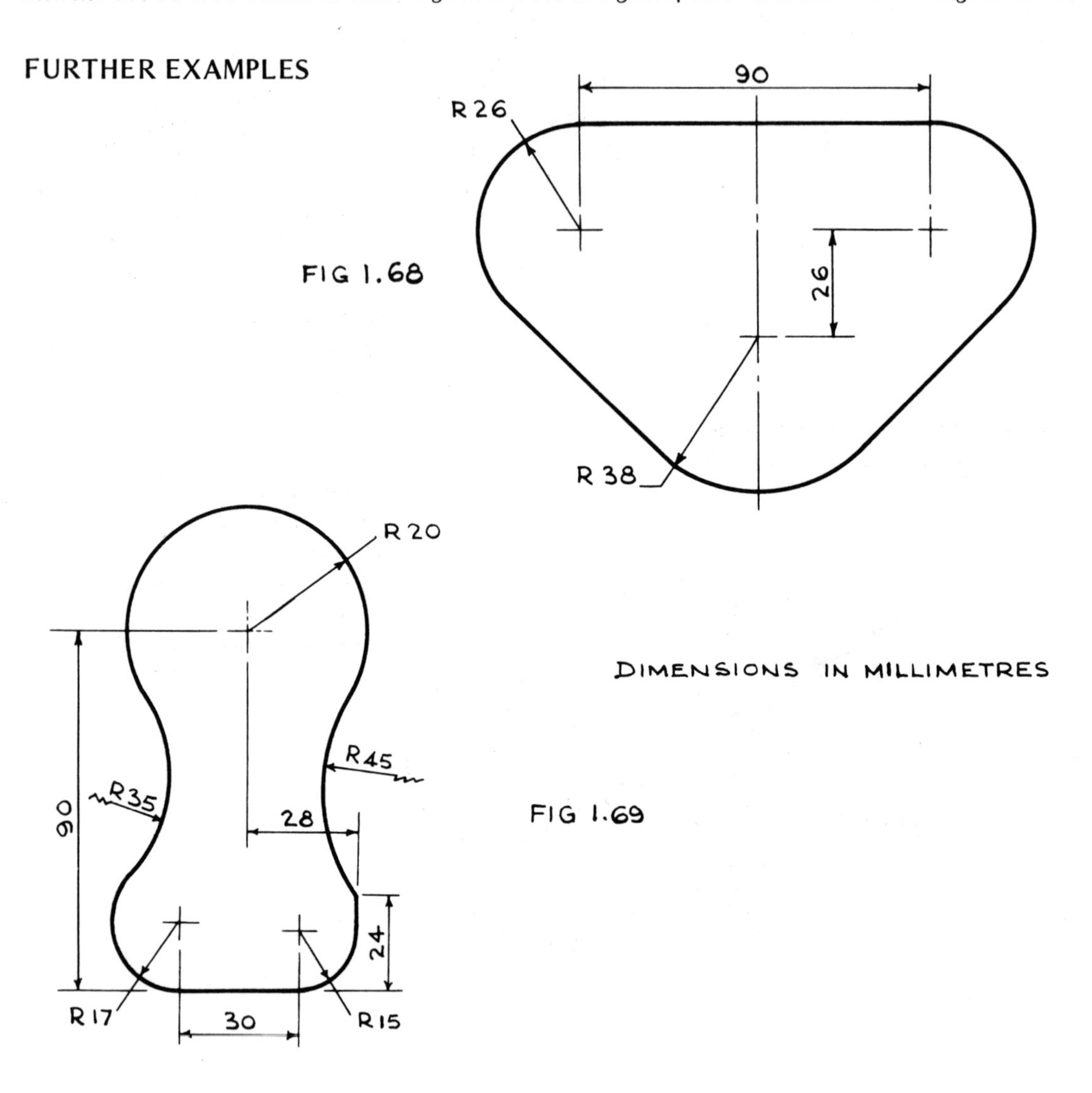

Example 1. Make a full-size drawing of the template shown in figure 1.68.
Example 2. Make a twice full-size drawing of the template shown in figure 1.69.
Example 3. Draw a parallelogram of 95 and 100 mm sides with contained angle 130°. Radius the obtuse-angled corners 25 mm and the acute-angled corners 12 mm.
Example 4. A circle 50 mm diameter is centred 50 mm from a straight line. Draw two circles to touch this circle and the straight line at a point P where P is 70 mm from the centre of the given circle. Measure and state the diameter of each circle.
Example 5. Two converging lines contain an angle of 30°. Draw any three circles to touch these lines and one another. How many such circles, in theory, could be drawn fulfilling these conditions?

EXERCISE 28 SOLUTION

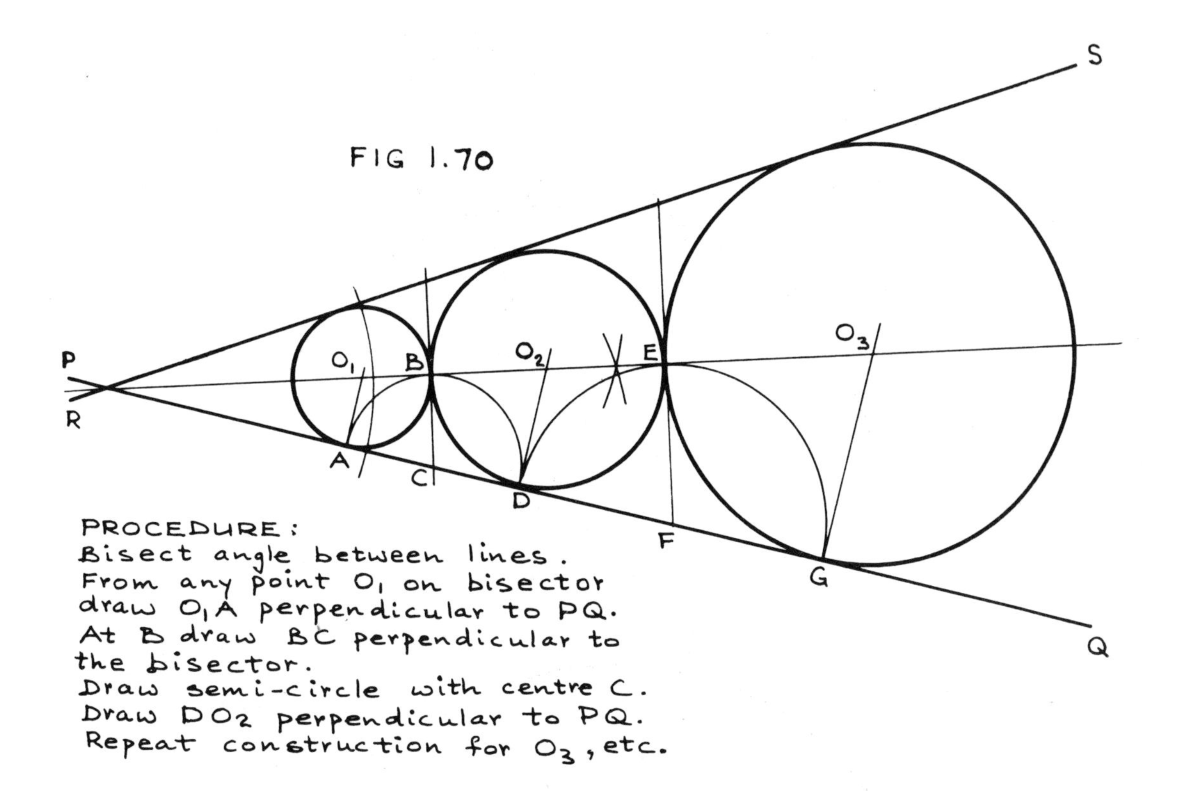

EXERCISE 29 SOLUTION

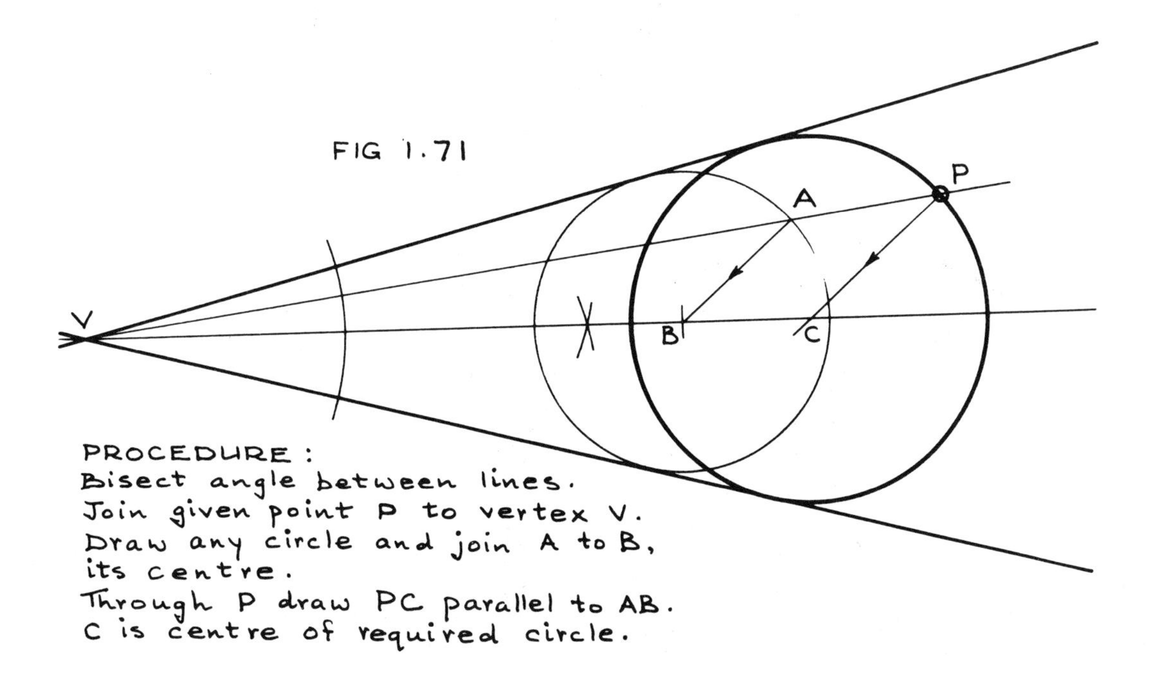

EXERCISE 30 SOLUTION

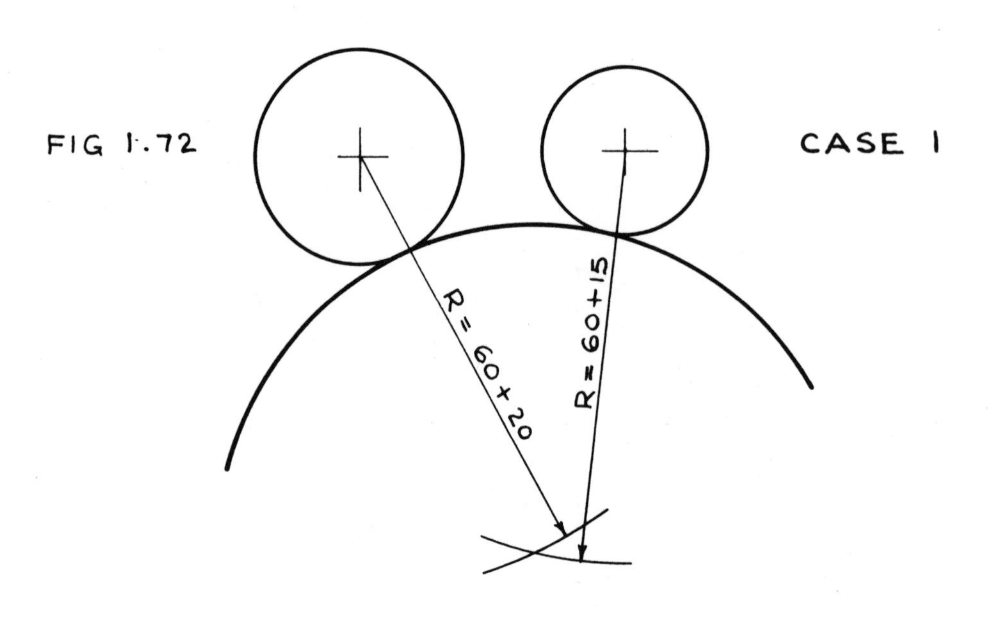

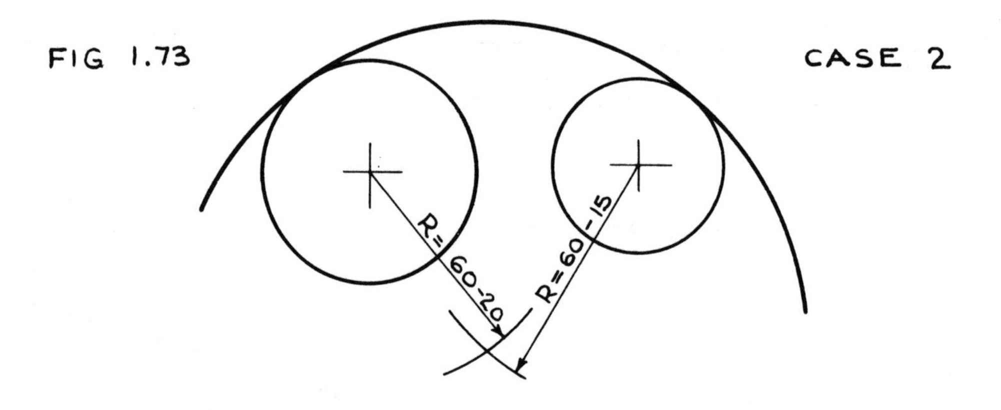

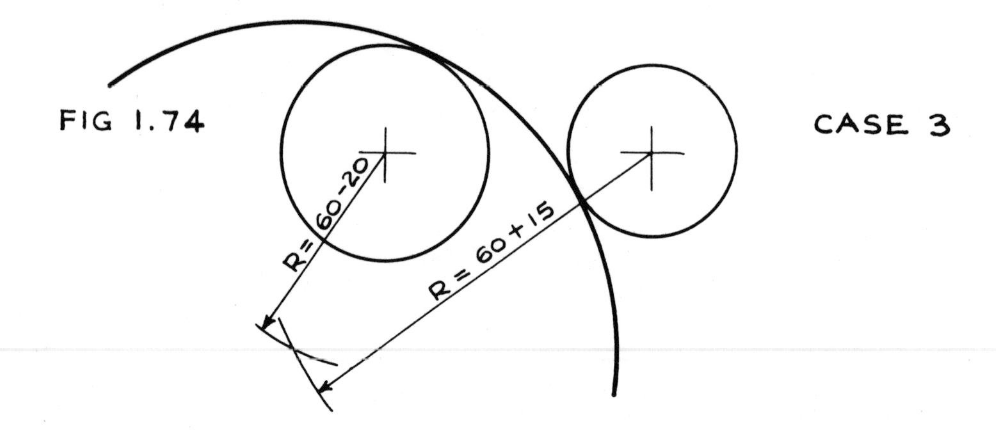

EXERCISE 31 SOLUTION

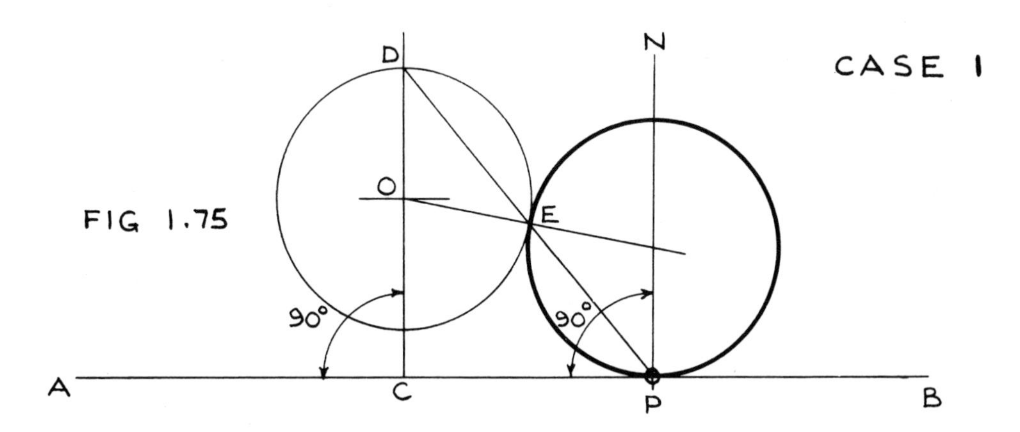

PROCEDURE:
Draw OC and PN perpendicular to AB.
Join DP and produce, if necessary.
Join OE and produce.

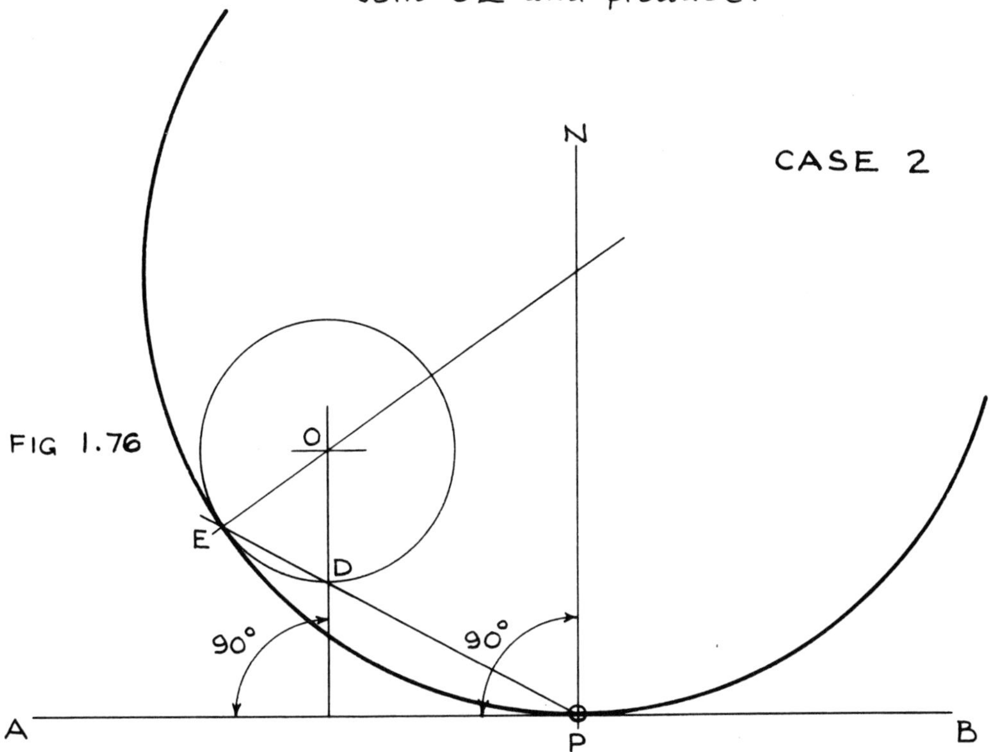

NOTE: The solution to problems of this type makes use of the following facts:

(i) Circles which touch have their contact point on the line joining their centres, produced if necessary;

(ii) The centre of a circle lies on the line at right-angles to a tangent.

2

ORTHOGRAPHIC PROJECTION

This is the projection system used by Engineers and Architects to define the shape of an object by means of a drawing.

There are two *principal planes* envisaged in this system of projection. One is the *vertical plane*, usually referred to simply as the *VP* and the other, a *horizontal plane* perpendicular to this *VP*, is usually referred to as the *HP*.

It is unfortunate that two slightly different types of projection have evolved under orthographic projection. There is the First Angle System, sometimes called the English or European projection, which would appear to be losing favour to the American system known as Third Angle projection.

Prior to the Second World War the First Angle System was used very largely in the United Kingdom but during the war, with interchange of drawings and technical information with the United States and Canada, more use began to be made of the Third Angle System. Neither system has any significant advantage over the other. The reasons for using one particular system in any organisation is purely a matter of tradition and is largely dependent on the firm's location or business interests – just as the question of which side of the road to drive a vehicle on depends upon the traffic legislation of the country in question.

Both First and Third Angle systems of orthographic projection will be explained and every effort will be made to encourage the student to use either system with confidence.

FIRST ANGLE PROJECTION

Figure 2.1 shows a pictorial isometric view of the principal planes intersecting to form four quadrants. An object is shown positioned in the first quadrant. By viewing this object in a *horizontal* direction, perpendicular to the *VP*, the view ABCDEFGH would be obtained. This view is called the *front view* of the object and is almost invariably the *most important view* of any object.

The reason why the Front View is the most important view is that other views depend upon it. The manner in which the Front View of an object is obtained *determines the remaining views in the system of projection being used.*

If the object shown in figure 2.1 is now viewed *vertically downwards*, perpendicular to the *HP*, the view IJKLMNOP would be obtained. This view is called a *plan view or top view* of the object.

The idea of a third plane, called an *auxiliary vertical plane,* is employed in order to obtain another elevation. This additional elevation is called a *side view* and is shown in figure 2.2. QRSTWX, shown projected on to the auxiliary *VP* is a *side view.*

The student should bear in mind that *all* elevations, that is, Front Views or Side Views are obtained as a result of viewing the object in a *horizontal* direction i.e. viewing *level*.

If one imagines that these planes are hinged as shown in figure 2.2 then when the planes are opened out on to a horizontal table as shown in figure 2.3 the relationship between the three views is clearly demonstrated. The student should carefully note this relationship between the views. For example, the manner in which the Tee square is used to project from the Front View to the Side View and how the 90° set-square is used to project from Front View to Plan View. Also notice the use of a 45° set-square to ensure that the plan depth of the object is equal to the Side View width.

Other views of an object may be obtained as demonstrated by figures 2.4 and 2.5. In some cases one view, properly dimensioned, may serve to describe a simple object but generally at least two views are required. However, it is not often that more than four views are necessary and the view in direction of

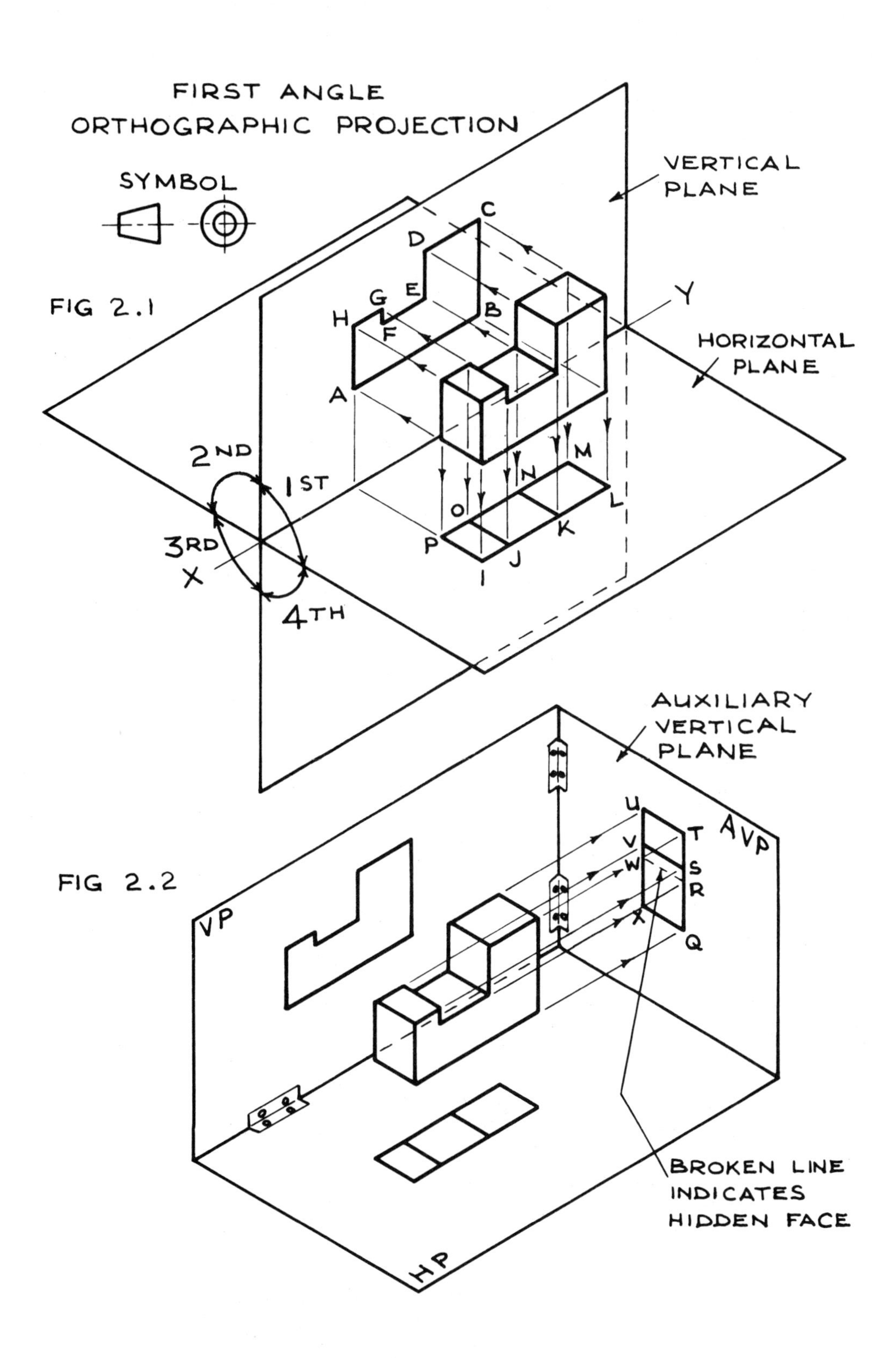
FIRST ANGLE
ORTHOGRAPHIC PROJECTION
SYMBOL
FIG 2.1
VERTICAL
PLANE
HORIZONTAL
PLANE
C
D
E
G
H
B
F
A
Y
M
N
L
O
K
P
J
I
2ND
1ST
3RD
X
4TH
FIG 2.2
AUXILIARY
VERTICAL
PLANE
AVP
U
T
V
W
S
R
X
Q
VP
HP
BROKEN LINE
INDICATES
HIDDEN FACE

arrow T, i.e. a view of the bottom of the object, is not often required. The student should carefully note that the Front View is the fundamental view on which all other views in the projection system depend. It is the 'hub' of the other views.

The basic rule for the First Angle System of orthographic projection may be summed up as follows.

Second and subsequent views are always drawn on the *opposite side* of the Front View to that in which the arrow showing the viewing direction is placed.

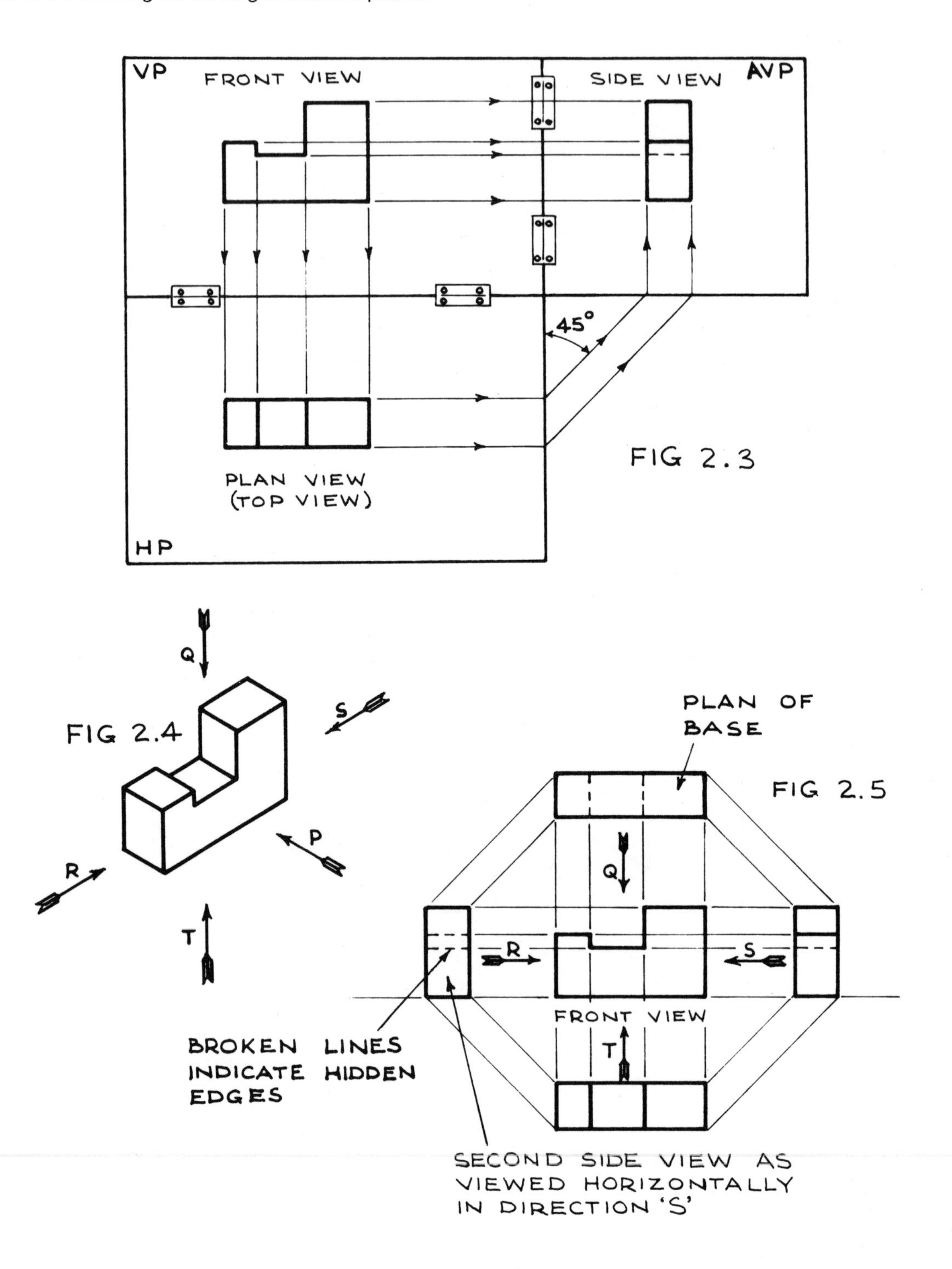

FIG 2.3

FIG 2.4

FIG 2.5

Note. Viewing arrows are not placed on drawings but are used here merely as an aid to understanding the principles. Also, projector lines are not shown so pronounced as in figure 2.3 and usually they are not shown at all.

The symbol for First Angle projection is a small truncated cone, shown in Front and Side Views as in figure 2.1. Suitable dimensions for this symbol are 10 mm and 5 mm for Side View diameters and 10 mm for the length of Front View. Thin chain centre lines should be indicated in each view.

THIRD ANGLE PROJECTION

To understand the philosophy behind the Third Angle System a little more imagination is required on the part of the student. The object, placed in the third quadrant, is imagined to be viewed through *transparent* principal planes and a transparent Auxiliary VP, as in figure 2.6. Thus the Plan would be projected *above* the Front View, as shown in figure 2.7. Similarly the Side View in the direction of arrow S would be drawn on the right, i.e. on the *same side* of the Front View as the viewing arrow.

Once again, five views are possible with the Front View as the basic view, as shown in figure 2.8.

The basic rule for the Third Angle System of orthographic projection may be summed up as follows.

Second and subsequent views are always drawn on the *same side* of the Front View as that in which the arrow showing the viewing direction is placed.

Notice that the symbol for Third Angle projection, the truncated cone shown in figure 2.6, obeys the given rule.

Note. The student who has read and understood the foregoing explanation of First and Third Angle orthographic projection may well ask himself what all the fuss was about! Transparent planes, four quadrants – two of which are not used and never will be! The author recommends that the student merely bears in mind these basic rule summaries, which are simple, and that he should not worry too much about transparent planes, etc., and all that is involved in the quadrant system philosophy.

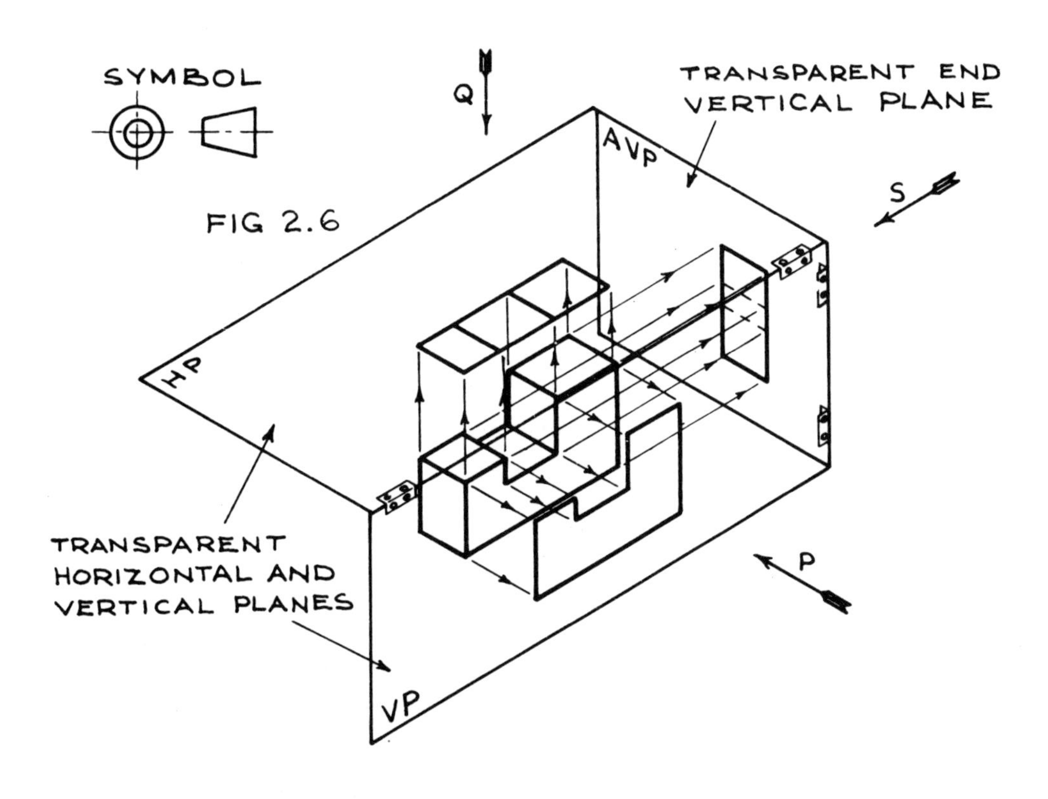

FIG 2.6

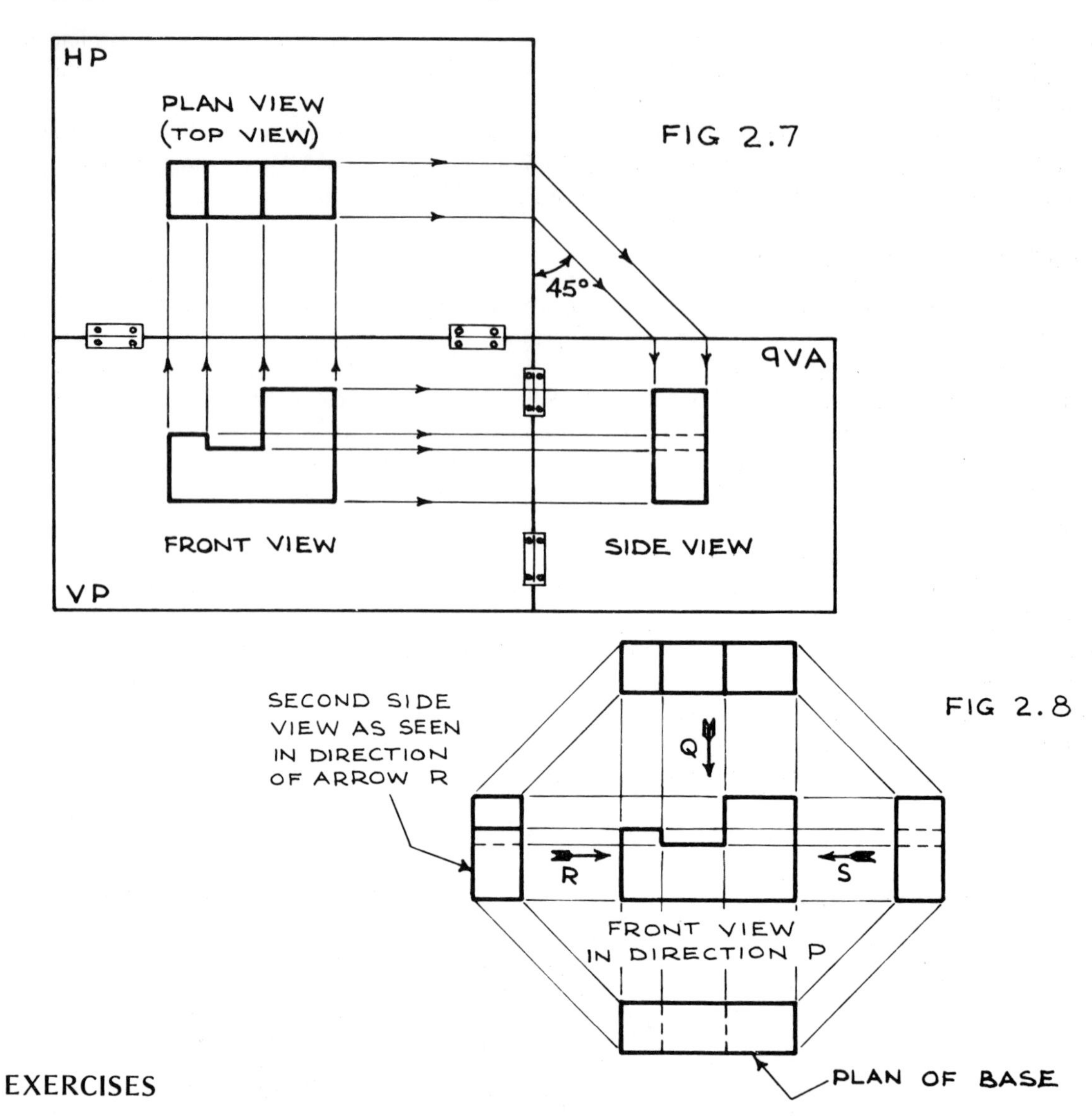

EXERCISES

Exercise 1. Figure 2.9 shows an isometric drawing of an object drawn on isometric graph paper, for which the triangle grid may be taken as 5 mm.

(a) Using First Angle orthographic projection draw the following views of the object:
 (i) A Front View as seen in the direction of the arrow P;
 (ii) A Side View as seen in the direction of the arrow Q;
 (iii) A Plan;
 (iv) A second Side View as seen in the direction of the arrow R.

(b) Draw similar views using Third Angle orthographic projection.

Note. Use plain drawing paper, A4 size, or 5 mm squares graph paper if available.

If graph paper is used the answers are revealed with greater clarity if finished in red, blue or green ball pen.

Solutions are given for *Exercise 3* as an example for the guidance of the student for the remainder of the *Exercises 1* to *6*.

Exercises 2, 3, 4, 5 and *6*.
Repeat the instructions as for *Exercise 1*, applied to figures 2.10, 2.11, 2.12, 2.13 and 2.14.

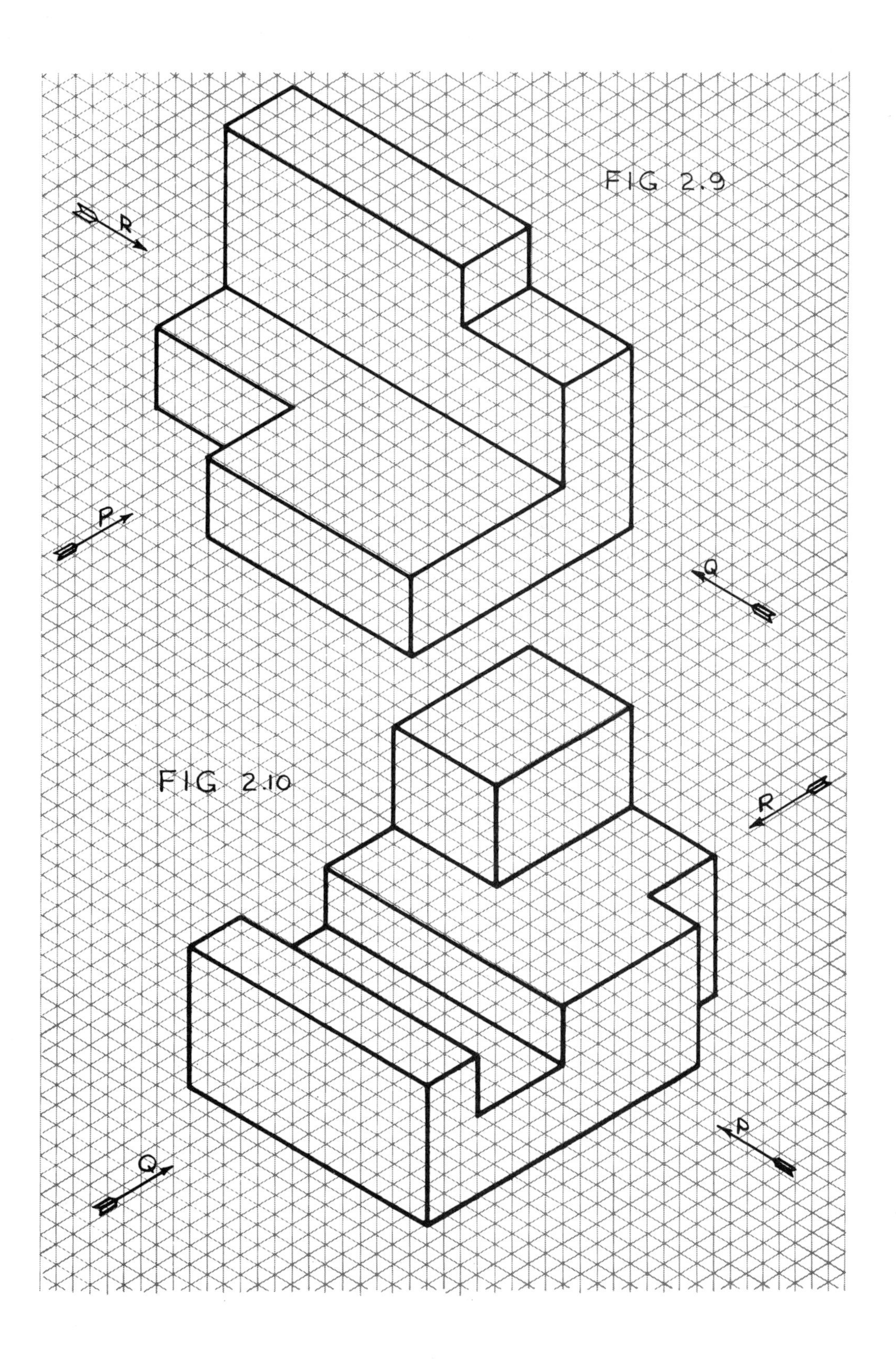
FIG 2.9
R
P
Q
FIG 2.10
R
P
Q

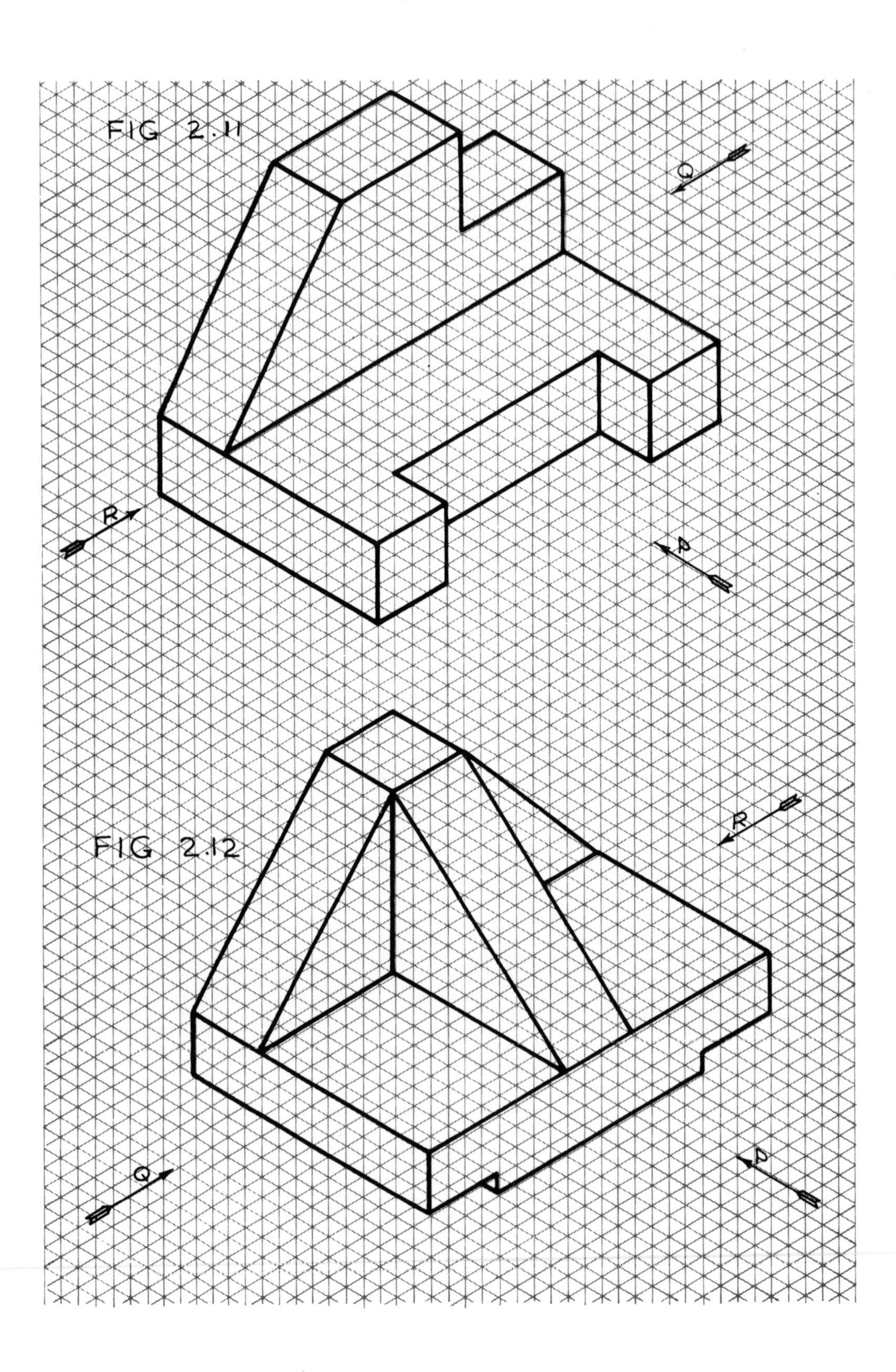
FIG 2.11
Q
R
P
FIG 2.12
R
Q
P

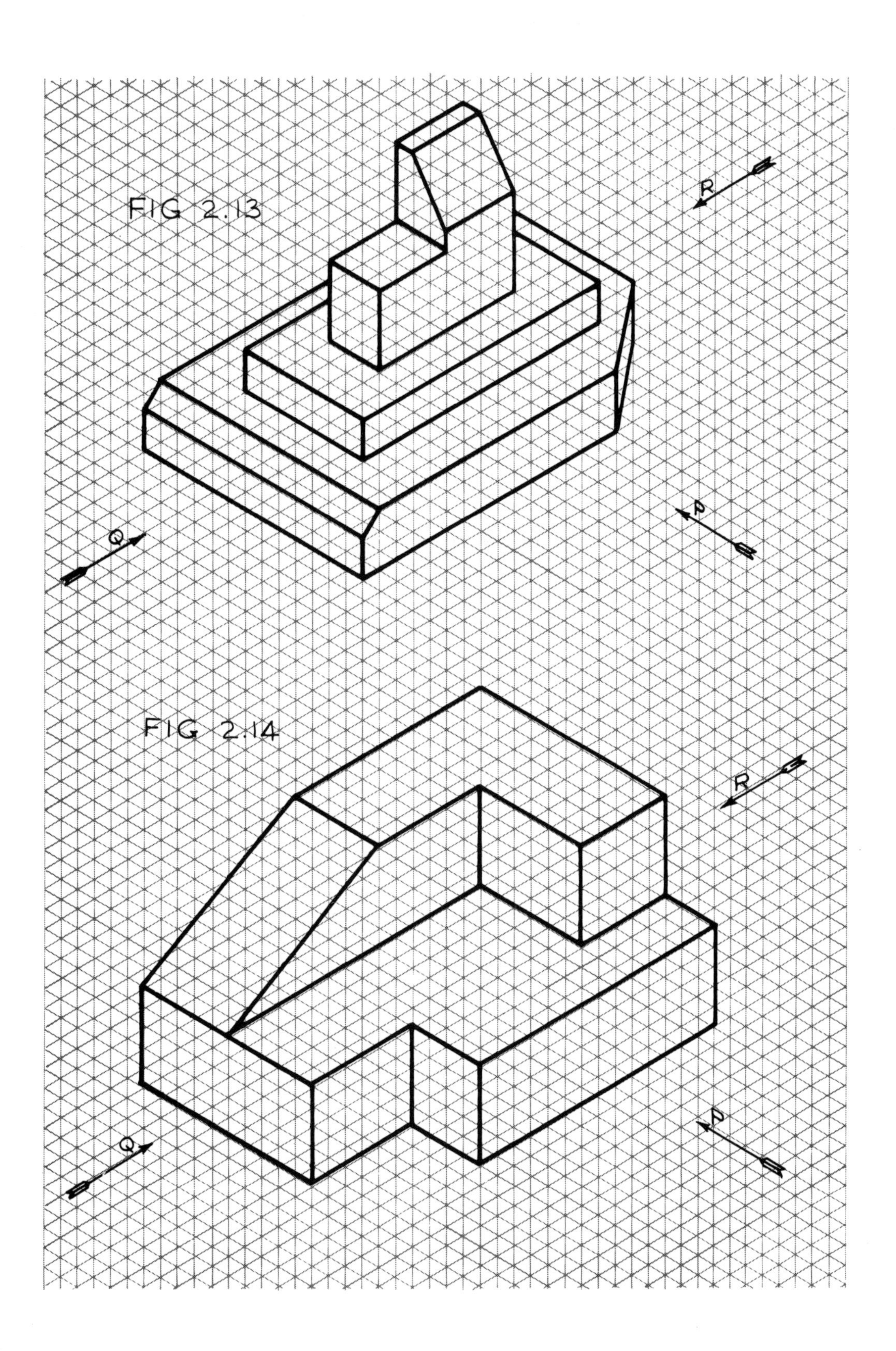
FIG 2.13
R
P
Q
FIG 2.14
R
P
Q

EXERCISE 3 SOLUTION

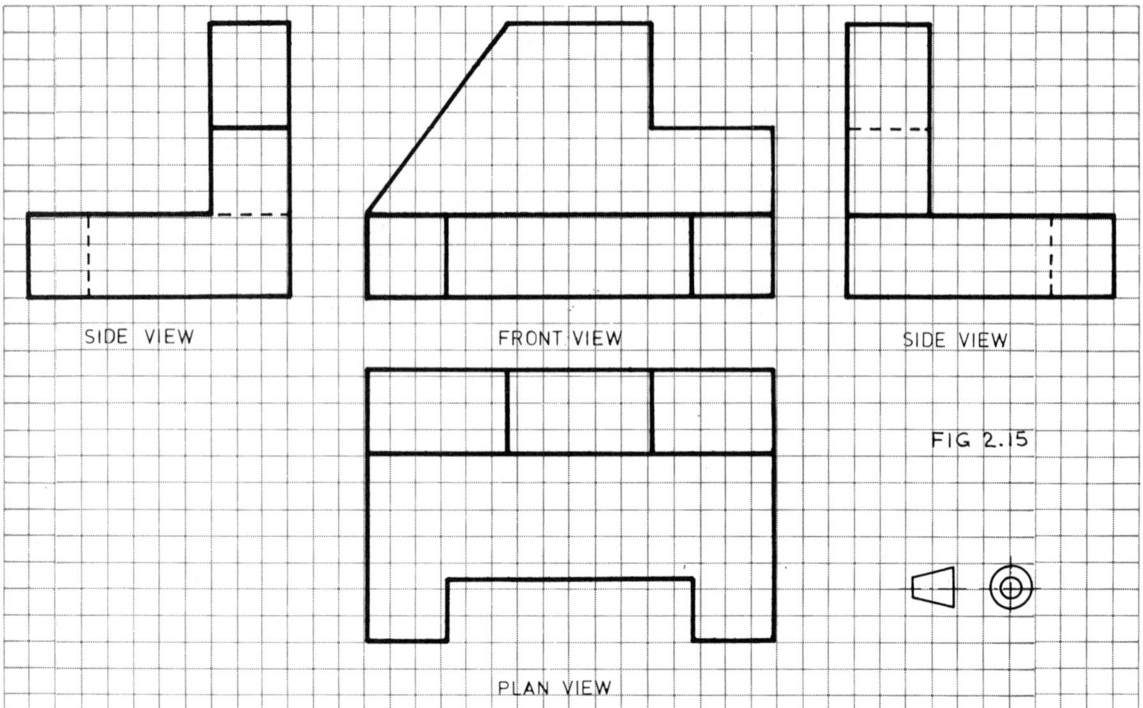

SIDE VIEW

FRONT VIEW

SIDE VIEW

FIG 2.15

PLAN VIEW

EXERCISE 3 SOLUTION

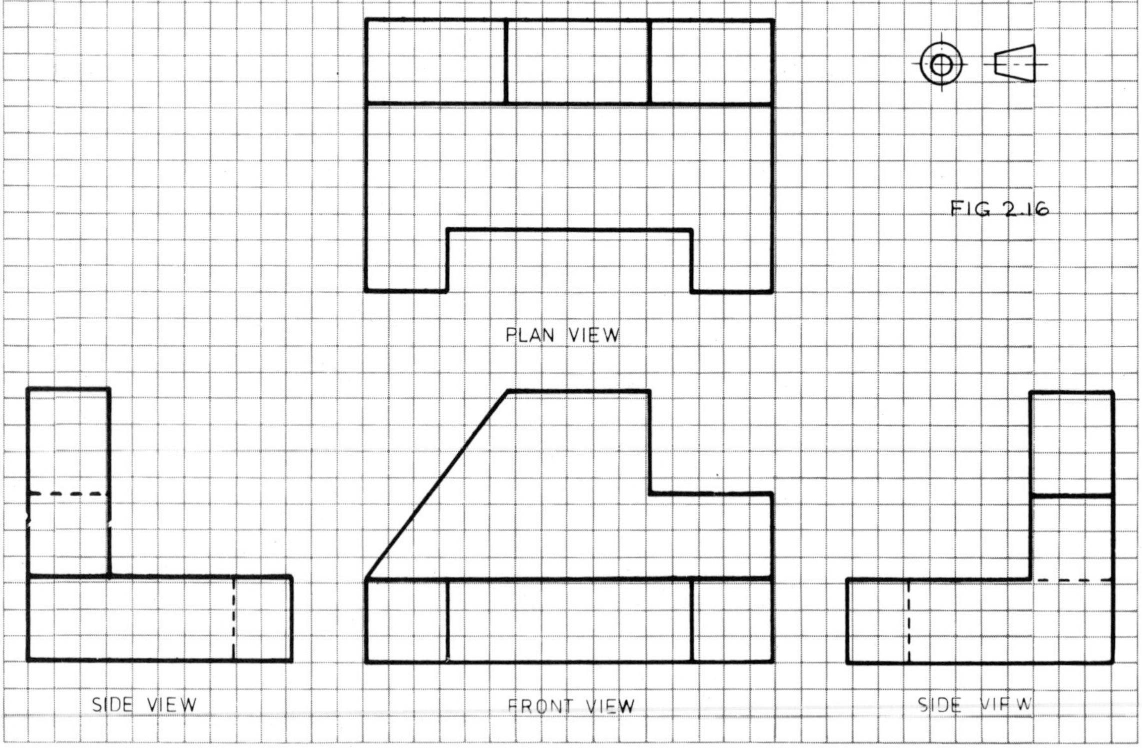

FIG 2.16

PLAN VIEW

SIDE VIEW

FRONT VIEW

SIDE VIEW

FILLETS

Generally, small fillet radii may be drawn in freehand, or with the aid of spring bows, or more usually with a radius aid. In engineering drawings it is not essential, and indeed it would be a waste of the draughtsman's time, to show a fillet radius accurately. The correct dimensioning is all important and the 'picture' just has to be reasonably correct. Occasionally a large fillet curve has to be shown and points for this may be plotted geometrically as described in chapter 5.

There may be occasions when the student is undecided about the drawing of a line where one curved surface blends with another as, for example, in figures 2.17 and 2.18. In fact, geometrically speaking, there is no definite line, but in engineering drawing office practice a line is shown, perhaps drawn less dense than other finished lines.

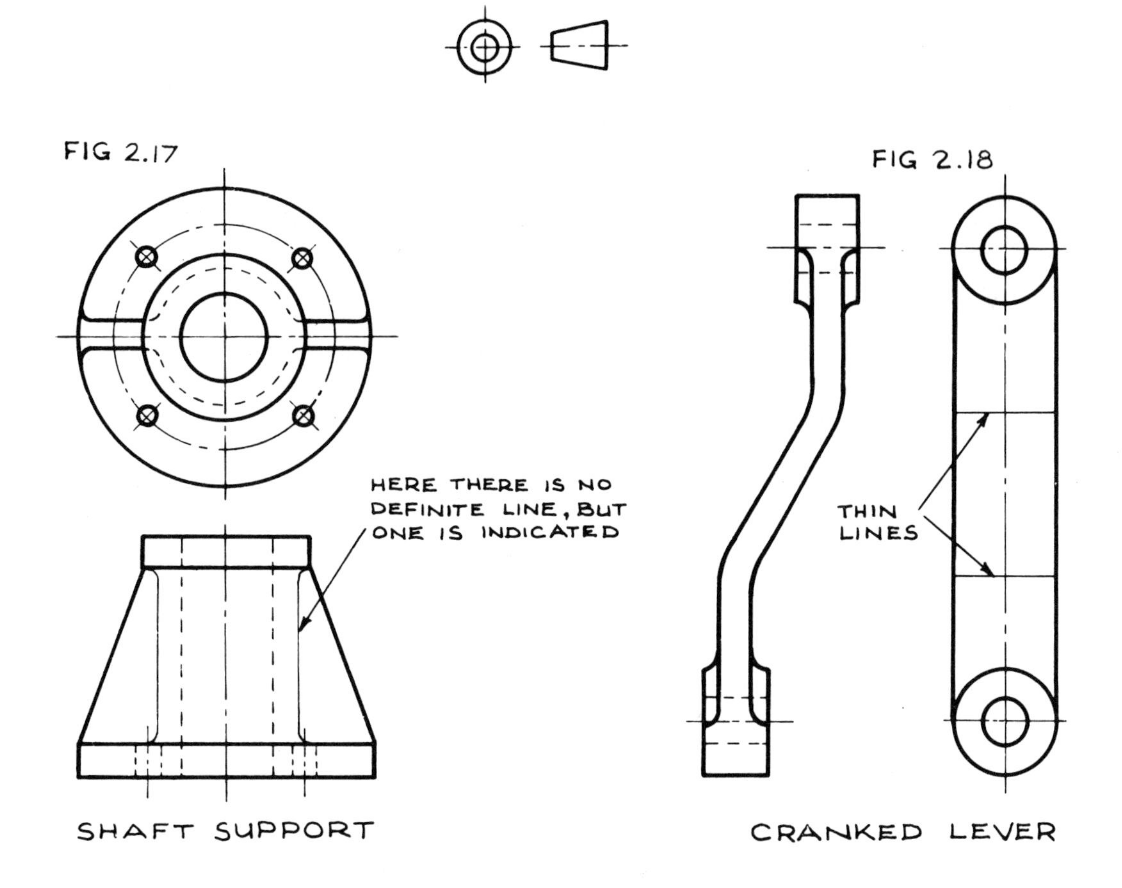

Sometimes shading is resorted to, i.e. parallel lines of variable spacing and of varying lengths, to indicate where surfaces meet when curved. This practice is not to be recommended, however, simply because an elementary drawing ought not to require shading whilst a complex drawing may be made even more difficult to interpret by the use of extra shading lines. The foregoing remarks apply to orthographic projection drawings and not to pictorial drawings where the use of shading lines may greatly aid interpretation.

PREPARATION OF THE DRAWING SHEET

Figure 2.19 shows one satisfactory method of preparing a drawing sheet, suitable for any type of finished work. A very definite border and a carefully spaced and lettered title serve to enhance the appearance of any drawing work. The dimensions given are suitable for any size of drawing sheet.

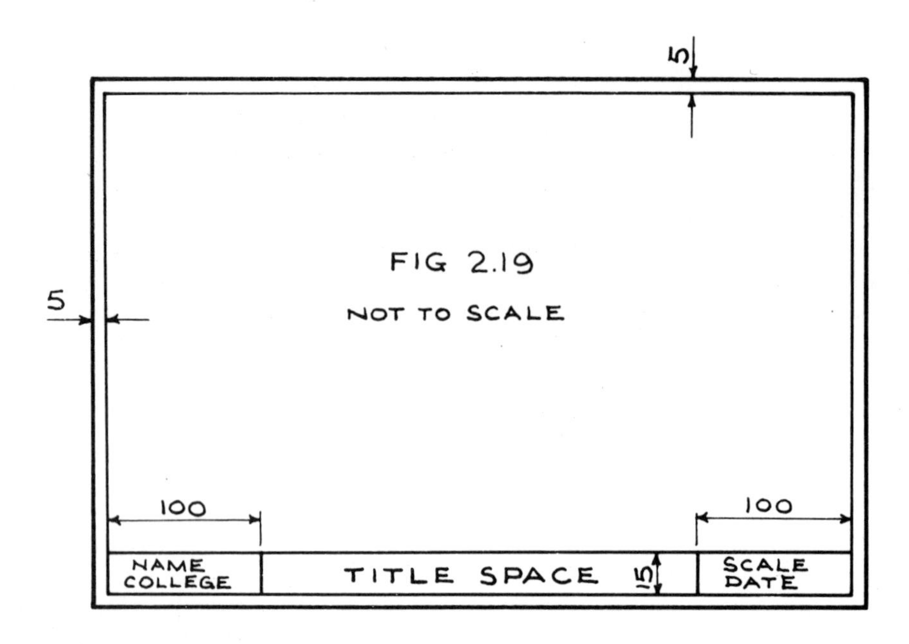

DIMENSIONING A DRAWING

The dimensioning of a drawing is most important and this is explained in detail in Engineering Drawing Notes, Chapter 8. However, for the present the student is advised to do *Exercises 7, 8, 9* and *10*, solutions to which are shown fully dimensioned, and he will thus gain valuable experience of the technique of dimensioning.

Important note. After centre lines and feint outlines of a new drawing have been laid out *it is advisable to draw the curves with a finished line before drawing any finished straight lines*. Straight lines, which are usually tangential to any curves, may then be finished by blending neatly with the finished curves.

DIMENSIONING AND FINISHING EXERCISES

Figures 2.20 to 2.23 show isometric drawings of simple engineering brackets. Fully dimensioned orthographic drawings are given in figures 2.24 to 2.27.

Exercises 7 and *8*. Instead of the views shown in First Angle projection by figures 2.24 and 2.25 draw the same views in Third Angle projection. Fully dimension the drawings and add a title, etc. Use A2 size drawing paper.

Exercises 9 and *10*. Instead of the views shown in Third Angle projection by figures 2.26 and 2.27 draw the same views in First Angle projection. Fully dimension the drawing and add a title, etc. Use A2 size drawing paper.

Note. Internal corners of cast pieces are usually rounded. In the isometric drawings, figures 2.20 to 2.23, no fillet radii are indicated. The student is asked to show fillet radii in the appropriate places in the orthographic drawing answers to *Exercises 7* to *10*.

Some points to observe on dimensioning

(1) Dimensions are kept well clear of the drawing and are adequately spaced.
(2) Dimension lines are thin.
(3) Projector lines are thin.
(4) Projector lines do not touch main lines of the drawing.
(5) Figures do not touch dimension lines.
(6) Arrow heads are small.

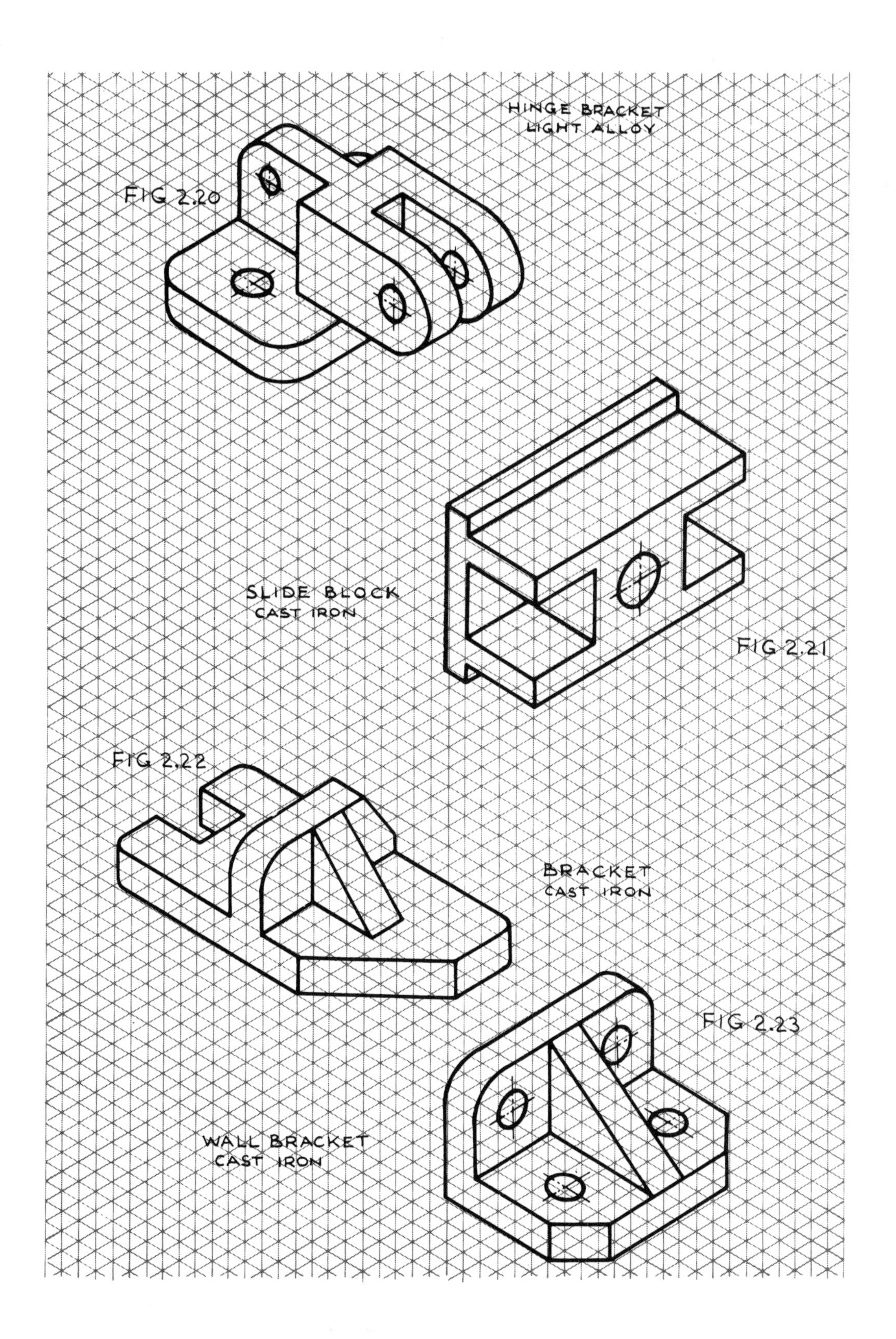
HINGE BRACKET
LIGHT ALLOY
FIG 2.20
SLIDE BLOCK
CAST IRON
FIG 2.21
FIG 2.22
BRACKET
CAST IRON
FIG 2.23
WALL BRACKET
CAST IRON

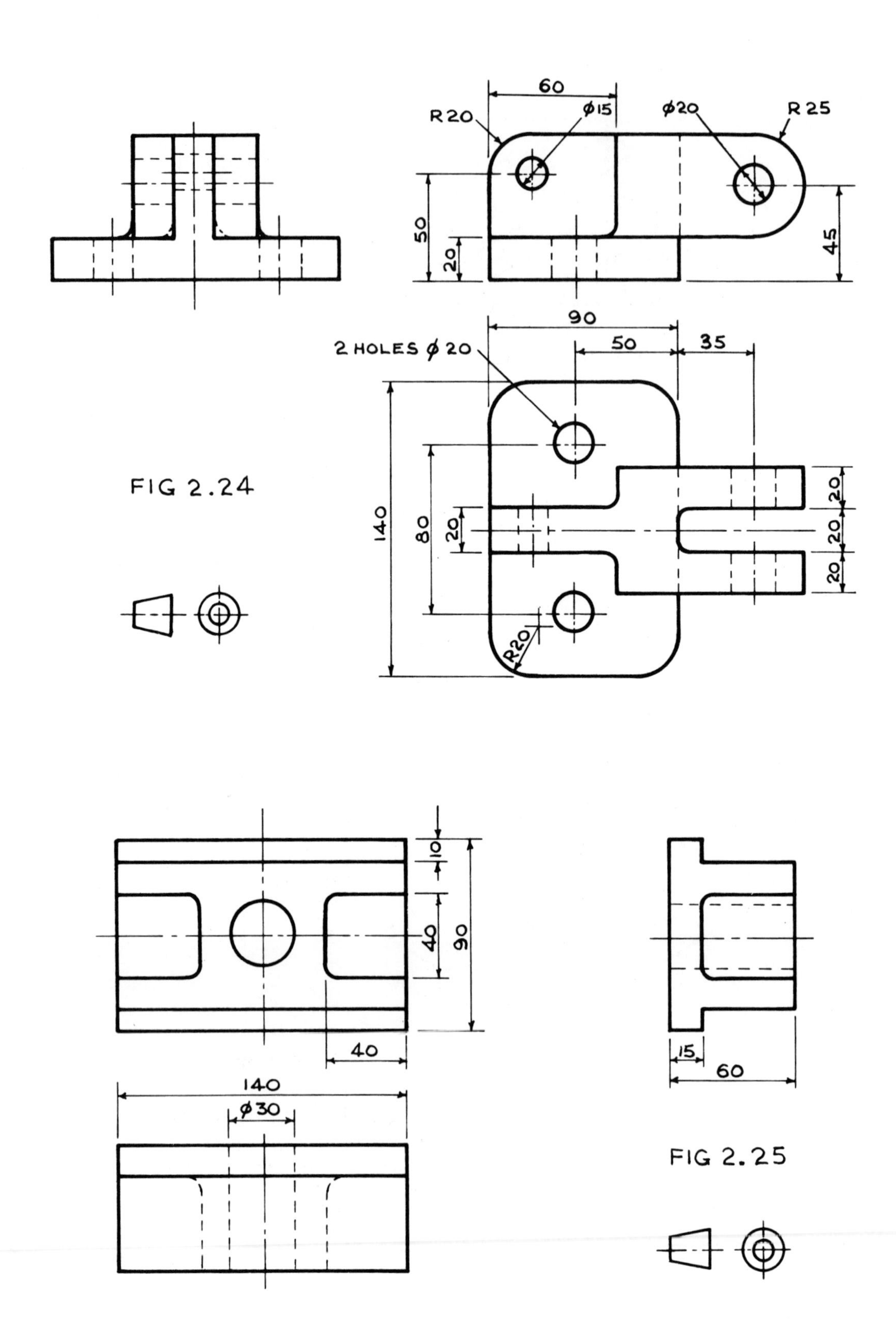

FIG 2.24

FIG 2.25

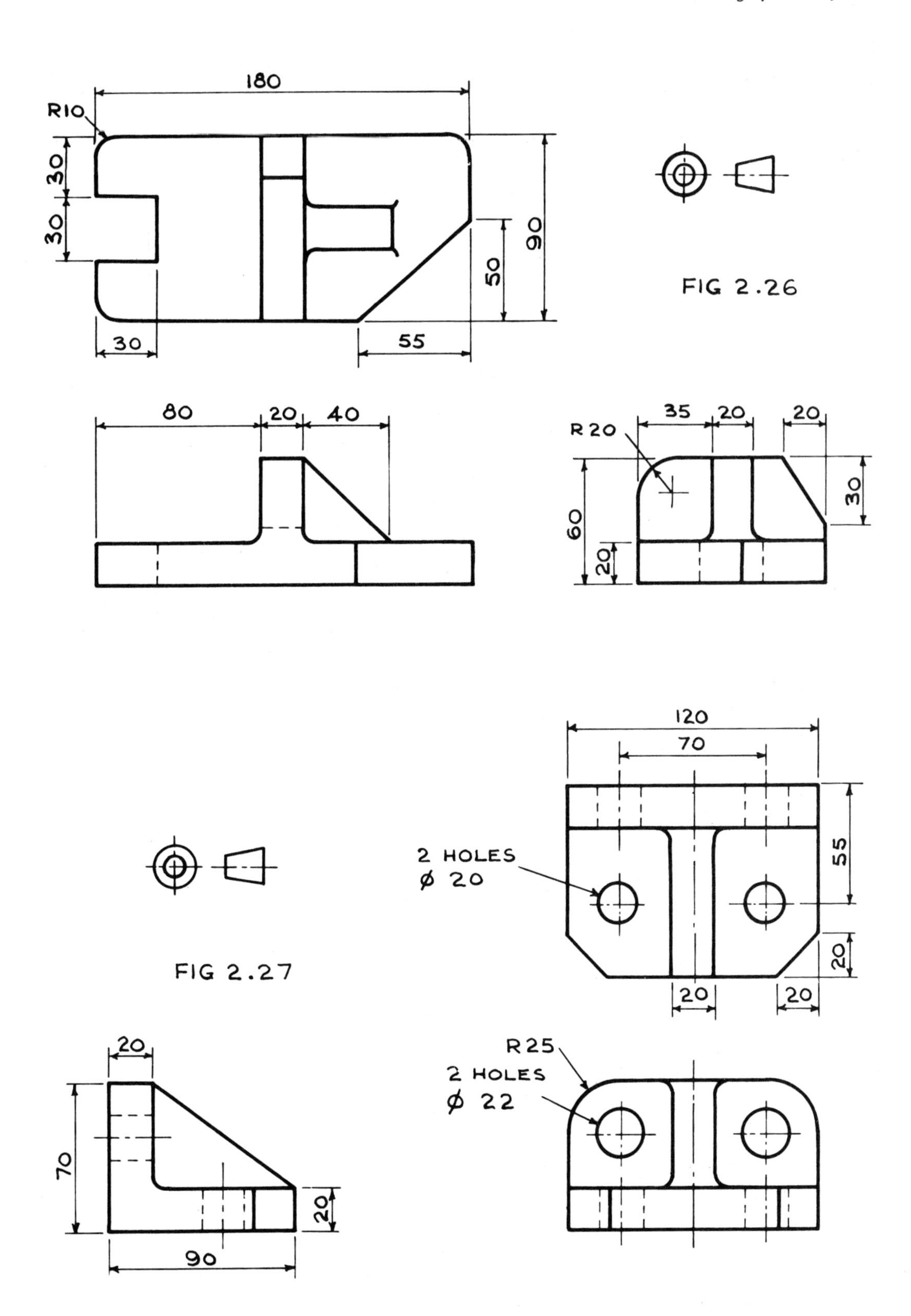

FIG 2.26

FIG 2.27

SECTIONING

Sectional views are used in orthographic projection in order to reveal the inside of an object, or to clarify an assembly of objects.

Obviously the simpler objects do not require to be shown in section but, nevertheless, the general principles of sectioning may be taught by reference to such simple objects.

Figure 2.28 shows a simple object pictured in isometric drawing.

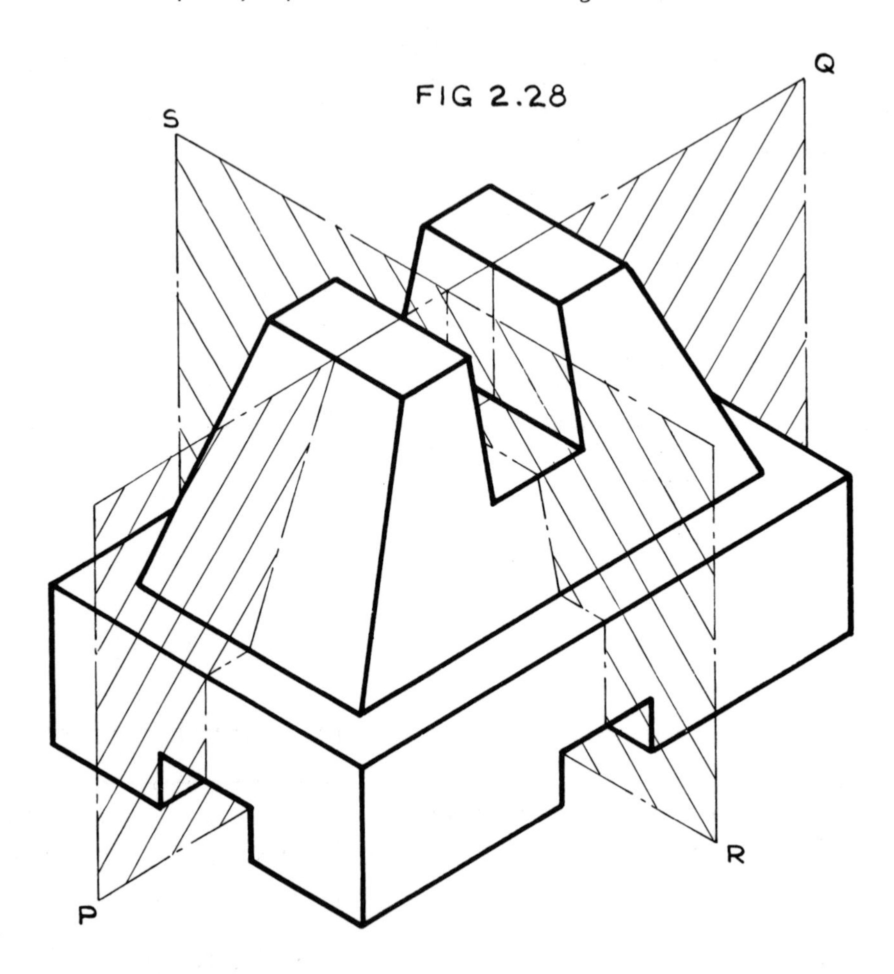

If one imagines a vertical cutting plane such as PQ to intersect the object shown in figure 2.28, the resulting *sectioned front view* will be as shown by *section* P-Q in figure 2.29. It is just as if the object were sawn in halves and one half removed to reveal the other. The 'saw cut' marks are drawn with a 45° set-square and are referred to as *hatching lines* on the sectioned view.

Again, another imaginary vertical cutting plane such as RS in figure 2.28 would give rise to *sectioned side view* R-S in orthographic projection shown in figure 2.29.

Sometimes other section cutting planes are employed, such as the horizontal plane A-A shown in figure 2.30 in Third Angle orthographic projection. The projection of *sectioned plan* A-A is clearly shown in figure 2.30.

In order to save time and space on paper *half-sectioned views* are sometimes used. Such a view is shown in figure 2.30 by *half section* B-B, where the right hand side is shown sectioned whilst the left hand side is an *outside view*. The student should carefully note that the dividing line between the outside half and the sectioned half is conventionally shown by a *thin chain line*.

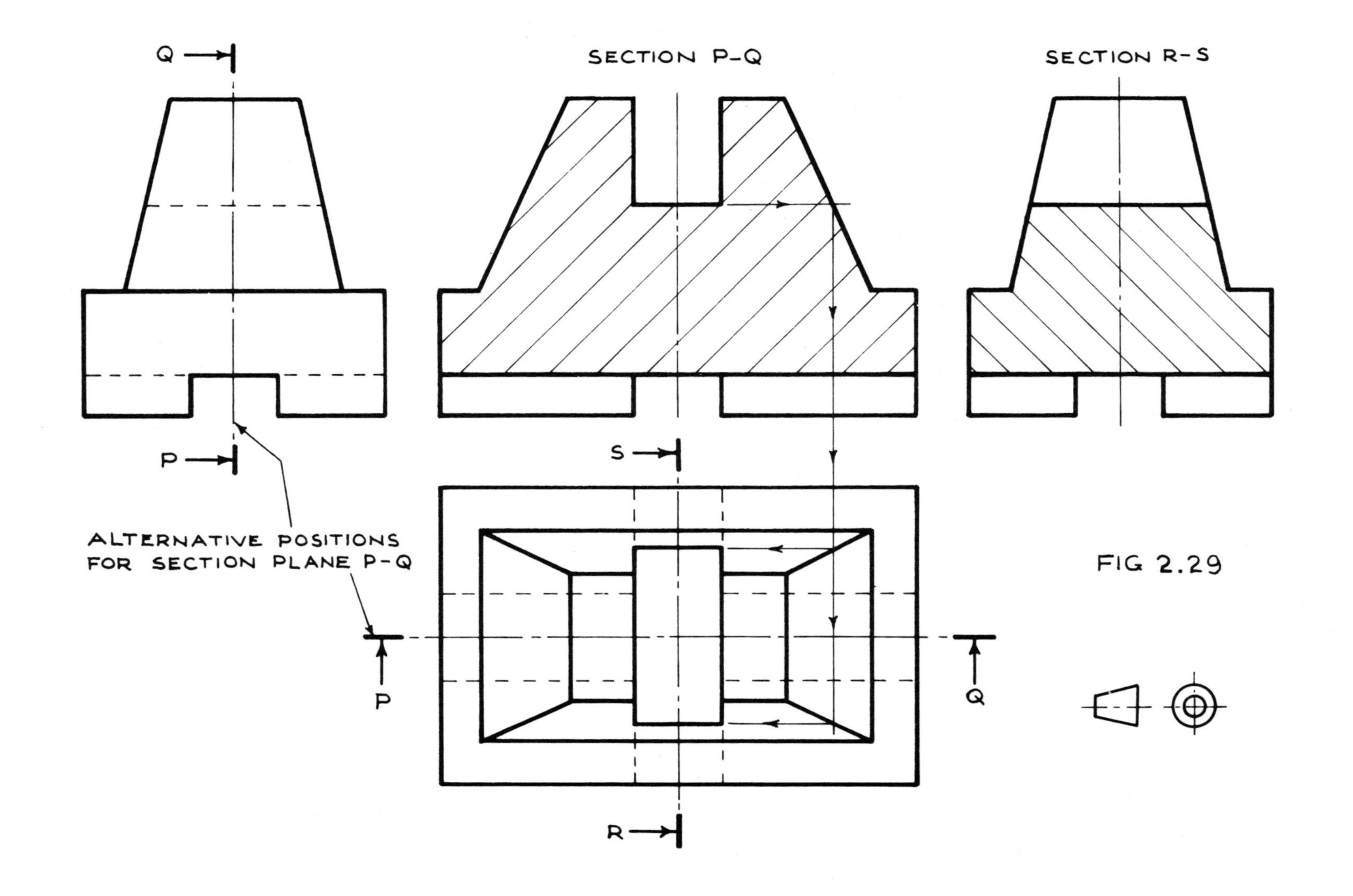
Q
SECTION P-Q
SECTION R-S
P
S
ALTERNATIVE POSITIONS
FOR SECTION PLANE P-Q
P
Q
R

FIG 2.29

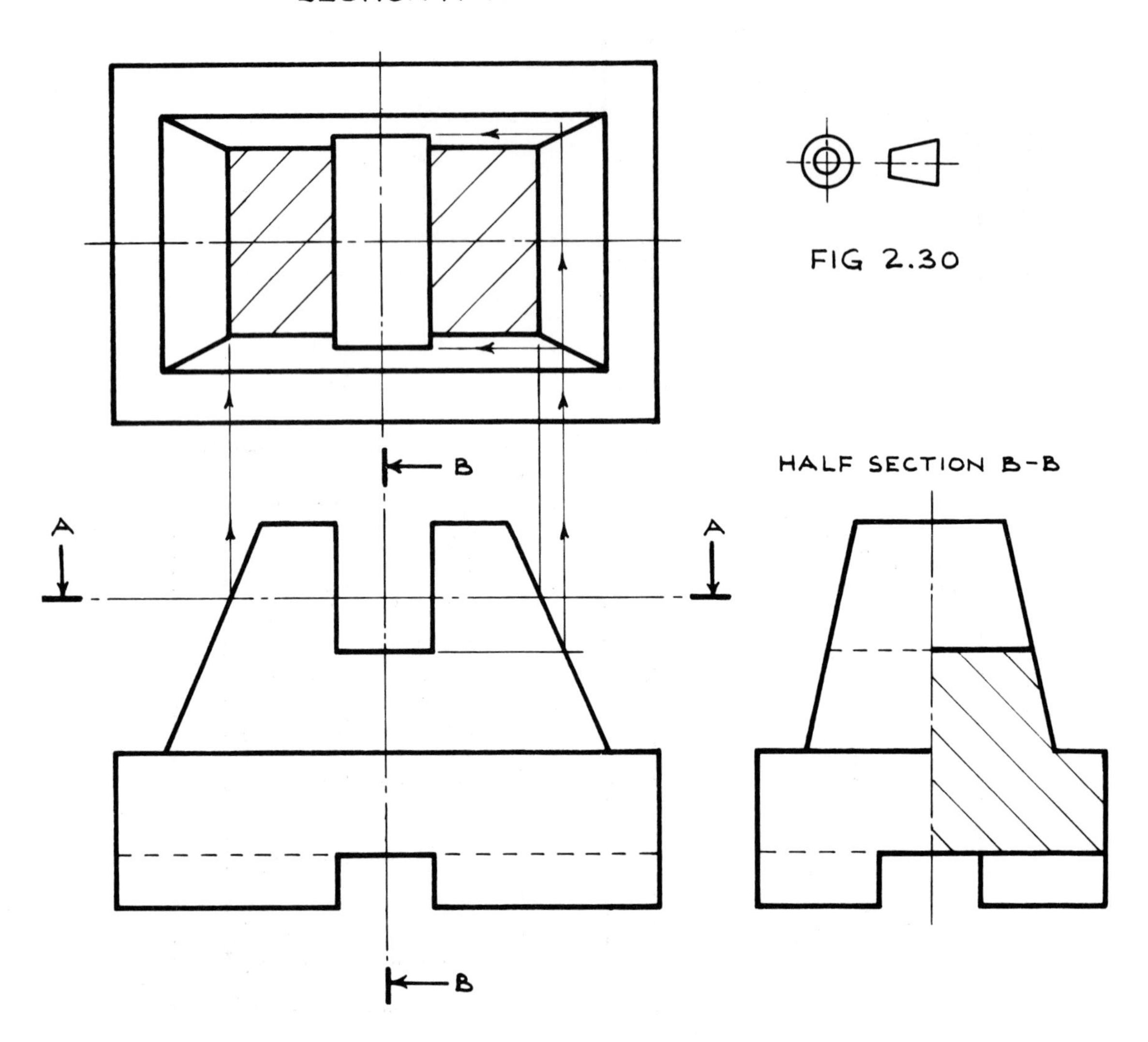

Careful note should be made of the following points with regard to sectioning conventions:

(i) Hatching lines, usually drawn with a 45° set-square, are *thin* lines and these are fairly *widely spaced*, in accordance with the latest standards regulations.

(ii) A section cutting plane is indicated in orthographic projection by a *thin chain line* having a *thick* portion at each end on which arrowheads are shown.

(iii) Hidden detail is not normally shown in a sectioned view.

Sectioning conventions are treated in detail in Engineering Drawing Notes, Chapter 8.

CENTRE LINES

Where appropriate, all views whether sectioned or not should have centre lines of symmetry indicated by a *thin chain line*.

Holes must always have centre lines shown, crossed lines or single lines, since dimensions are always given to hole centres.

SECTIONING AND OTHER EXERCISES

Exercises 11, 12 and *13* should be answered in both First Angle and Third Angle orthographic projection.
Exercise 11. Figure 2.31 shows an isometric drawing of a Vee Block. Draw the following orthographic views of the Block:

(i) A sectioned Front View on the centre line as viewed in the direction of arrow P;
(ii) A sectioned Side View on the centre line as viewed in the direction of arrow Q;
(iii) A Side View as viewed in the direction of arrow R;
(iv) A Plan.

Exercise 12. Refer to figure 2.32, Bearing Brass, and answer the question as for *Exercise 11.*
Exercise 13. For the Slide Block shown in figure 2.33 draw the following orthographic views:

(i) A Front View as viewed in the direction of arrow P;
(ii) A sectioned Side View on the centre line as viewed in the direction of arrow Q;
(iii) A Side View as viewed in the direction of arrow R;
(iv) A Plan.

Exercise 14. Draw a sectioned Front View, by a vertical plane through the axis and two flange holes, of the Shaft Support shown in figure 2.34. Add a Plan. Use Third Angle projection.
Important note. The Front View being sectioned in *Exercise 14* does *not* mean that only *half* of a Plan is to be drawn. This is a very common error among students at an early stage.

Sometimes half views of symmetrical objects are drawn to save space but generally the *entire object* must be drawn whenever non-sectional views are called for.

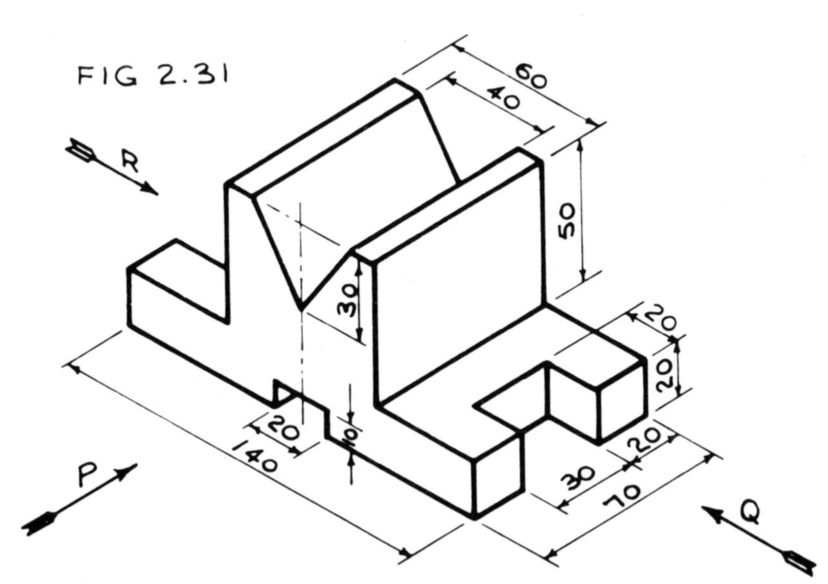

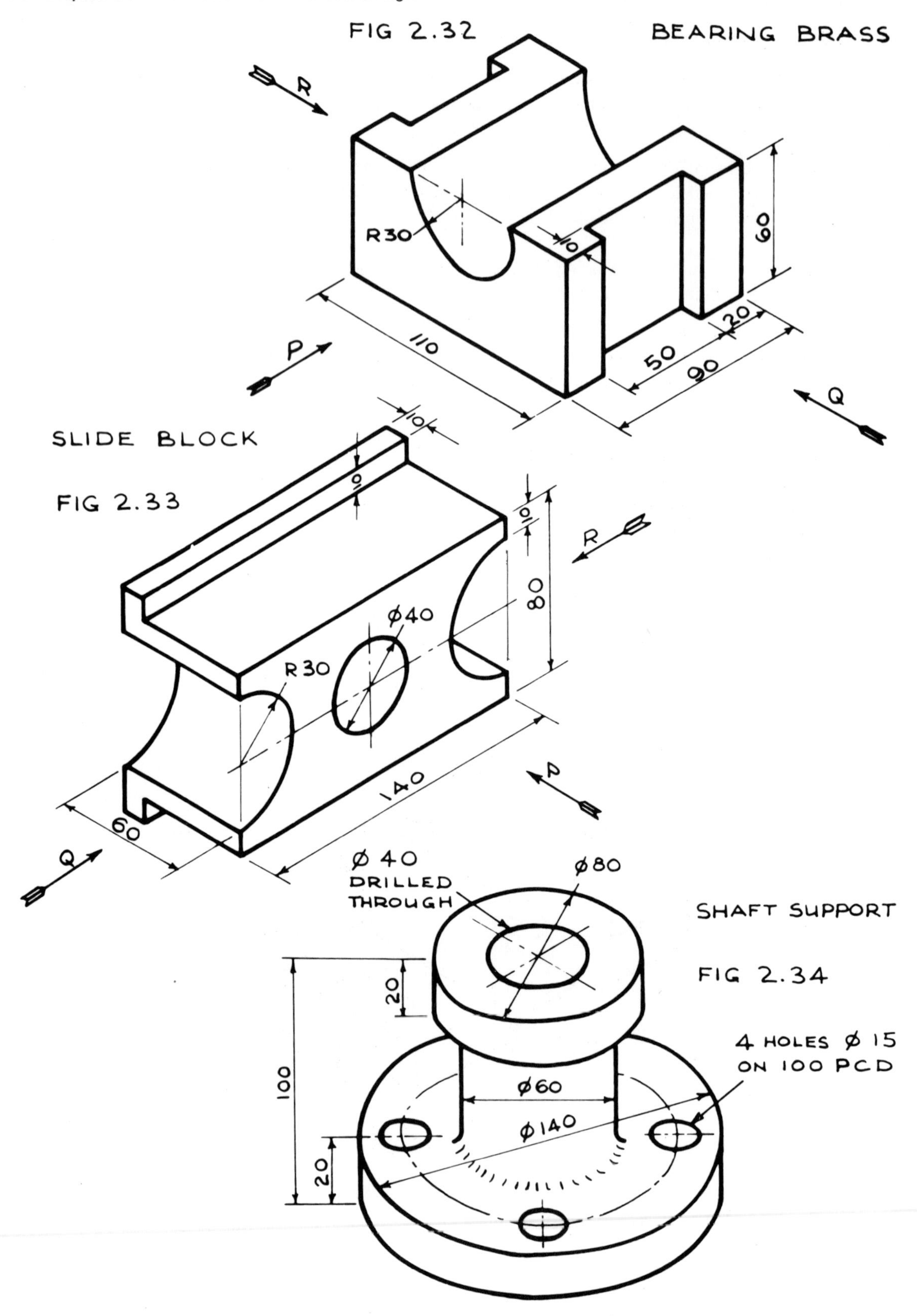
FIG 2.32
BEARING BRASS
R
R30
10
60
20
50
90
110
P
Q
SLIDE BLOCK
FIG 2.33
10
10
10
80
R
ϕ40
R30
140
P
60
Q
ϕ 40
DRILLED
THROUGH
ϕ80
SHAFT SUPPORT
FIG 2.34
20
100
4 HOLES ϕ 15
ON 100 PCD
ϕ60
ϕ140
20

WEB SECTIONING CONVENTION

The Shaft Support shown in figure 2.35 has two triangular Webs which are used to strengthen the casting without adding undue weight. Webs are usually relatively thin portions of a casting.

Figure 2.36 shows a sectioned Front View by a cutting plane through the centre line which 'splits' the webs. This view gives the impression of solidity and although correct geometrically is not recommended by standards organisations.

Figure 2.37 shows the recommended method of treating a sectioned web whenever the cutting plane splits the web longitudinally.

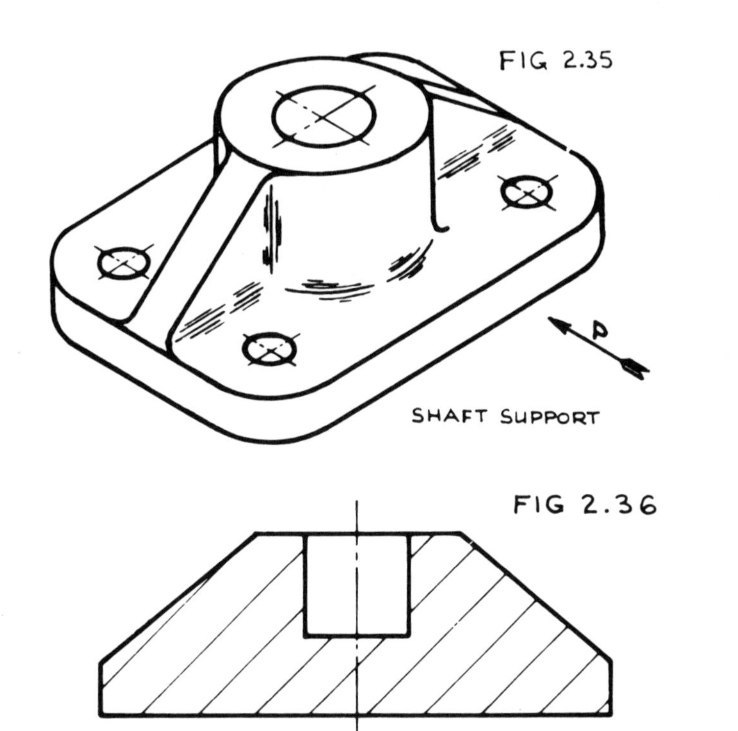

GEOMETRICALLY CORRECT — BUT NOT ACCEPTABLE

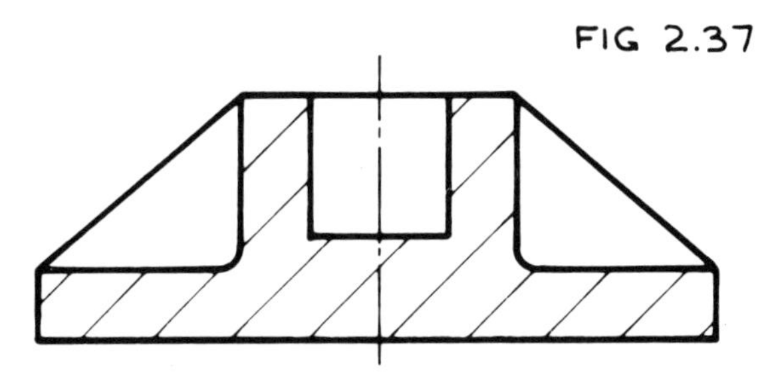

GEOMETRICALLY INCORRECT STANDARD CONVENTION

NOTE: THIS CONVENTION FOR THE SECTIONING OF A WEB APPLIES ONLY TO A LONGITUDINAL SECTION AS IN FIG 2.37. A CROSS-SECTION OF A WEB IS TREATED IN THE NORMAL GEOMETRICAL MANNER AND IS HATCHED ACCORDINGLY.

Exercise 15. With reference to the Shaft Support shown in figure 2.35 draw the following views, using the 5 mm triangular grid as the basis for measurements:

(i) A Sectioned Front View as viewed in the direction of arrow P, with the section plane splitting the webs longitudinally along the centre line;
(ii) A Side View;
(iii) A Plan View.

Use either First Angle or Third Angle projection and completely dimension and finish the drawing with title, etc.

Exercise 16. For the Bar Support shown in figure 2.38 draw the following views full size in First Angle projection, using A2 size paper:

(i) A sectioned Front View on the centre line of the object as viewed in the direction of arrow P;
(ii) A Side View as viewed in the direction of arrow Q;
(iii) A Side View as viewed in the direction of arrow R;
(iv) A Plan.

Exercise 17. Using Third Angle projection draw the following views of the Simple Bearing shown in figure 2.39.

(i) A Sectioned Front View on the centre line as viewed in the direction of arrow P;
(ii) A Side View;
(iii) A Plan View.

Exercise 18. A Shaft Bracket is pictured in figure 2.40. Draw, full size, using First Angle projection, the following views of the Bracket:

(i) A Front View as viewed in the direction of arrow P;
(ii) A Sectioned Side View on the centre line as viewed in the direction of arrow Q;
(iii) A Plan.

Exercise 19. Draw the following views, in Third Angle projection, of the Adjustable Bearing Block shown in figure 2.41.

(i) A Front View as viewed in the direction of arrow P;
(ii) A Side View as viewed in the direction of arrow Q;
(iii) A Plan View.

BAR SUPPORT

FIG 2.38

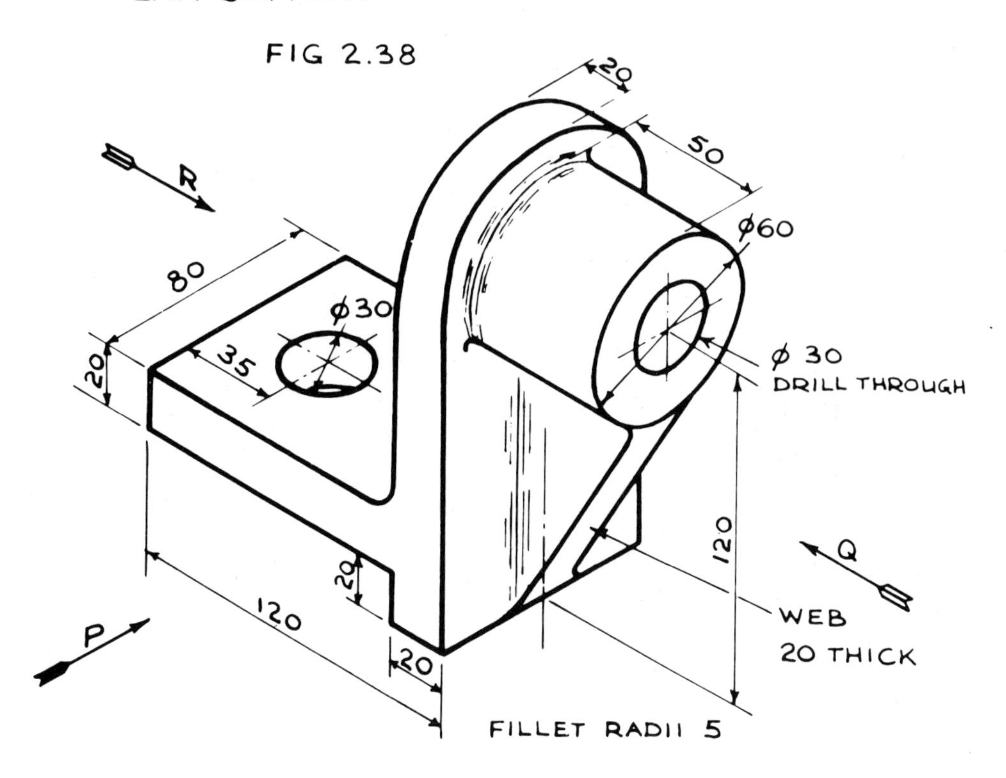

SIMPLE BEARING

FIG 2.39

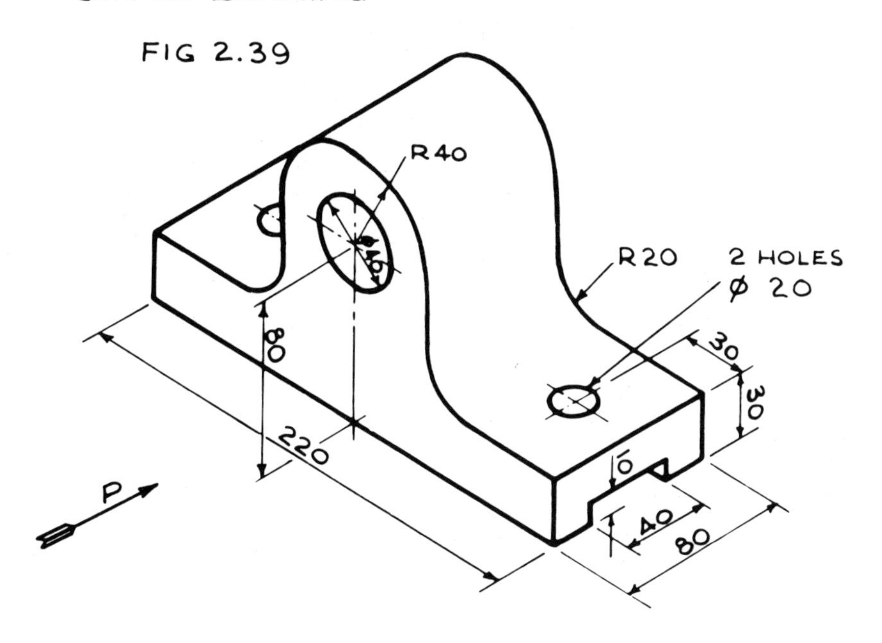

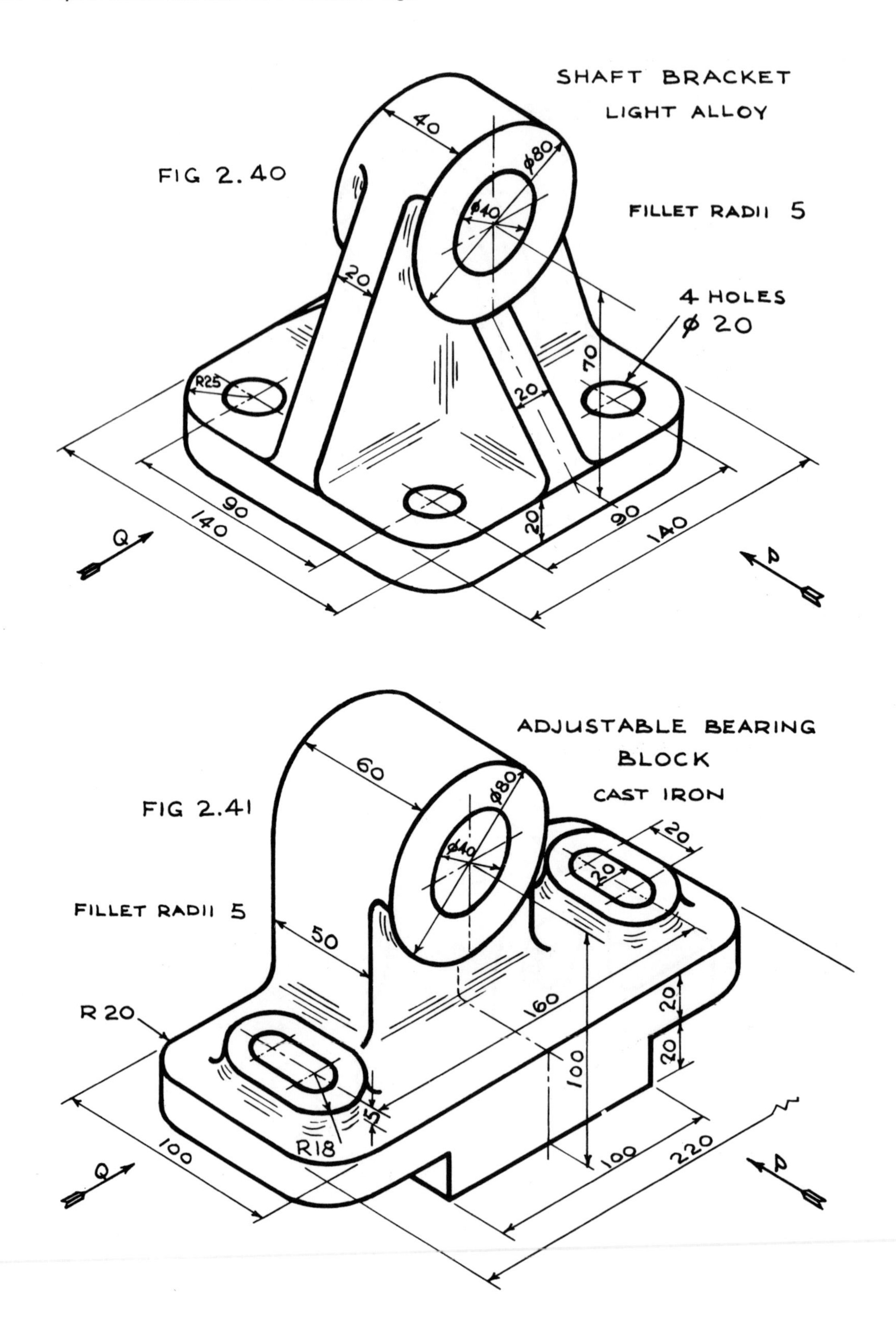
SHAFT BRACKET
LIGHT ALLOY
FIG 2.40
FILLET RADII 5
4 HOLES
ϕ 20
40
ϕ80
ϕ40
20
70
R25
20
90
140
20
90
140
Q
P
ADJUSTABLE BEARING
BLOCK
CAST IRON
FIG 2.41
FILLET RADII 5
60
ϕ80
ϕ40
20
20
50
R 20
160
20
20
100
5
R18
100
100
220
Q
P

3
ISOMETRIC AND OBLIQUE DRAWING

An *isometric drawing* is used to portray a three-dimensional object by means of a single diagram. A 30° set-square is used as shown in figure 3.1.

In *conventional isometric drawing* the full dimensions of the object are set off in each of the three directions OA, OB and OC. This results in a 'larger-than-life' size pictorial view of the object, since no account has been taken of foreshortening of sloping edges. The correction for this foreshortening effect is dealt with in chapter 7 where the use of isometric scale is explained.

FIG 3.1

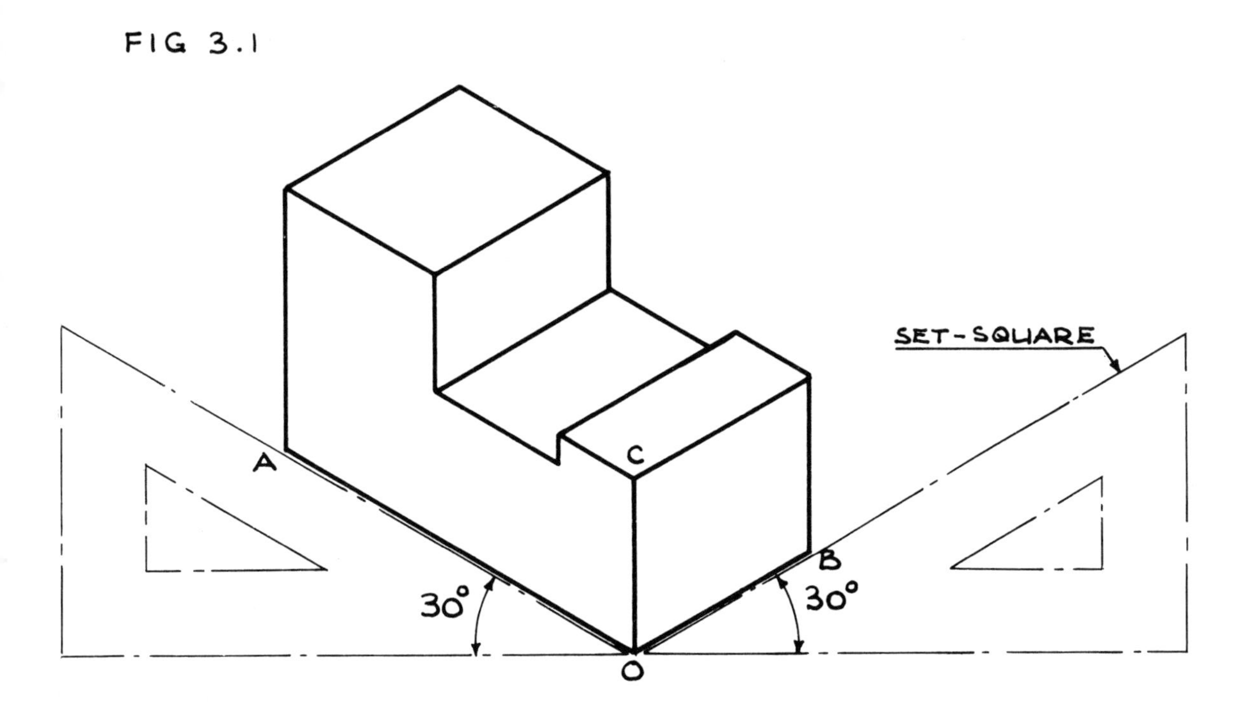

The directions OA, OB and OC are sometimes referred to as ISOMETRIC AXES

Figure 3.2 shows some of the different isometric drawings which it is possible to make of the object shown in figure 3.1. (Figure 3.2 viewed upside down results in five equally valid isometric views of the object, from a low-level view point.)

Figure 3.1, figure 3.2(a), figure 3.2(c) and figure 3.2(d) show very clearly what the object is like and any of these views would serve as an isometric drawing.

Figure 3.2(b) and particularly figure 3.2(e) are less suitable as isometric drawings.

When attempting to portray an object isometrically care should be exercised in choosing an appropriate view.

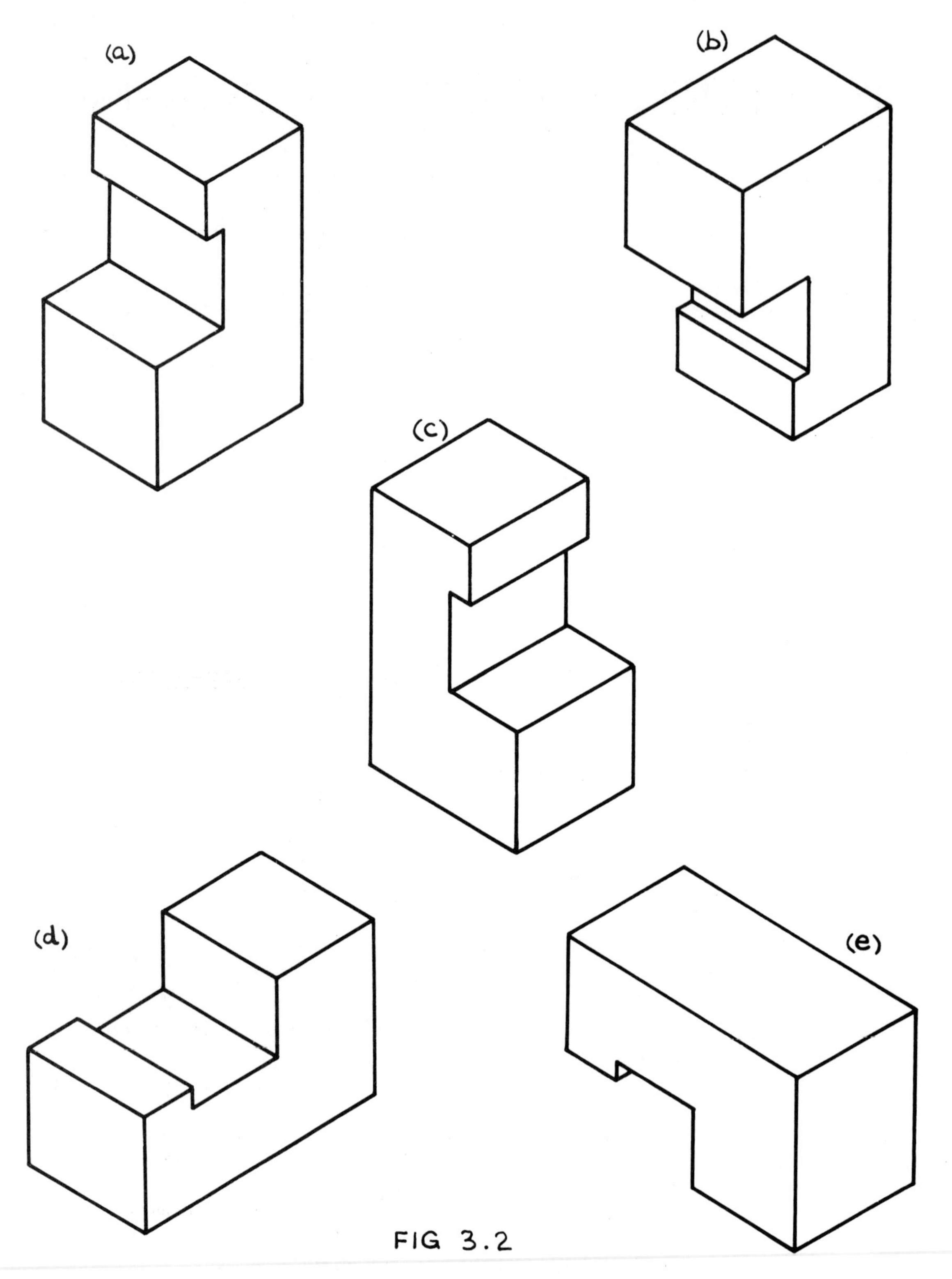

FIG 3.2

DIFFERENT ISOMETRIC DRAWINGS OF FIGURE 3.1

Exercise 1. Using the dimensions given in figure 3.3 make three or four different isometric drawings of the given object.
Exercise 2. Using plain drawing paper make isometric drawings of the objects shown in figures 3.4, 3.5, 3.6, 3.7 and 3.8. Make at least two different drawings of each of these objects.
Exercise 3. Figures 3.9 and 3.10 show isometric drawings of figure 3.6 and 3.8 respectively drawn on isometric graph paper. Using isometric graph paper make two different isometric drawings of each of the objects given in figures 3.3, 3.4, 3.5, 3.6, 3.7 and 3.8.
Note. Assume that the equilateral triangular grid of the isometric graph paper is of 5 mm unit side. In fact the grid unit is less than 5 mm, but it is made thus in order to make the isometric drawing more true to scale by allowing for the foreshortening effect of isometric drawing.

A red or blue coloured ball pen, or a fibre tipped pen, shows up best on the isometric graph paper.
Exercise 4. Using plain drawing paper make isometric drawings of a regular hexagonal prism 30 mm side of hexagon and 60 mm long.
Exercise 5. Using plain drawing paper make an isometric drawing of a regular hexagonal pyramid 50 mm side of hexagon and altitude 100 mm.

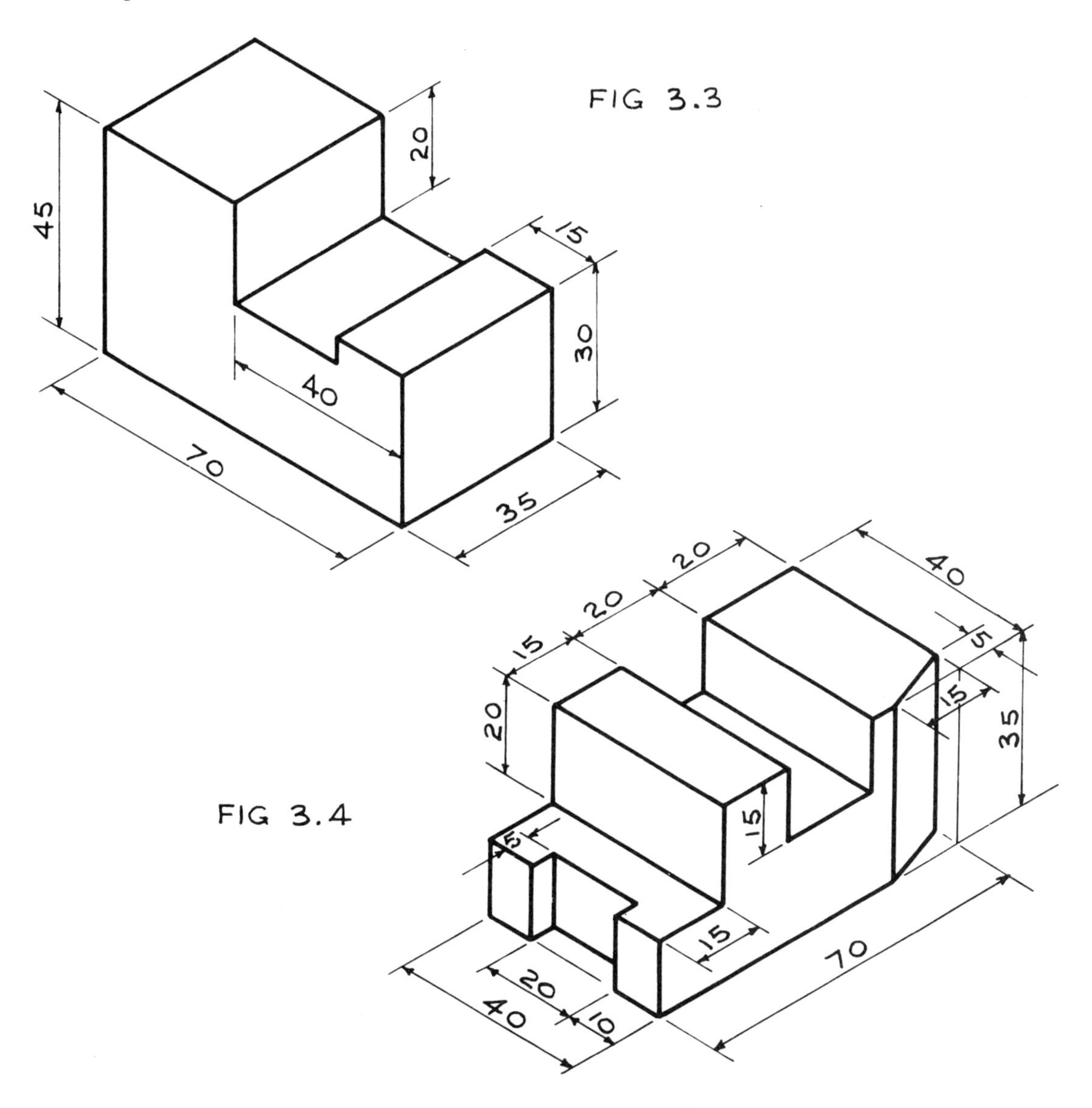

FIG 3.3

FIG 3.4

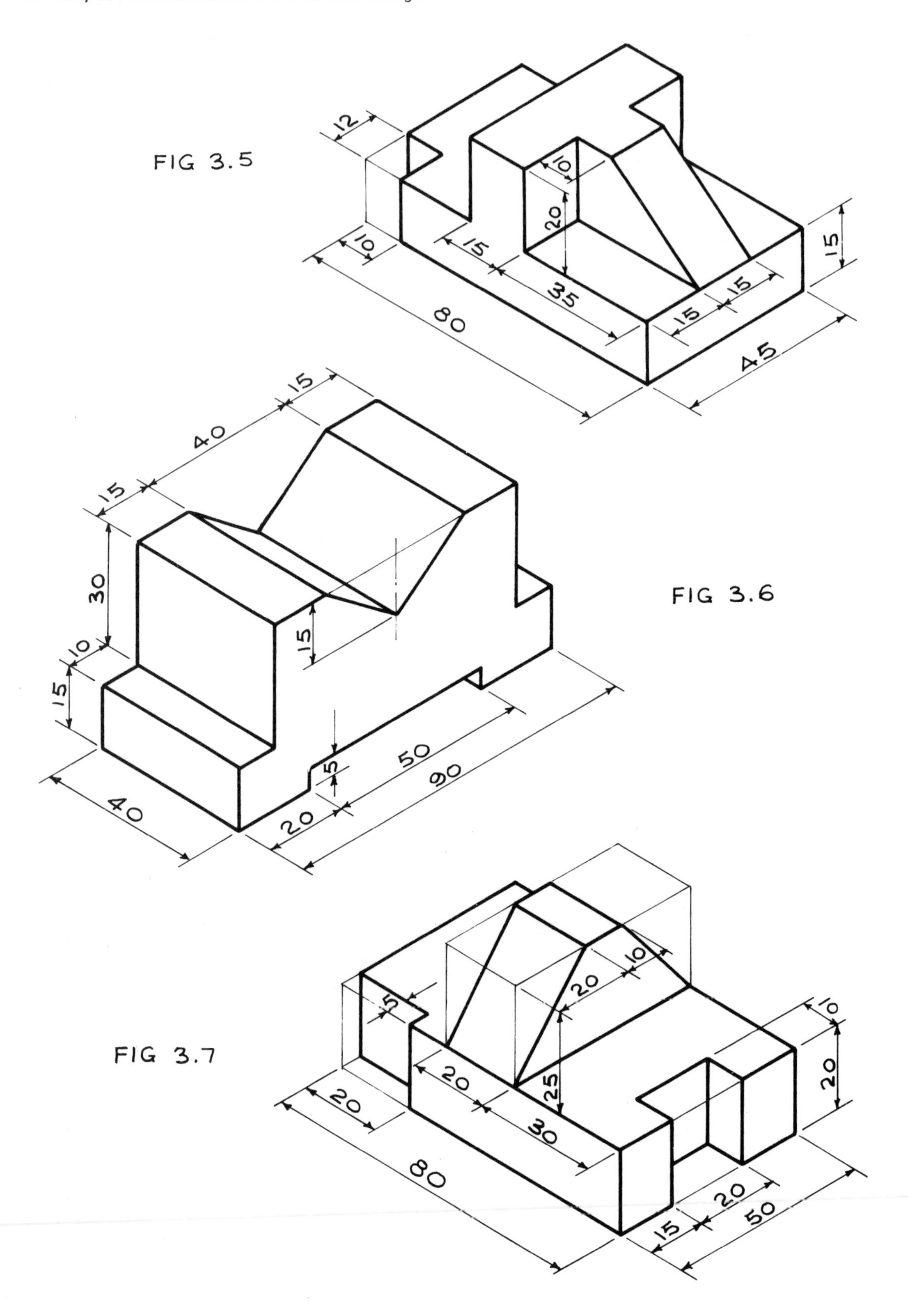
FIG 3.5
12
10
20
10
15
35
15
15
80
45
15
FIG 3.6
15
40
15
30
10
15
15
5
50
90
20
40
FIG 3.7
10
20
5
25
20
20
30
80
10
20
15
20
50

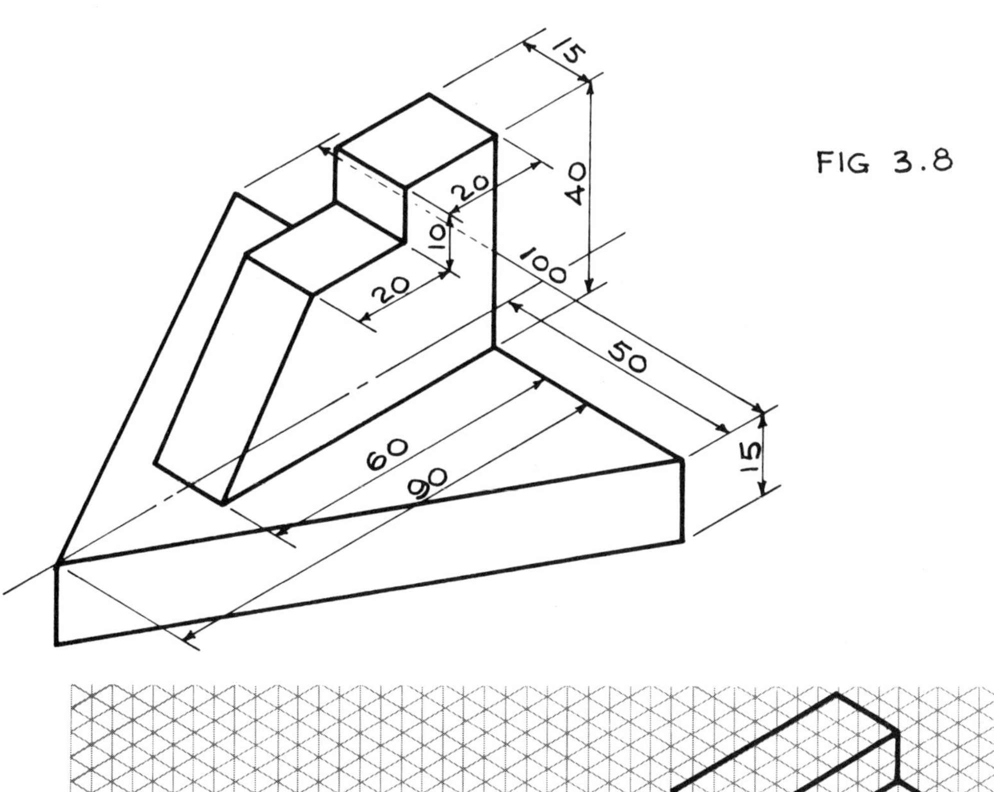

FIG 3.8

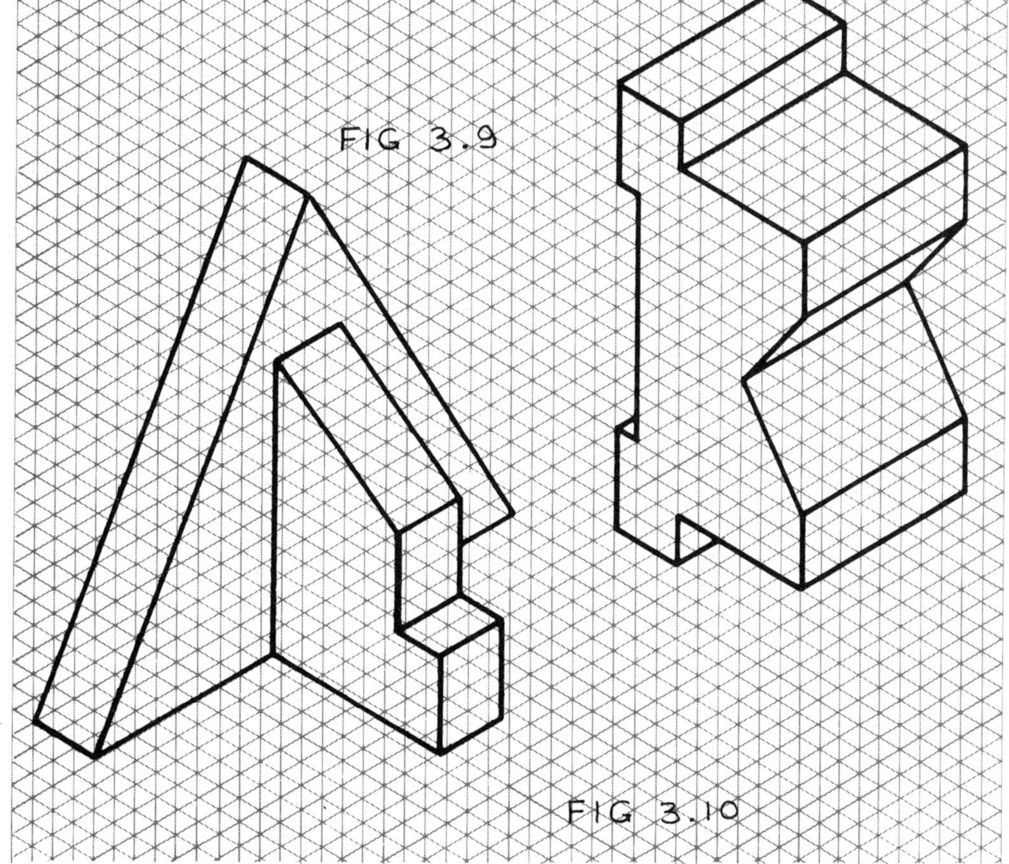

FIG 3.9

FIG 3.10

EXERCISE 4 SOLUTION

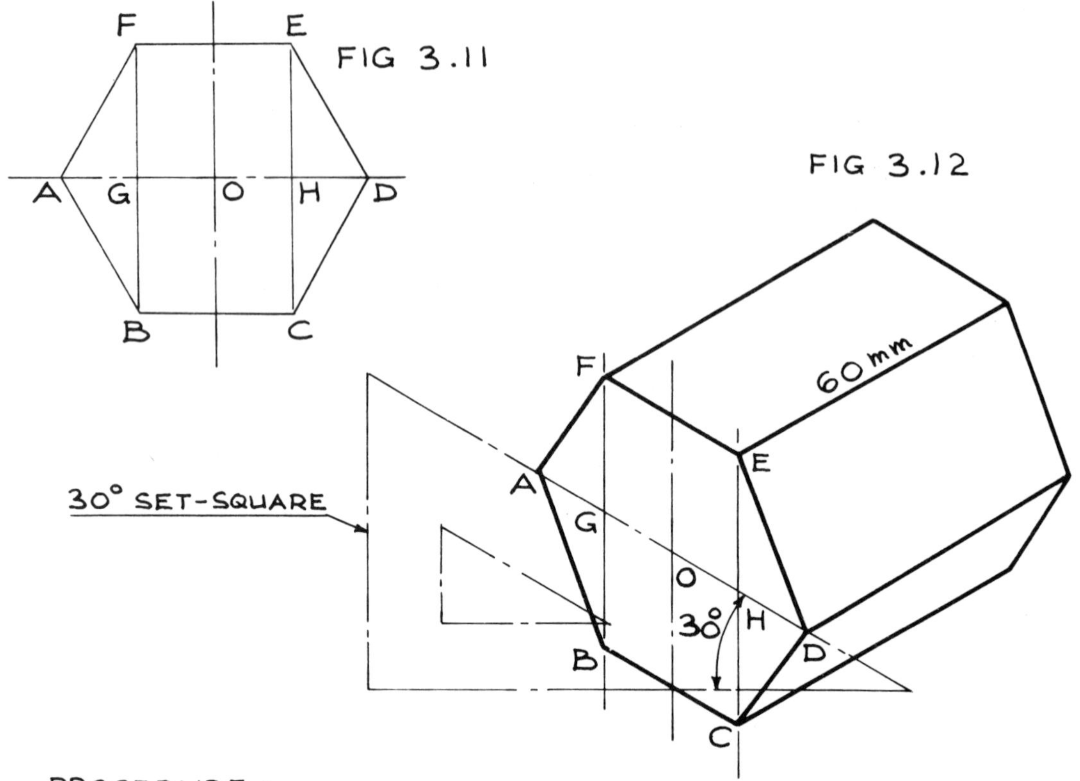

PROCEDURE:

First draw a regular hexagon 30 mm side and draw diagonals, as a grid, for offsets.

Ordinate positions from centre O are marked with compasses, i.e. OG, OH; OA, OD are taken from the regular hexagon, Fig. 3.11, and set off along the 30° diagonals AD in Figs. 3.12 and 3.13.

GF, GB; HE and HC are then marked off, on vertical ordinates in Fig. 12 and along 30° ordinates in Fig. 3.13.

Points A, B, C, D, E and F are joined and then length of prism, 60 mm, is marked with compasses along the edges.

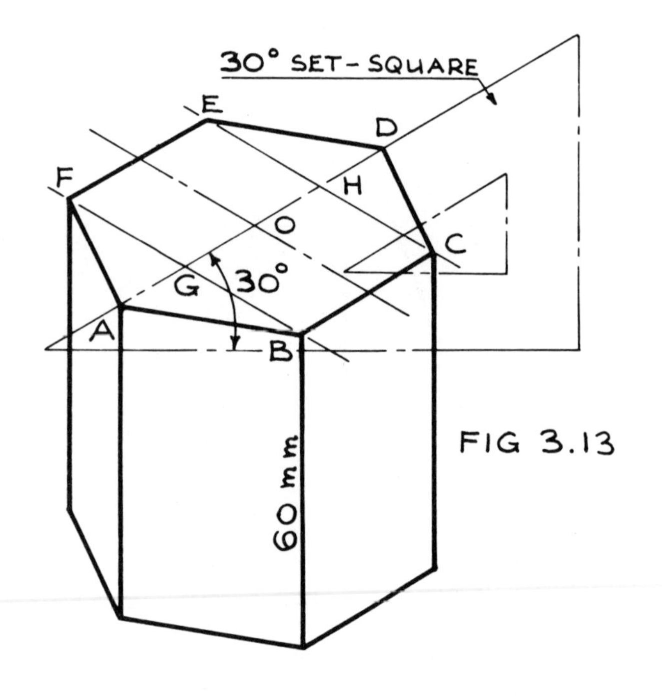

EXERCISE 5 SOLUTION

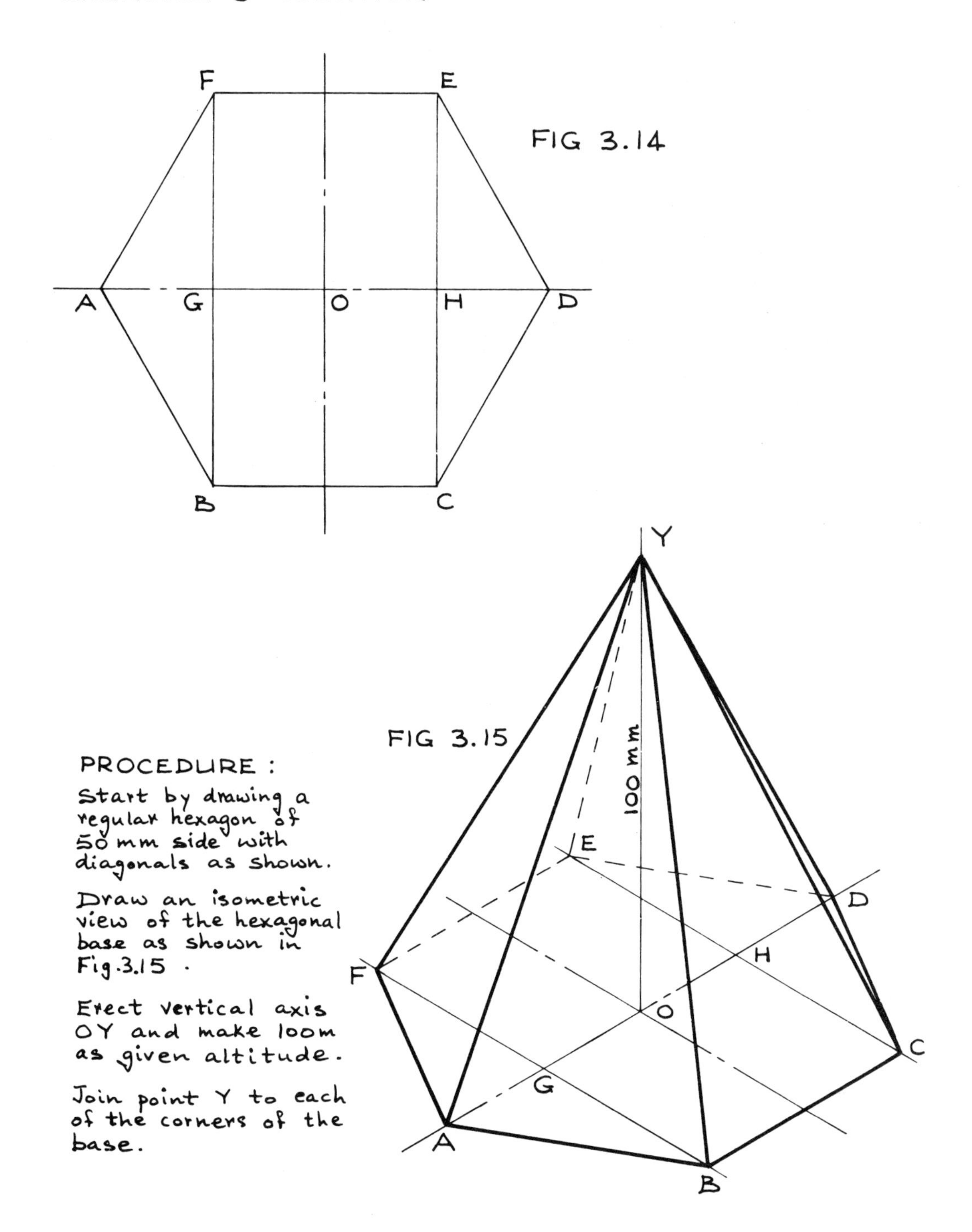

FIG 3.14

FIG 3.15

PROCEDURE:

Start by drawing a regular hexagon of 50 mm side with diagonals as shown.

Draw an isometric view of the hexagonal base as shown in Fig. 3.15.

Erect vertical axis OY and make 100m as given altitude.

Join point Y to each of the corners of the base.

NOTE:

Hidden edges, though not usually shown on isometric drawings, are indicated here by the broken lines.
Notice also that if altitude OY had been greater, say 120 mm, then edge YD would also have been hidden.

For *Exercises 6* to *12* use plain drawing paper.

Exercise 6. Make an isometric drawing of a circle 100 mm in diameter.

Exercise 7. Make an isometric drawing of a cube of 80 mm side and inscribe an 80 mm diameter circle in each of the three visible faces.

Exercise 8. A cylinder has a diameter of 90 mm and a length of 45 mm. Make isometric drawings of the cylinder:

(i) with a plane circular end vertical;
(ii) resting on a plane circular end.

Exercise 9. Make an isometric drawing of a hollow cylinder with a plane face vertical. Outside diameter 100 mm, inside diameter 70 mm and thickness (length) 20 mm.

Exercise 10. A cylinder 120 mm diameter and 30 mm long has a small hole 30 mm diameter drilled along the axis. Make an isometric drawing with the axis of the cylinder vertical.

Exercise 11. A regular hexagonal prism 50 mm side of hexagon and 15 mm long has an axial hole drilled 50 mm in diameter. Make isometric drawings:

(i) with a plane face vertical;
(ii) with the axis vertical.

Exercise 12. A corner is cut from a plank of wood 20 mm thick and is shown in figure 3.26. Make a drawing to your own measurements, something like the irregular shape given, and then make an isometric drawing of the piece.

EXERCISE 6 SOLUTION

FIG 3.16

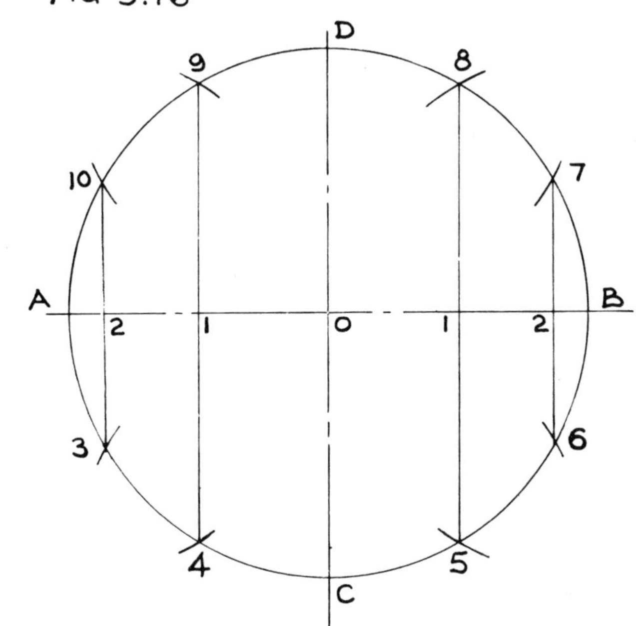

PROCEDURE:

First draw a circle 50 mm radius and two right-angled diameters AB and DC, Fig. 3.16.

Using compasses set at the same radius divide the circumference into 12 equal parts.

Draw ordinates 3, 10; 4, 9 etc.

FIG 3.17

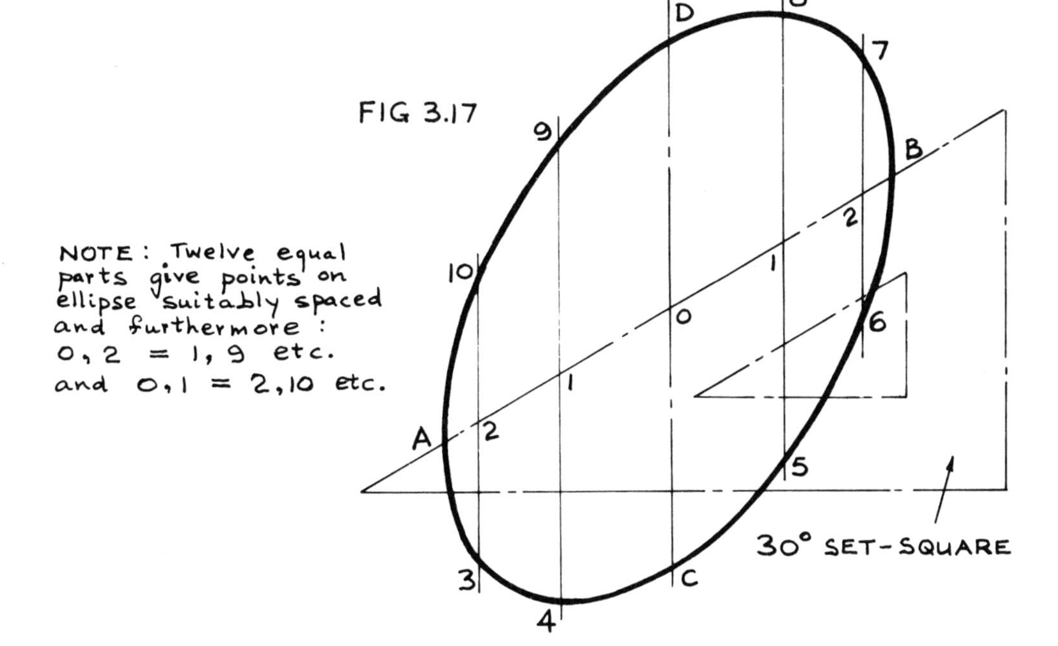

To complete the isometric drawing:

Using compasses at 50 mm radius mark off OA, OB, OC and OD.

Mark off O2 and O1 on either side and erect ordinates.

Using half-ordinate length on compasses mark points 2, 10; 2, 3; 2, 7; 2, 6 and similarly for 1, 9; 1, 4; 1, 8; 1, 5, all compass settings being taken from figure 3.16.

EXERCISE 7 SOLUTION

FIG 3.18

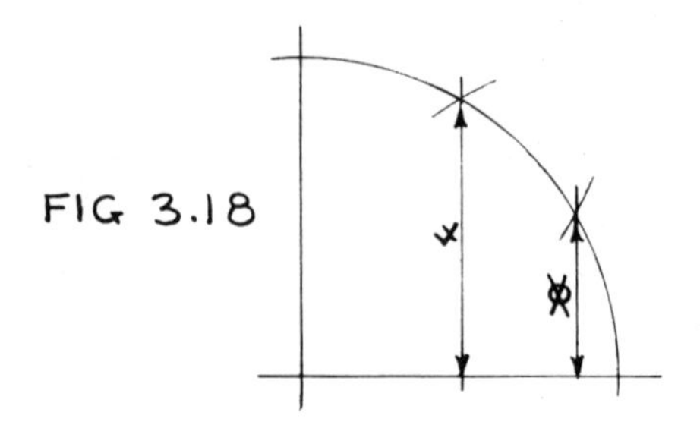

FIG 3.19

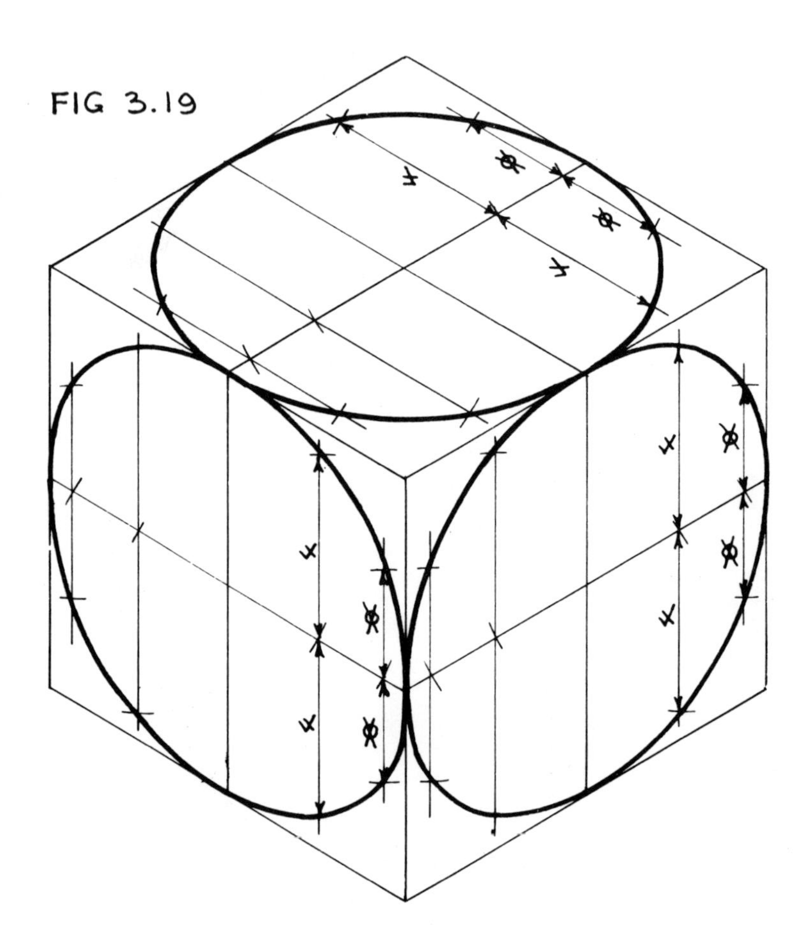

NOTE: A quarter circle only is required, Fig.3.18, to obtain half-ordinates for ellipse construction as shown in Fig.3.19.

EXERCISE 8 SOLUTION

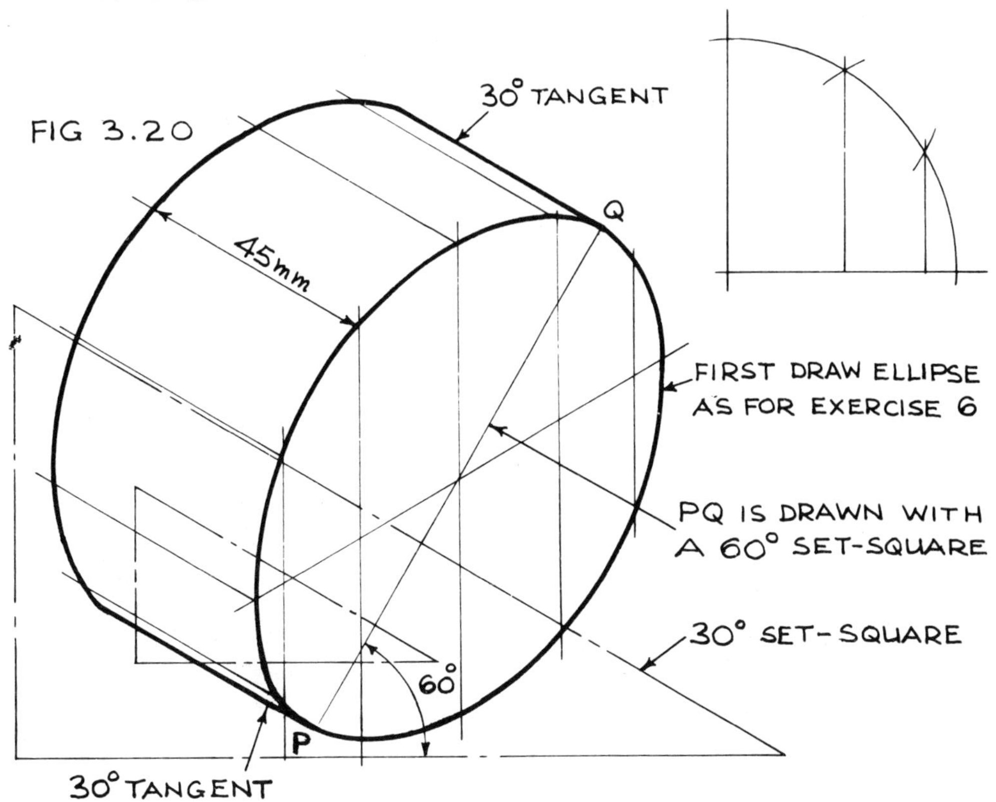

Set compasses to length 45 mm and mark along 30° parallels, Fig.3.20 and vertical parallels, Fig.3.21 .

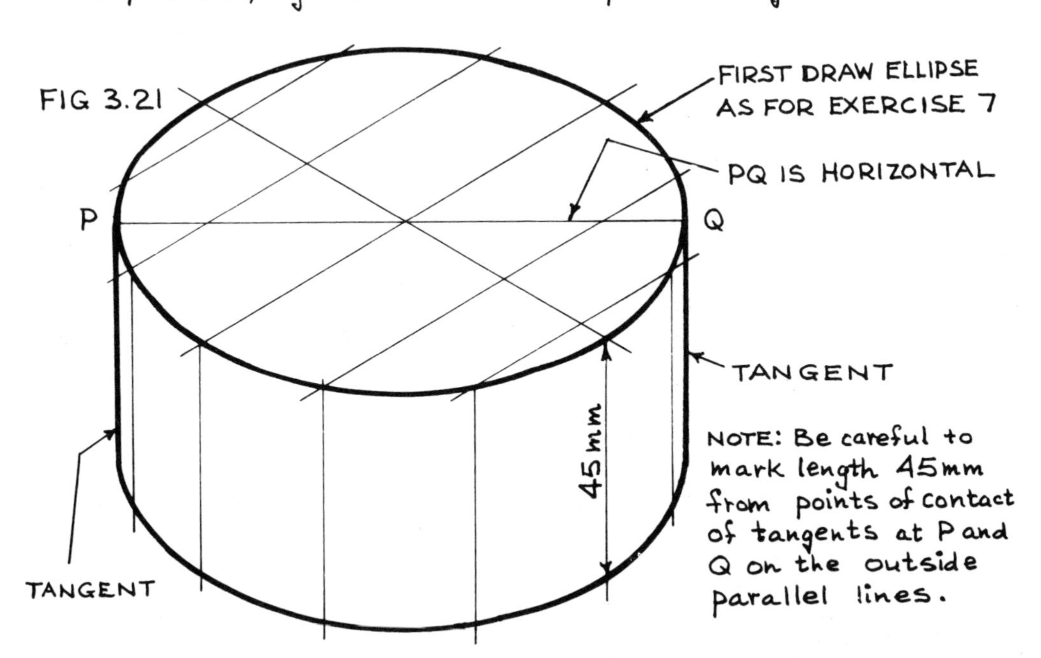

EXERCISE 9 SOLUTION

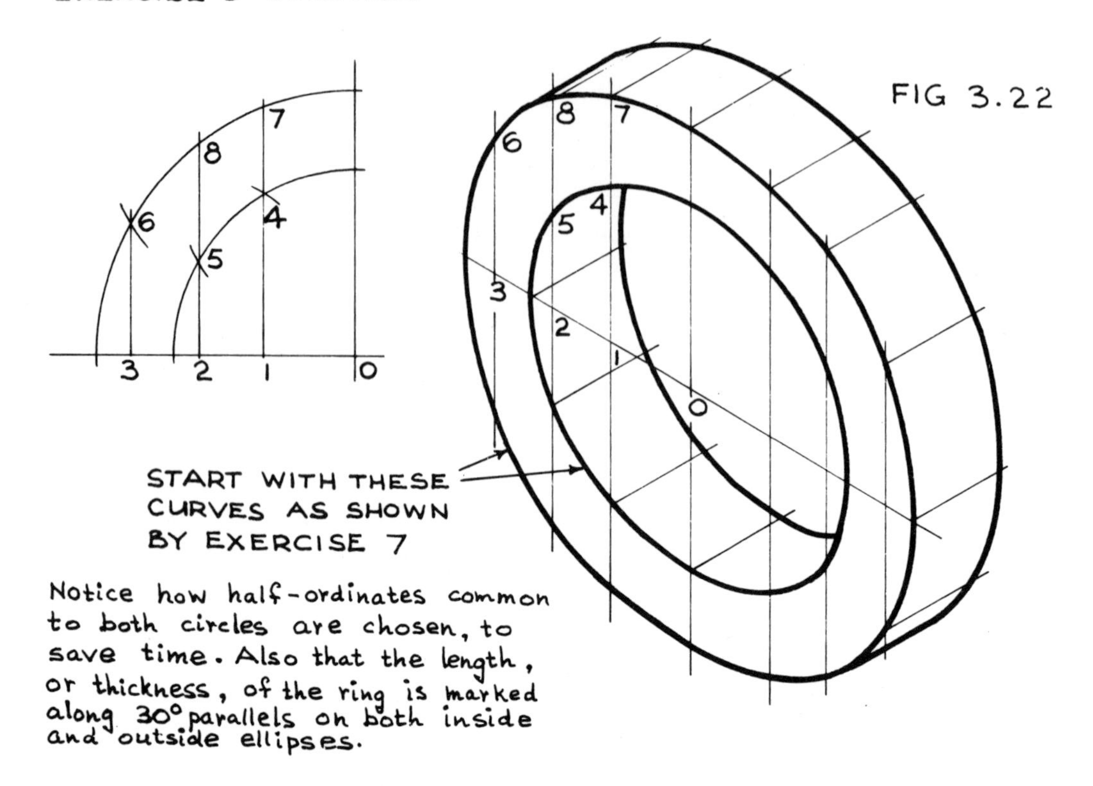

EXERCISE 10 SOLUTION

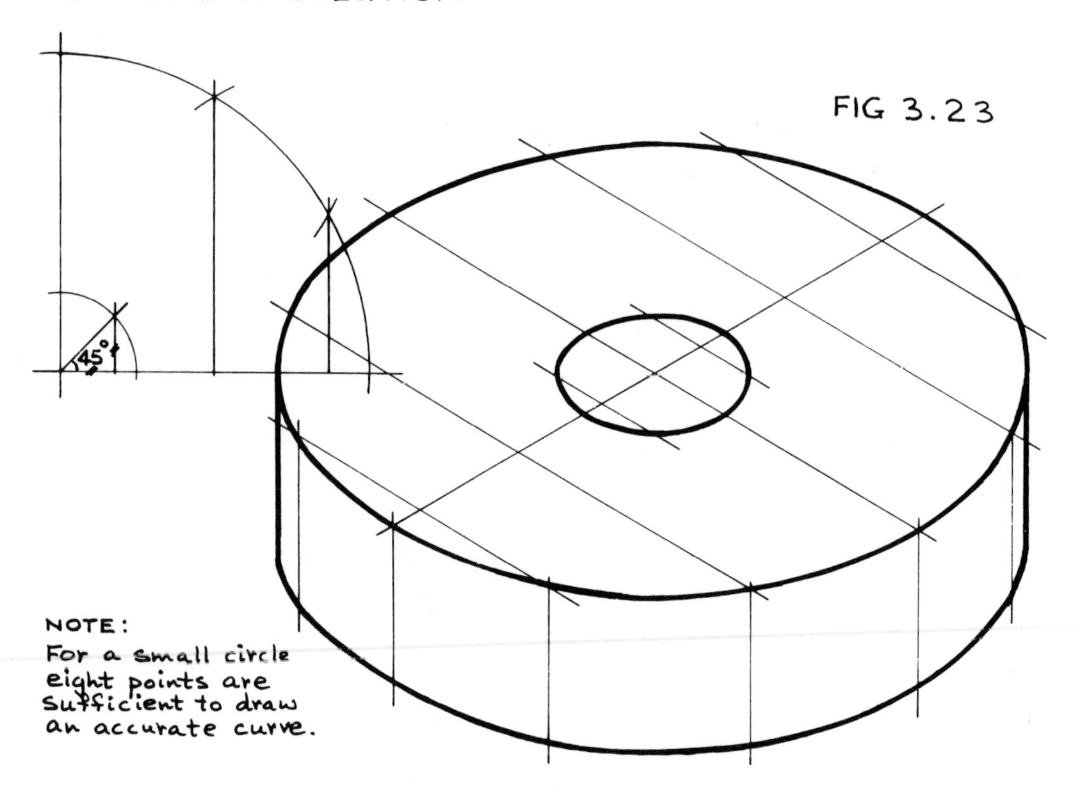

EXERCISE 11 SOLUTION

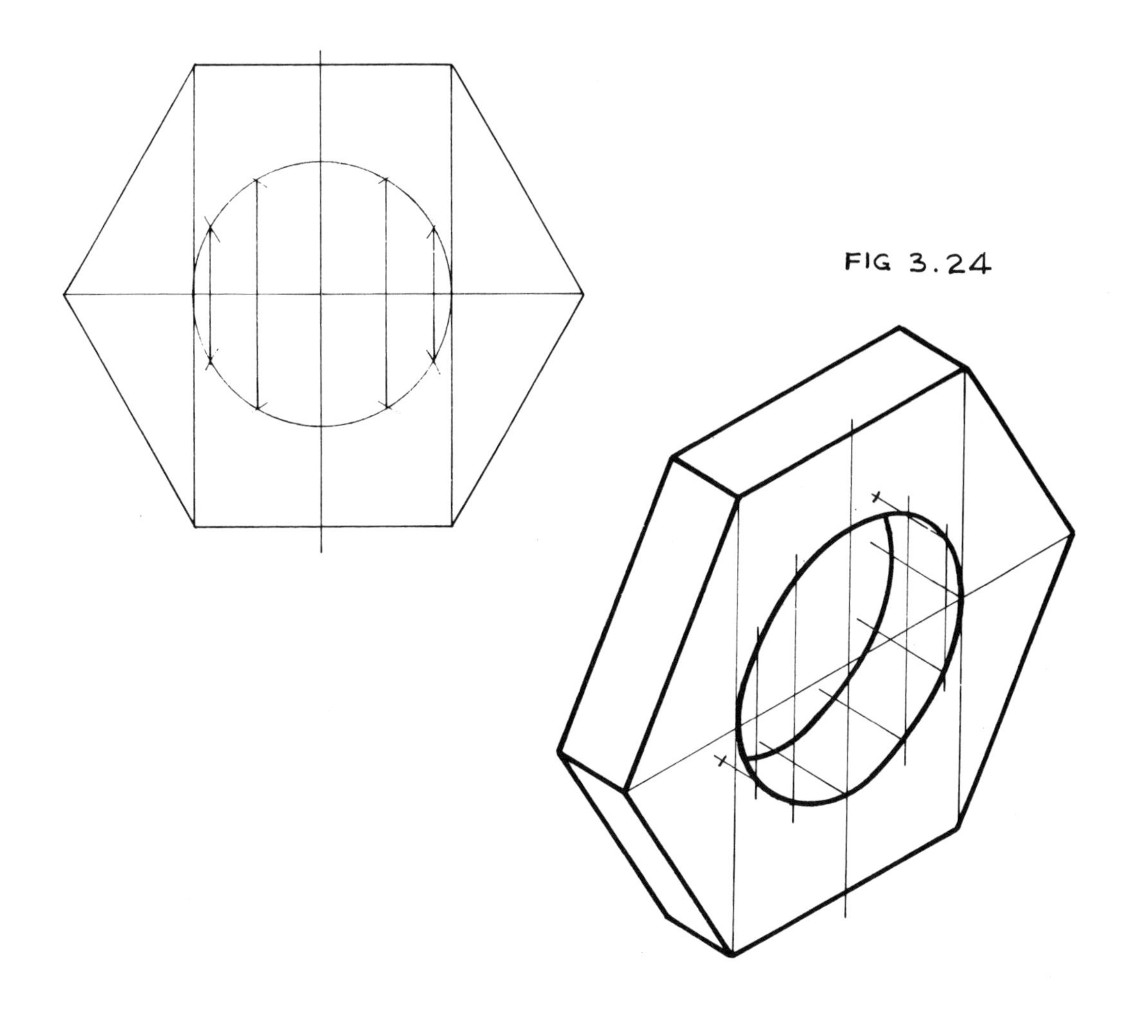

FIG 3.24

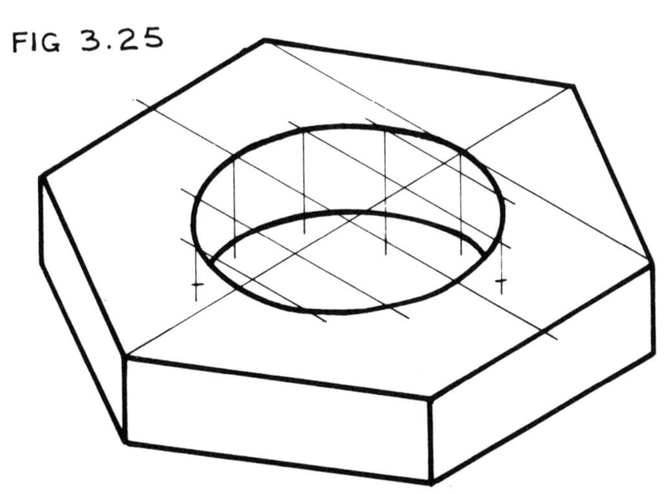

FIG 3.25

EXERCISE 12 SOLUTION

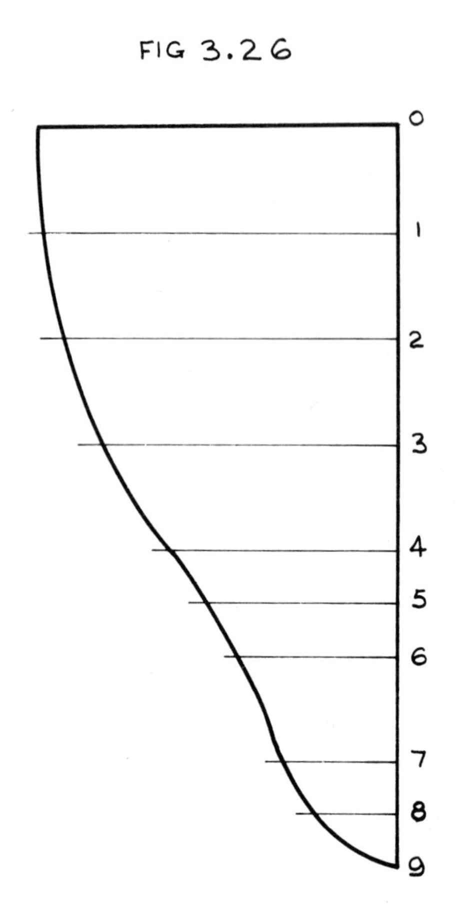

FIG 3.26

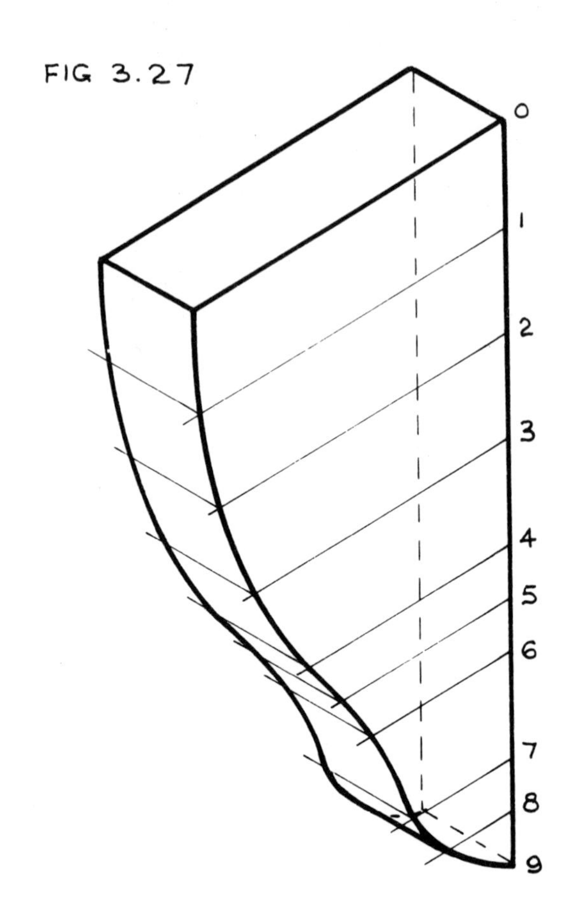

FIG 3.27

NOTE: USE ANY SUITABLY SPACED HORIZONTAL LINES AS SHOWN, FIG.3.26. LINES SHOULD BE DRAWN CLOSER TO OBTAIN THE SHAPE AT CURVATURE CHANGE.

Exercise 13. A Shaft Support, figure 3.28, is sketched on isometric graph paper. Using plain drawing paper make an isometric drawing of the object. Assume the triangle grid of the graph paper to be 5 mm.
Exercise 14. Repeat the task of the previous exercise but refer to the Double Eye Bracket shown sketched in figure 3.29.
Exercise 15. Referring to figures 3.30 and 3.31 make compass drawn approximations to ellipses, to show isometric drawings of a circle 100 mm in diameter. Use plain drawing paper.
Exercise 16. Using plain drawing paper and the compass method make isometric drawings to represent a cylinder 50 mm diameter and 90 mm long.
Exercise 17. Using plain drawing paper and the compass method make an isometric drawing of a hollow cylinder with a plane face vertical. Outside diameter 120 mm, inside diameter 80 mm and length 45 mm.
Exercise 18. Using isometric graph paper repeat *Exercises 16* and *17*.

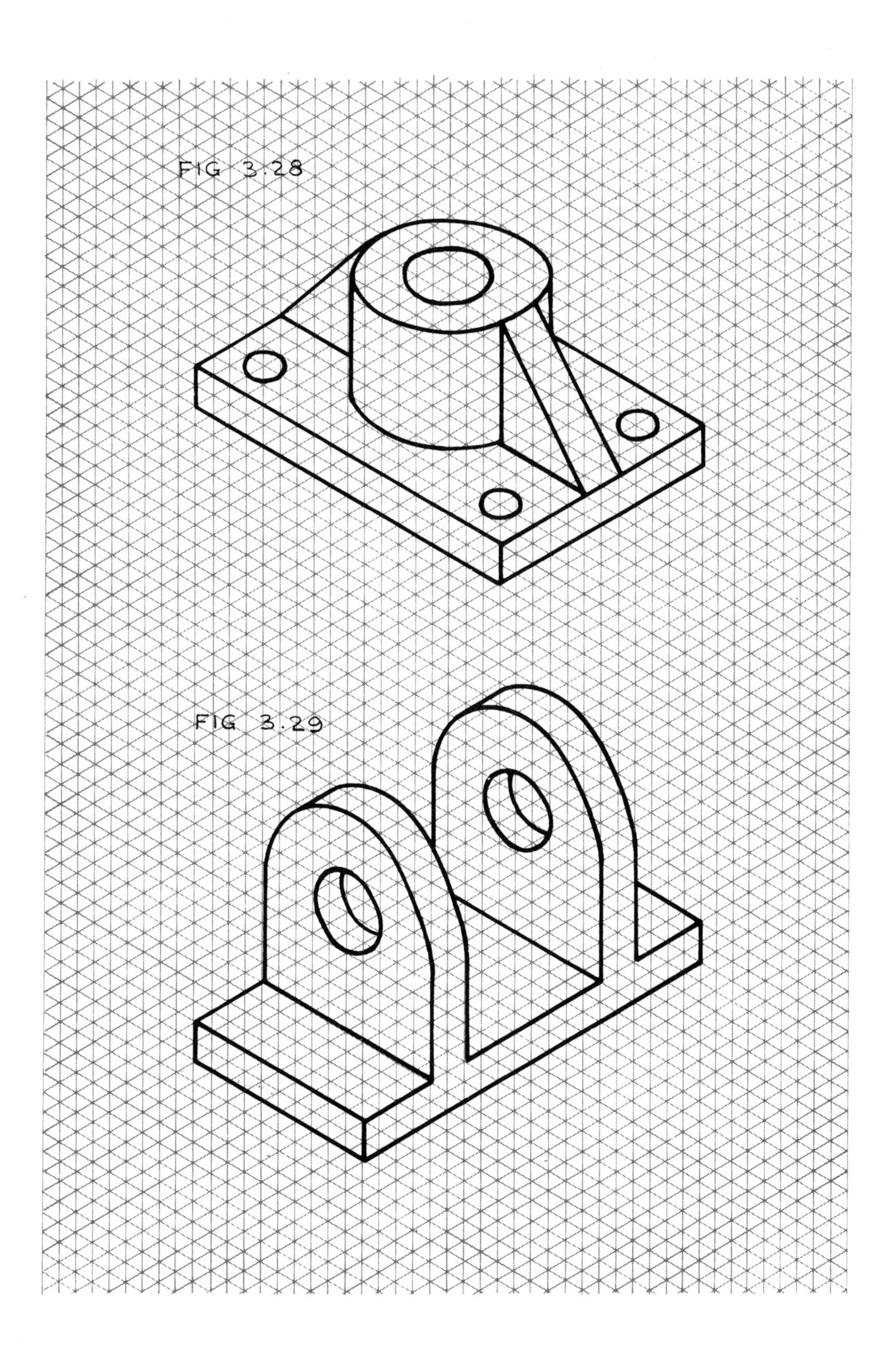
FIG 3.28
FIG 3.29

APPROXIMATE ELLIPSES FOR ISOMETRIC DRAWING

PROCEDURE :

First draw a rhombus, that is, a parallelogram with equal sides, using a 30° set-square. MAKE THE LENGTH OF SIDE OF THE RHOMBUS EQUAL TO THE DIAMETER OF THE REQUIRED CIRCLE.

Centres for the smaller compass drawn arcs are found on the longer diagonal of the rhombus.

To find these centres draw lines, making 60° with the longer diagonal, from each obtuse-angled corner.

Alternatively, draw Tee-square lines from corners, Fig.3.30, or 60° set-square lines, Fig. 3.31 .

Also notice that these lines through the obtuse-angled corners bisect the opposite sides of the rhombus.

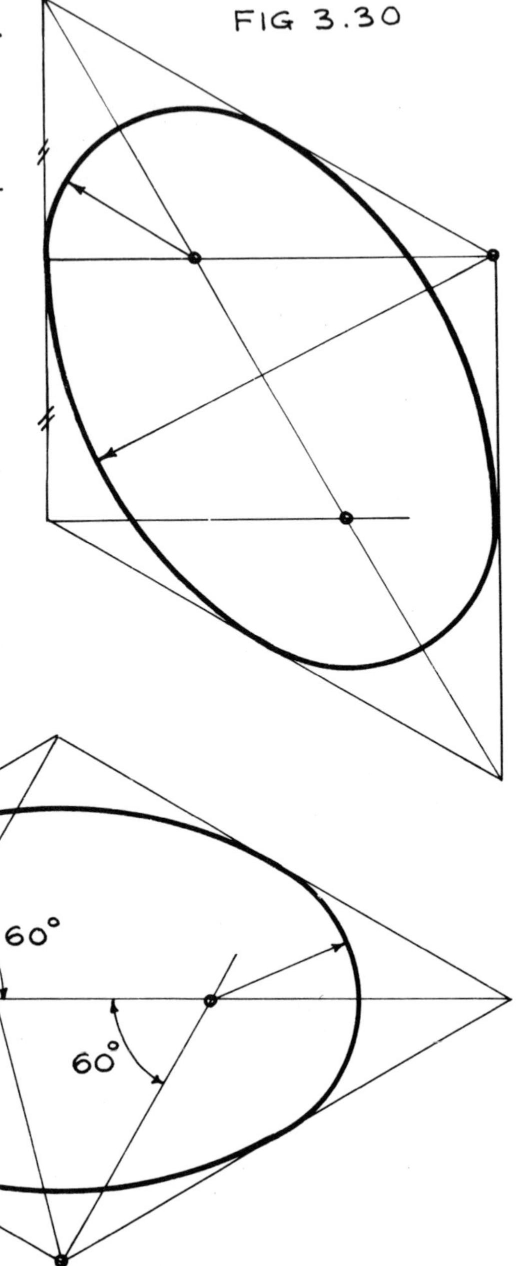

FIG 3.30

FIG 3.31

NOTE: This approximate construction is not generally acceptable for examination purposes. Never-the-less it is a simple, neat and quick construction for circles occurring in isometric drawing.

COMPARISON OF ORDINATE AND COMPASS METHODS

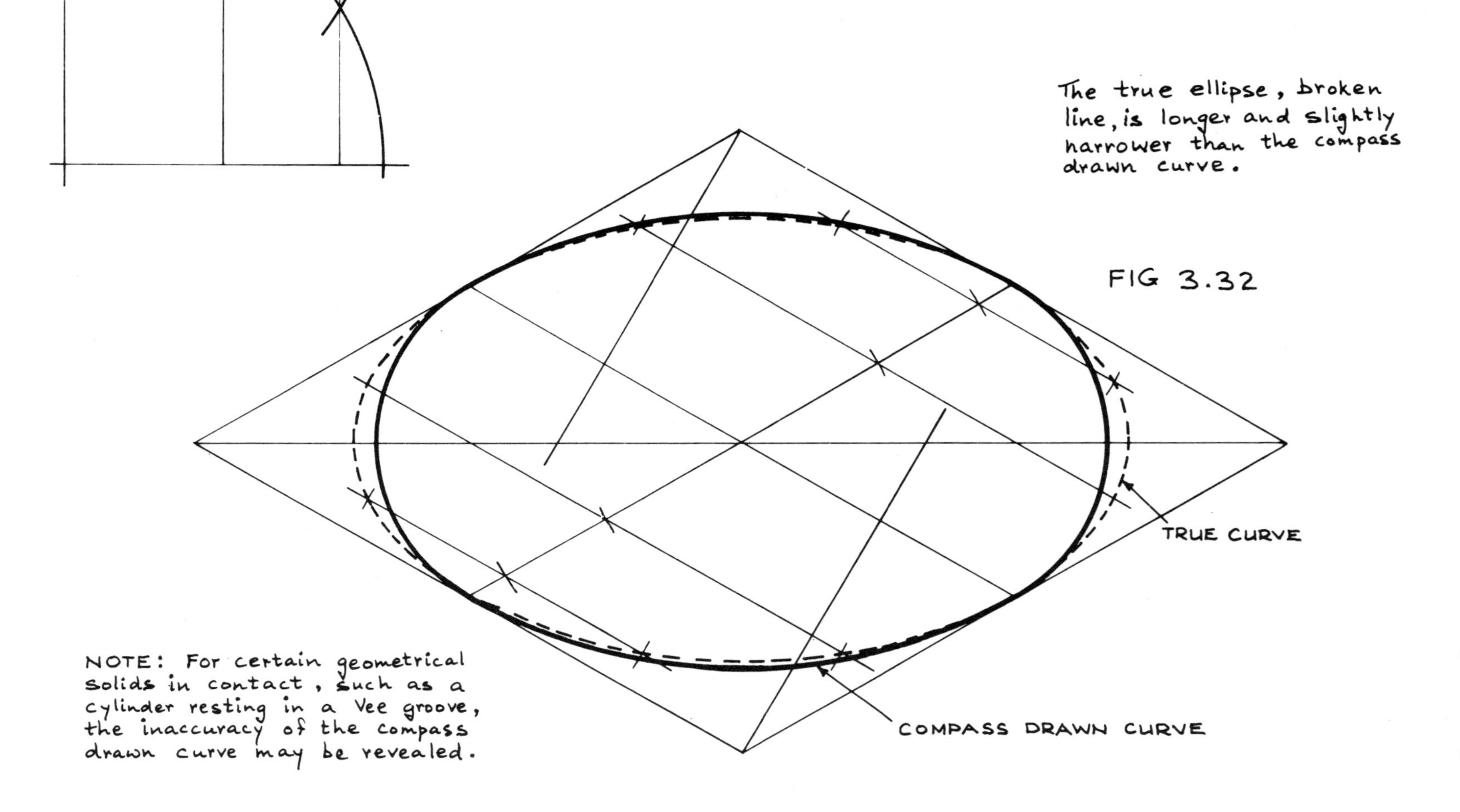

FIG 3.32

EXERCISE 16 SOLUTION

FIG 3.33

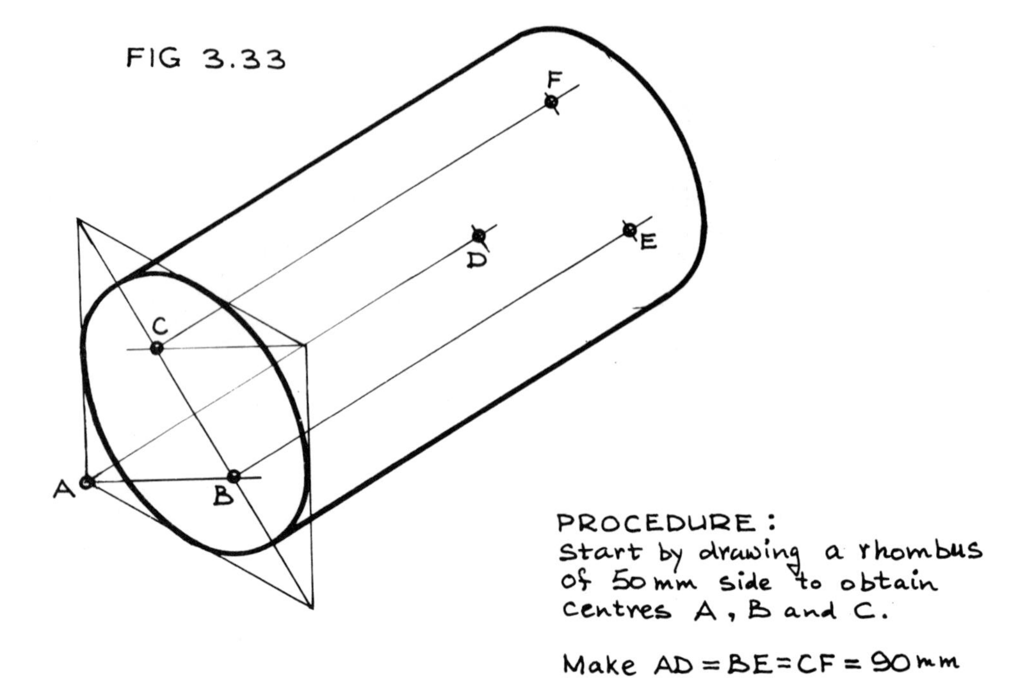

PROCEDURE:
Start by drawing a rhombus of 50 mm side to obtain centres A, B and C.

Make AD = BE = CF = 90 mm

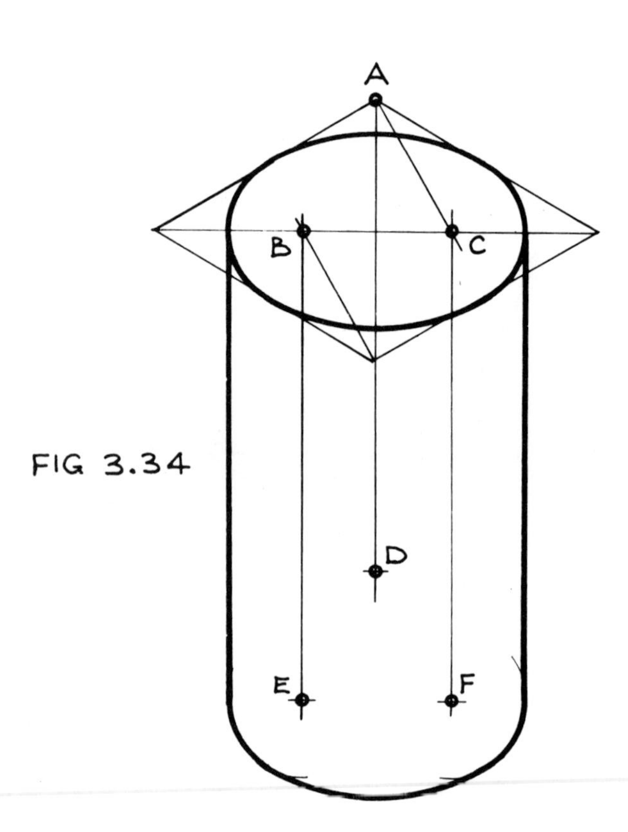

FIG 3.34

EXERCISE 17 SOLUTION

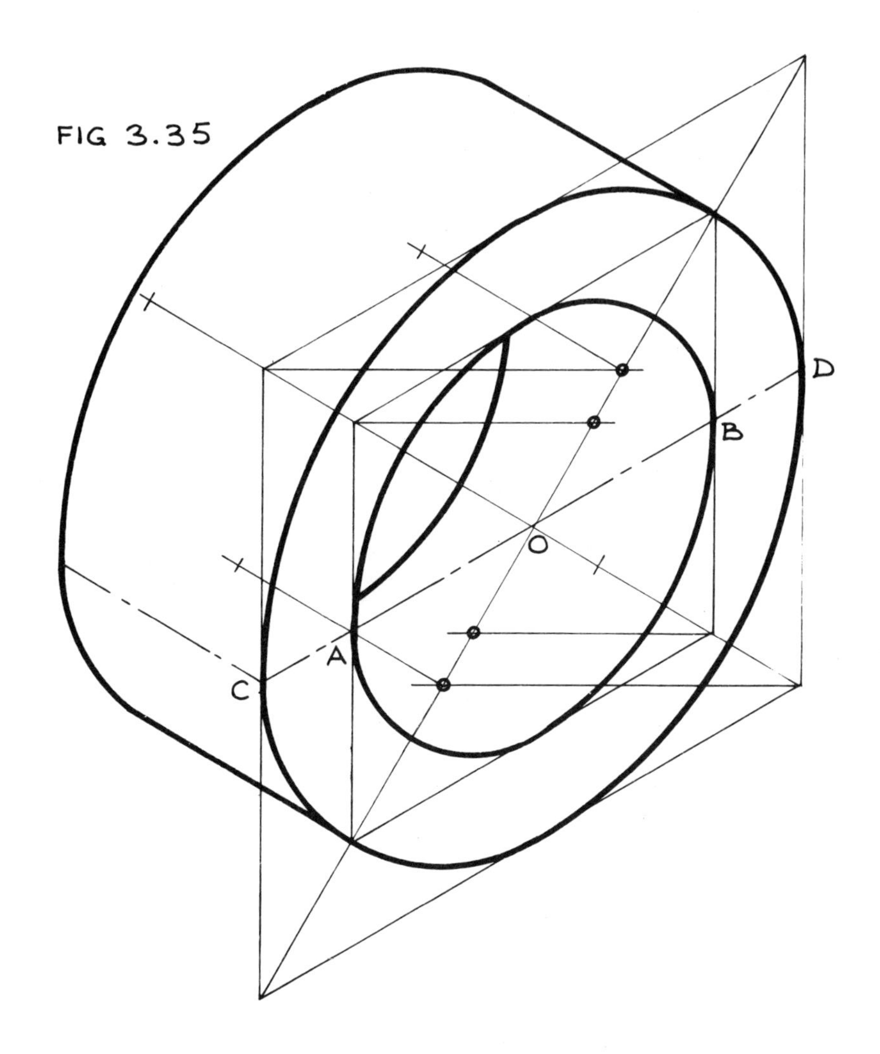

NOTE:

A separate rhombus is required for the inside 'circle'.

This rhombus is most easily drawn using the geometrical centre O and marking half the length of side of this smaller rhombus along OA and OB.

To draw an isometric 'semi-circle', a complete construction rhombus is still necessary.

Isometric graph paper may conveniently be used for this construction method.

Note. Use a scale of twice full size for the following exercises.

Exercise 19. Using the *ordinate* method, make an isometric drawing of the Open Bearing shown on 5 mm squares graph paper, figure 3.36 (Third Angle projection). Make point A the centre for isometric axes.

Exercise 20. Make an isometric drawing of the Open Bearing, figure 3.36, using isometric graph paper, with line PQ a vertical line in the isometric drawing.

Exercise 21. Using compasses to draw approximate ellipses, make an isometric drawing of the Open Bearing, figure 3.36, in any position of your choice.

Exercises 22, 23 and *24*. Follow the instructions for *Exercises 19, 20* and *21* respectively but refer to the Slide Block shown in Third Angle projection in figure 3.37.

Exercise 25. Make an isometric drawing of the Shaft Support shown in Third Angle projection in figure 3.38. Use the *ordinate* method and position the axis XX at 30° to the horizontal in the isometric view.

Exercise 26. Using isometric graph paper make an isometric drawing of the Shaft Support, figure 3.38, with a large circle diameter as a horizontal major axis of the ellipse in the isometric view.

Exercise 27. By compass method make an isometric drawing of the Shaft Support, figure 3.38, corresponding to the view shown in the previous exercise.

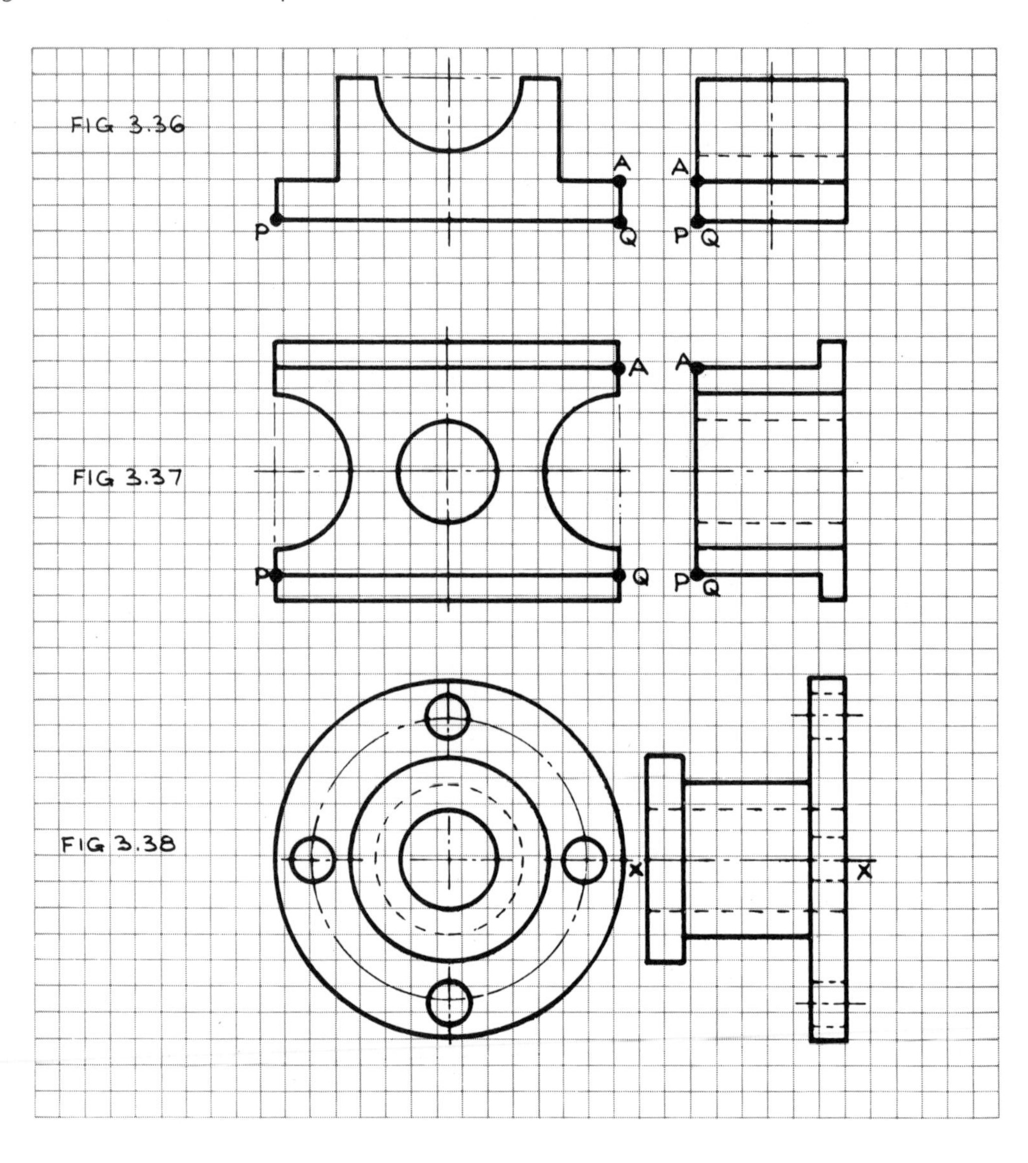

OBLIQUE DRAWING

Oblique drawing, sometimes called Oblique Projection, is another useful method of making a pictorial drawing of an object.

Cavalier projection or Cabinet projection are other names sometimes applied to oblique projection.

Oblique projection is most useful for portraying an object which has curves, probably circles, in one particular view. This front view with the curves is first of all drawn as in orthographic projection (treated in chapter 2) and then the side edges are drawn by parallel lines making an angle with the horizontal.

If the same scale is used along the inclined parallel edges as for the front face; it is known as *cavalier* projection. However, use of the same scale on both side and front face views leads to a much distorted effect so to correct for this in some measure, side measurements are made a certain fraction of the front face scale. Parallel side measurements of scale $\frac{1}{3}$, $\frac{1}{2}$ or $\frac{2}{3}$ of the front face measurements would use angles of 60°, 45° or 30° respectively. In general, however, an oblique drawing is made *full-scale on front face and half-scale on sides at an angle of 45°* with the horizontal.

Obviously oblique projection is not a true system of projection, since if a front face view is drawn as in orthographic projection then it is not possible to see the sides of the object if they are perpendicular to the front face. This fact is overlooked in the interests of simplicity and convenience.

Exercise 28. Using either plain drawing paper or 5 mm squares graph paper make oblique drawings of figures 3.3, 3.6, 3.5 and 3.8.

Note. In the solution to *Exercise 28*, shown in figures 3.39, 3.40, 3.41 and 3.42, the shaded face in each case shows the full-size front face view with which the oblique drawing is commenced. A 45° set-square is used to draw the parallel edges on sides and top and the scale is half-size in this direction.

Exercise 29. Make an oblique drawing of a hollow cylinder 80 mm outside diameter, 60 mm inside diameter and 100 mm long.

Exercise 30. A regular hexagonal prism of 50 mm side of hexagon and 15 mm thickness has an axial hole 50 mm in diameter drilled through. Make an oblique drawing of the object.

Exercise 31. Figure 3.45 shows a Simple Bearing in Third Angle orthographic projection, drawn on 5 mm squares graph paper. Make an oblique drawing of the bearing. Use plain drawing paper.

Exercise 32. A Shaft Support Bracket is shown in Third Angle orthographic projection in figure 3.46. Taking the squares grid as 5 mm, make an oblique drawing of the bracket. Use plain drawing paper.

Note. When dealing with the holes in the base of the Simple Bearing, figure 3.47, the student should bear in mind that the horizontal ordinates are full-size but the spacing of these ordinates on the 45° axis is *half* full-size.

SOLUTION TO EXERCISE 28

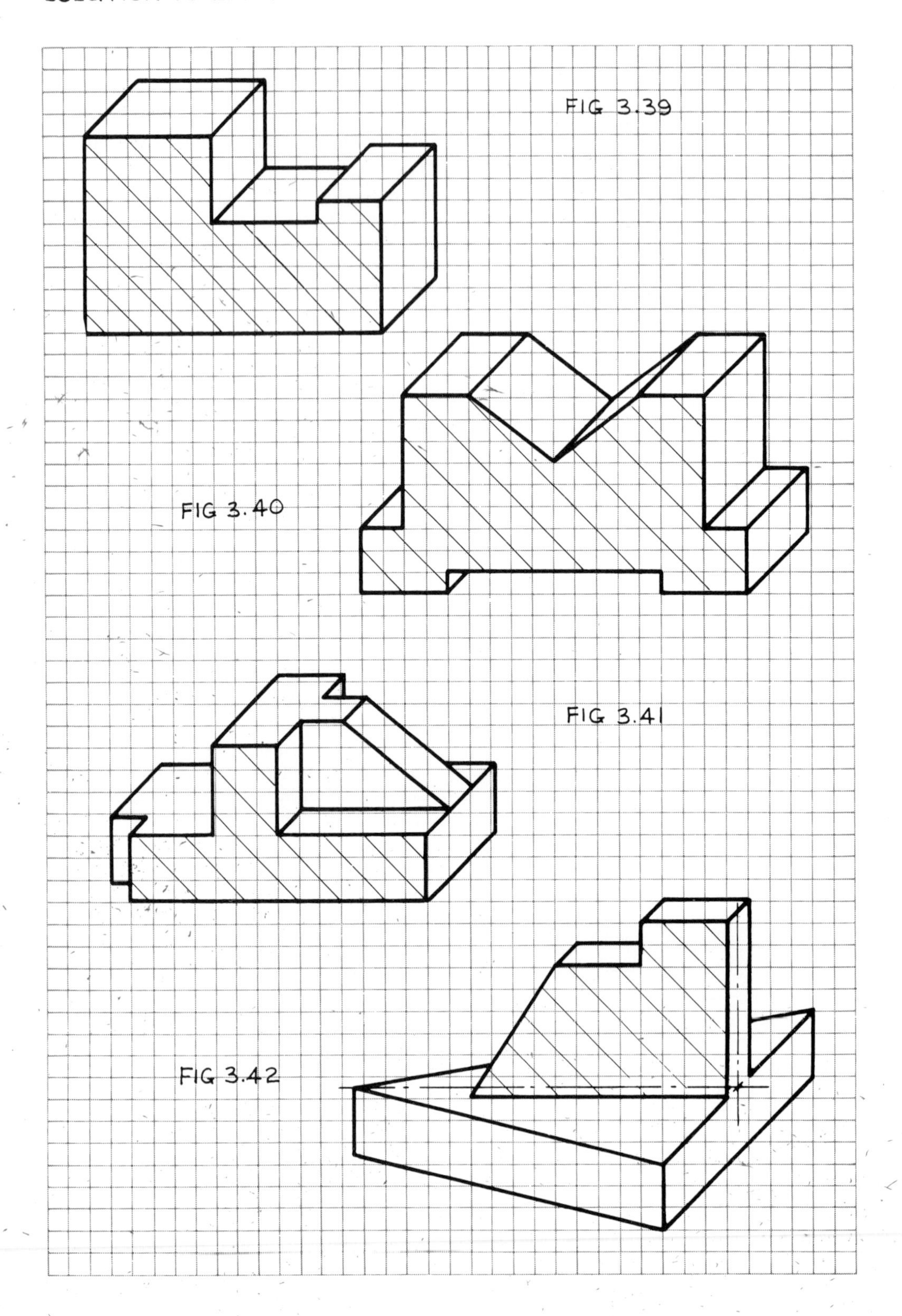

EXERCISES 29 AND 30 SOLUTIONS

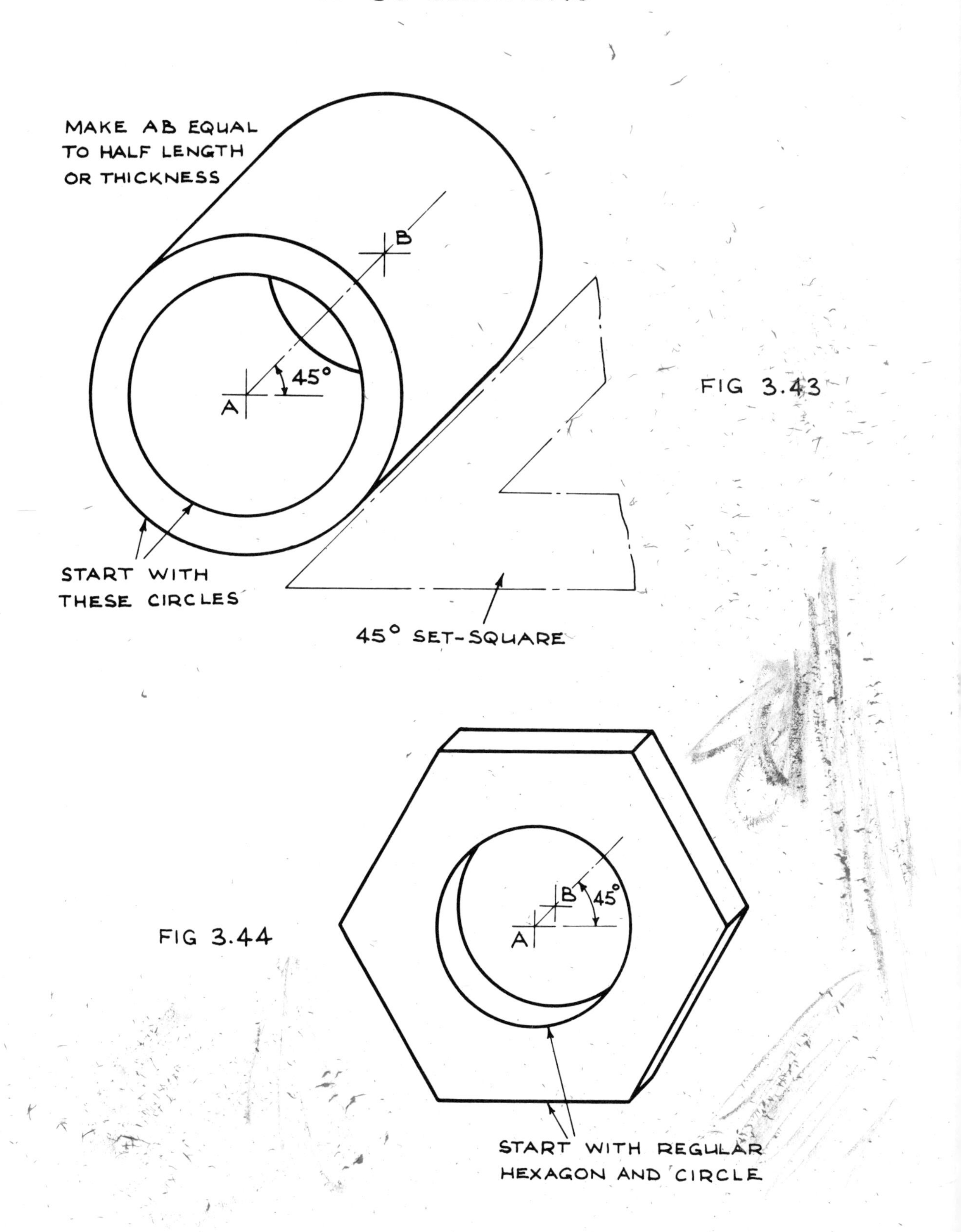

FIG 3.43

FIG 3.44

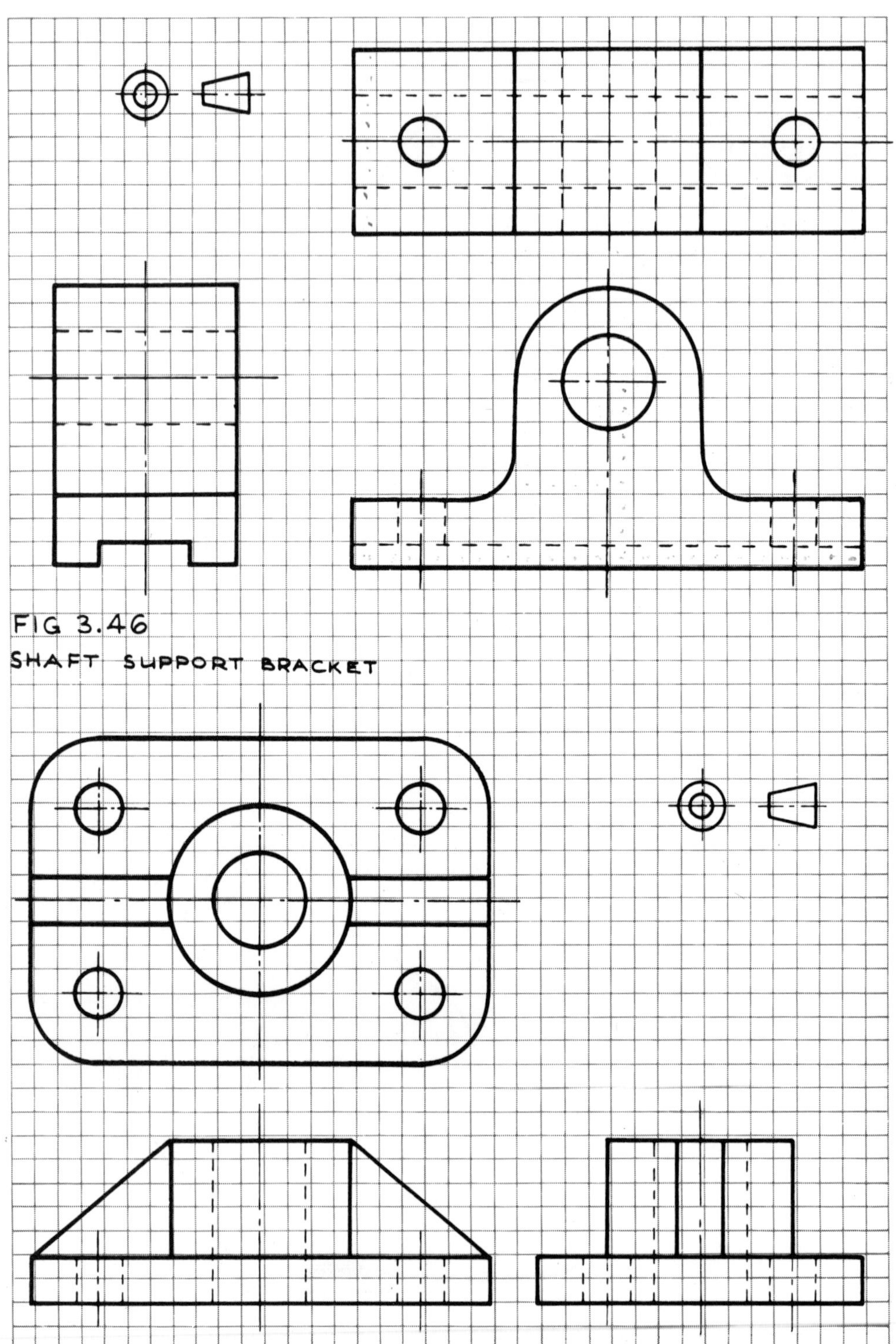
FIG 3.45
SIMPLE BEARING
FIG 3.46
SHAFT SUPPORT BRACKET

EXERCISES 31 AND 32 SOLUTIONS

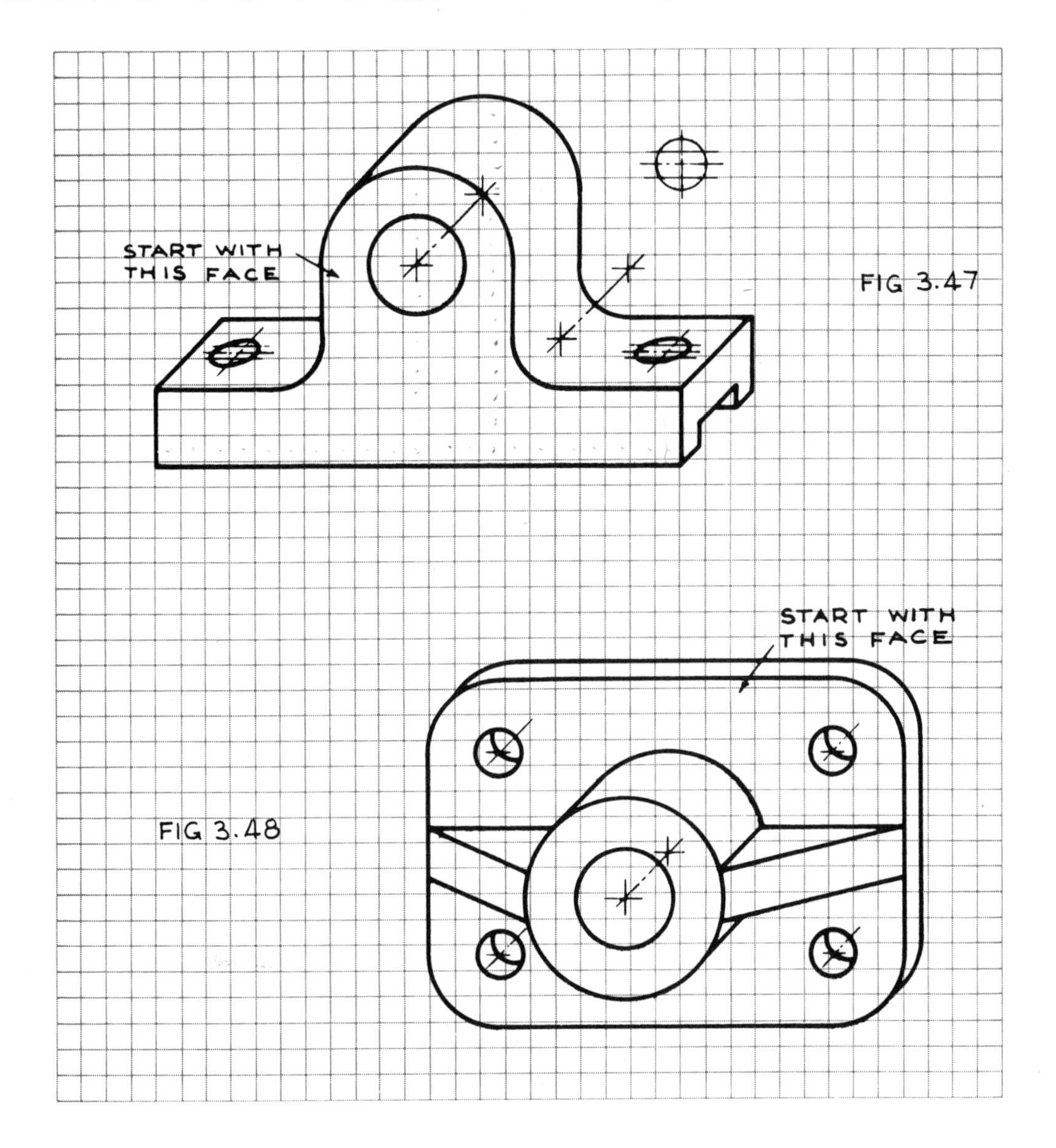

4
AUXILIARY PROJECTION

This is an extension of orthographic projection to obtain additional views, elevations or plans, of any given object. The student is referred to chapter 2 in which the principles of orthographic projection are dealt with in detail.

First to be considered here are additional elevations, called *auxiliary* elevations, of an object. Sometimes these additional elevations are referred to simply as *new* elevations.

Figure 4.1 shows a pictorial view of a simple object placed in the first quadrant, with the principal planes VP and HP. In First Angle projection the Front Elevation is obtained by viewing in the direction of arrow A and this is drawn on the principal vertical plane. The idea of an End Elevation is explained in chapter 2 where an auxiliary vertical plane perpendicular to the two principal planes is employed. But in the case of auxiliary projection of new elevations an auxiliary vertical plane such as is shown in figure 4.1 is used. This plane is perpendicular to the HP but may be inclined at any given angle to the VP. The horizontal trace of this auxiliary vertical plane is the X_1Y_1 line in figure 4.1.

If the object shown in figure 4.1 is now viewed in the direction of the arrow B, which is perpendicular to the X_1Y_1 line, then a *first auxiliary elevation* is obtained as shown.

The pictorial drawing of figure 4.1 is translated into orthographic projection as shown in figure 4.2. The student should carefully note the simple principles for obtaining auxiliary elevations, which are given as follows:

(i) Always project at right angles to the new XY line, which is sometimes called the new ground line and denoted by X_1Y_1.

(ii) Make the height measurements for the auxiliary elevation the same as for the given elevation.

Note on use of First or Third Angle projection

As noted in chapter 2 there is still no agreement on the exclusive use of either First Angle or Third Angle orthographic projection. There is a trend in favour of the Third Angle system, particularly with larger firms having world wide trading interests. However, the author finds that even the most fanatical advocates of the Third Angle system often revert to the First Angle system when dealing with auxiliary projection. The idea of a ground line (XY line) with the object 'resting' on the line perhaps makes changes of ground line more easily envisaged. However, for an object placed in the third quadrant the idea of a ground line is not really relevant since the object would in fact be below the ground!

From the educational view point both First and Third Angle systems of projection will be used and for this particular topic more emphasis will be given to First Angle.

The equivalent Third Angle orthographic projection solution to the First Auxiliary Elevation of the object shown pictorially in figure 4.1, is given in figure 4.3.

Note. The various 'XY' lines involved may be regarded simply as datum lines. In fact, when using First Angle projection, as in figure 4.2, the object need not necessarily be drawn on the XY line, but may be drawn above it. In some cases it may even be found convenient to dispense with XY lines and work with centre lines of views instead.

Two different First Auxiliary Elevations are shown in figure 4.4, of the object pictured in figure 4.7. First Angle projection is used for these views. Similarly, figure 4.5 shows two First Auxiliary Elevations of the object pictured in figure 4.8, but in this case the Third Angle system is used.

FIG 4.1

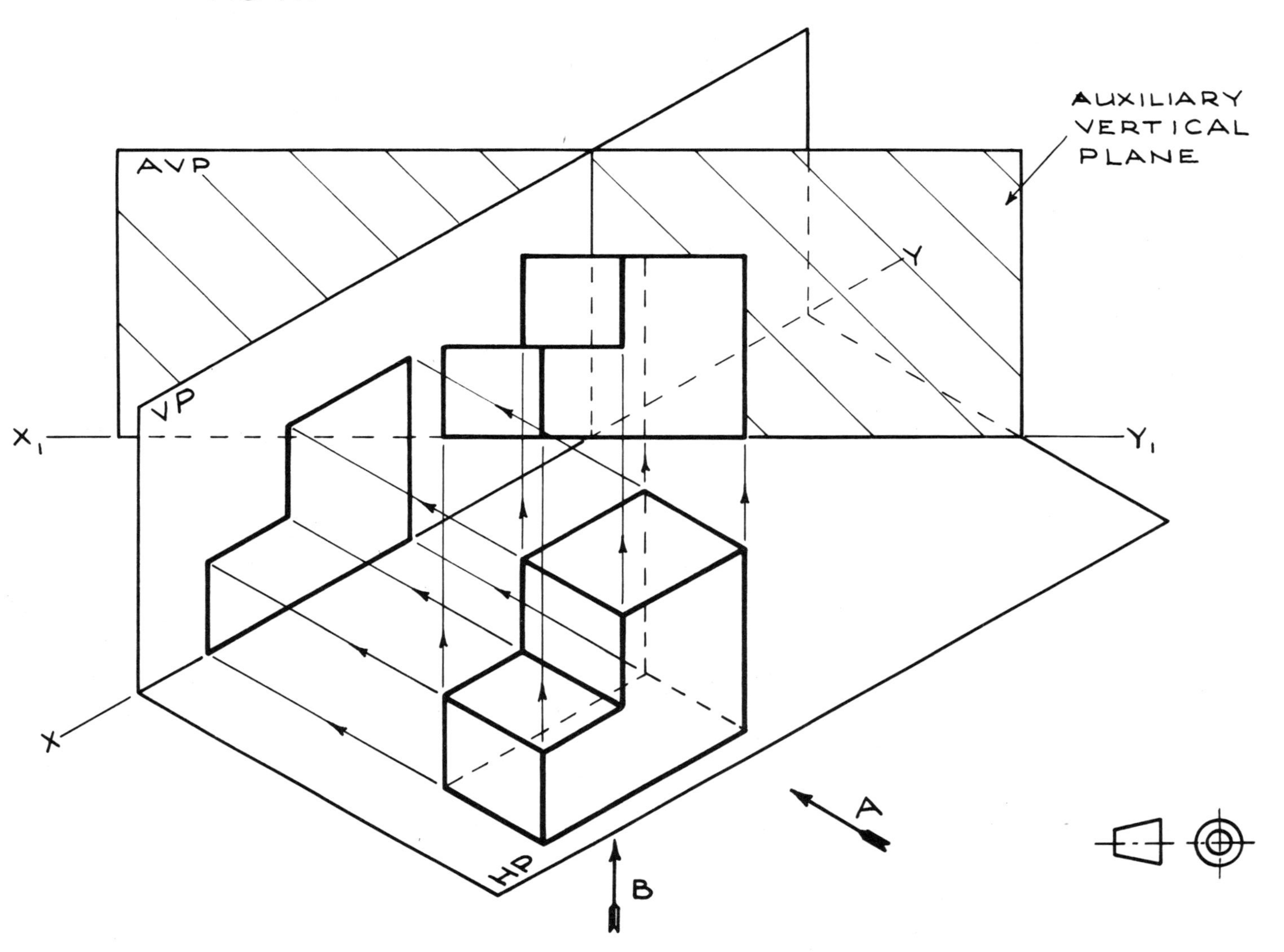

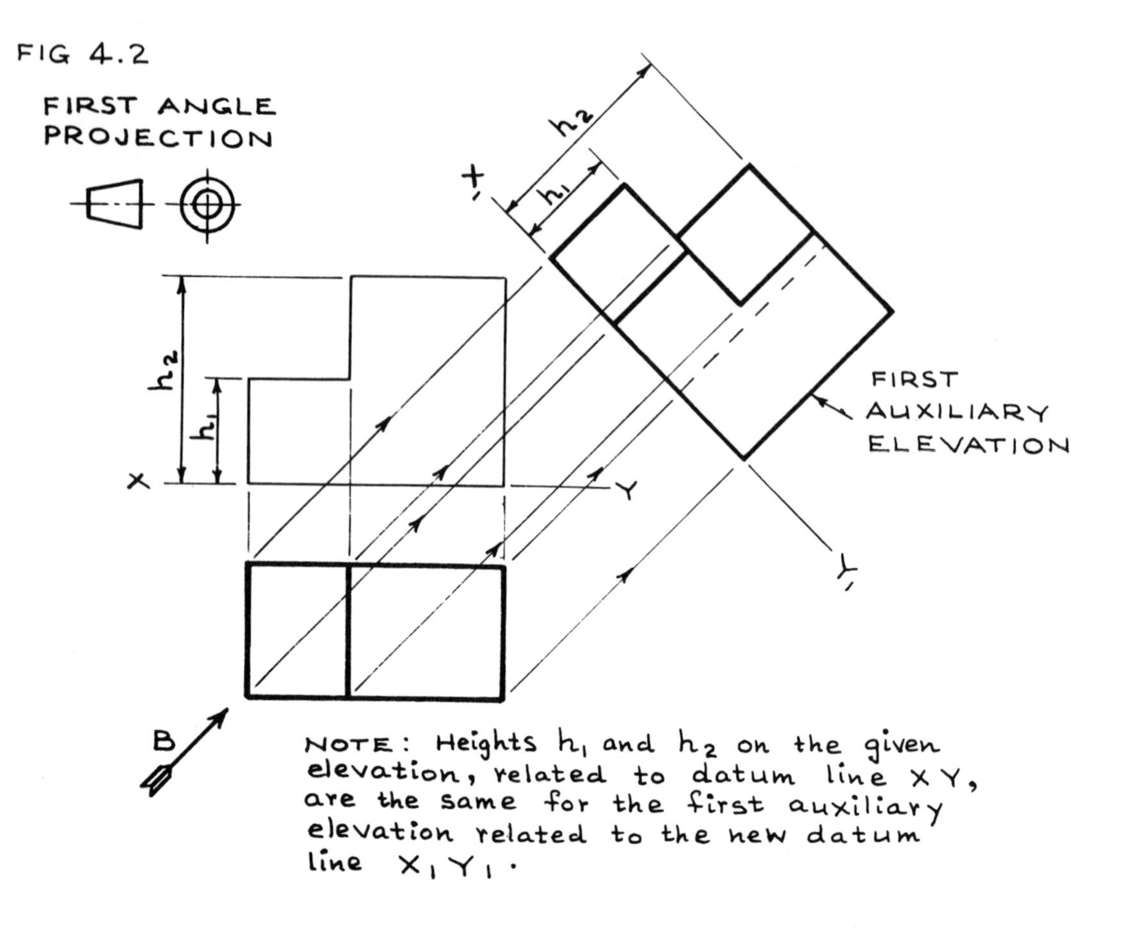

NOTE: Heights h_1 and h_2 on the given elevation, related to datum line XY, are the same for the first auxiliary elevation related to the new datum line X_1Y_1.

FIG 4.3

THIRD ANGLE PROJECTION

X_1

X

Y

d

B

h_2

h_1

Y_1

FIRST AUXILIARY ELEVATION

FIG 4.4

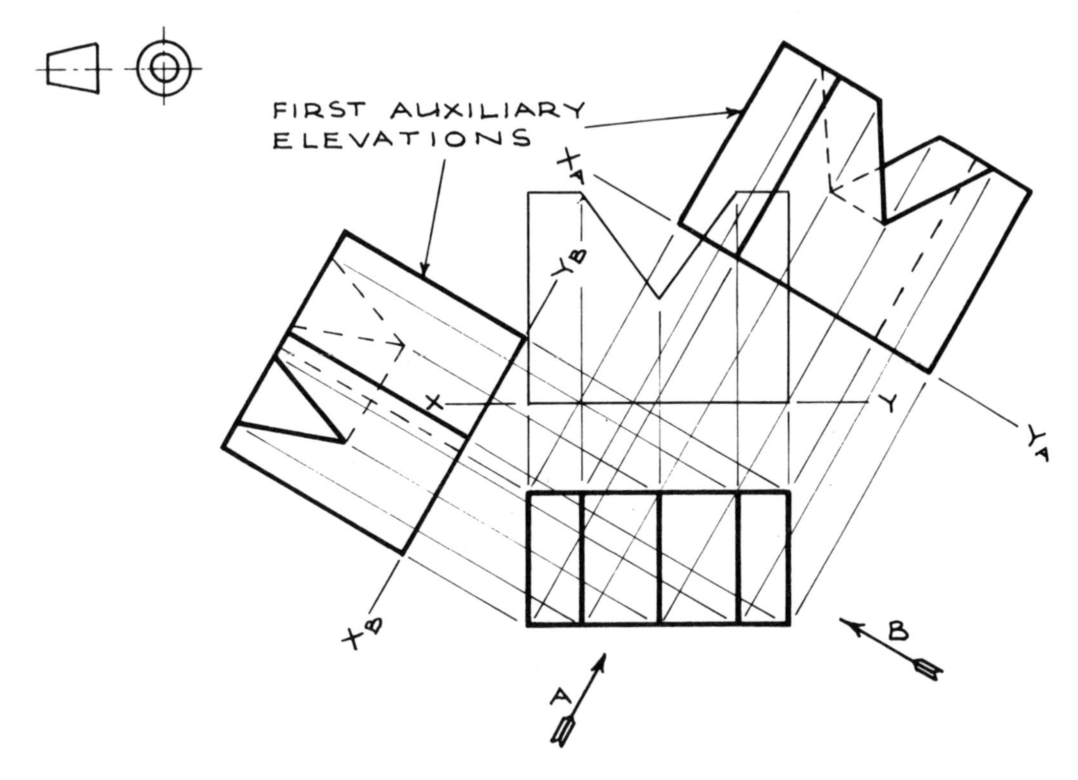
FIRST AUXILIARY
ELEVATIONS
X
Y
X_A
Y_A
X_B
Y_B
A
B

FIG 4.5

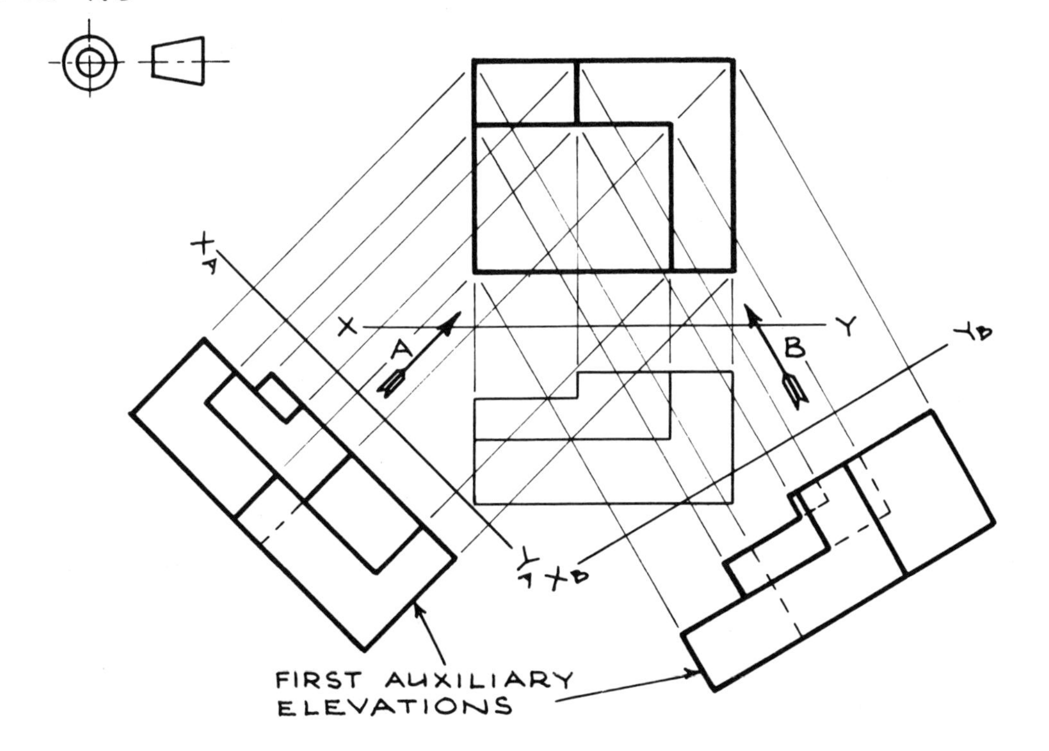
X
Y
X_A
Y_A
X_B
Y_B
A
B
FIRST AUXILIARY
ELEVATIONS

Exercise 1. Using First Angle orthographic projection draw a Front Elevation and Plan of the object pictured in figure 4.6. Add two separate First Auxiliary Elevations of the object using new datum lines (ground lines) inclined at any convenient angles to the horizontal datum line (XY line).

Note. Angles of 30°, 45°, or 60° are convenient angles unless the student is fortunate enough to possess an adjustable set-square, in which case any angles are convenient.

Exercises 2 to *6*. Repeat the instructions as for *Exercise 1* but refer to figures 4.7 to 4.11 respectively.

Exercises 7 to *12*. Repeat the instructions as for *Exercise 1* but use Third Angle orthographic projection and refer to figures 4.6 to 4.11 respectively.

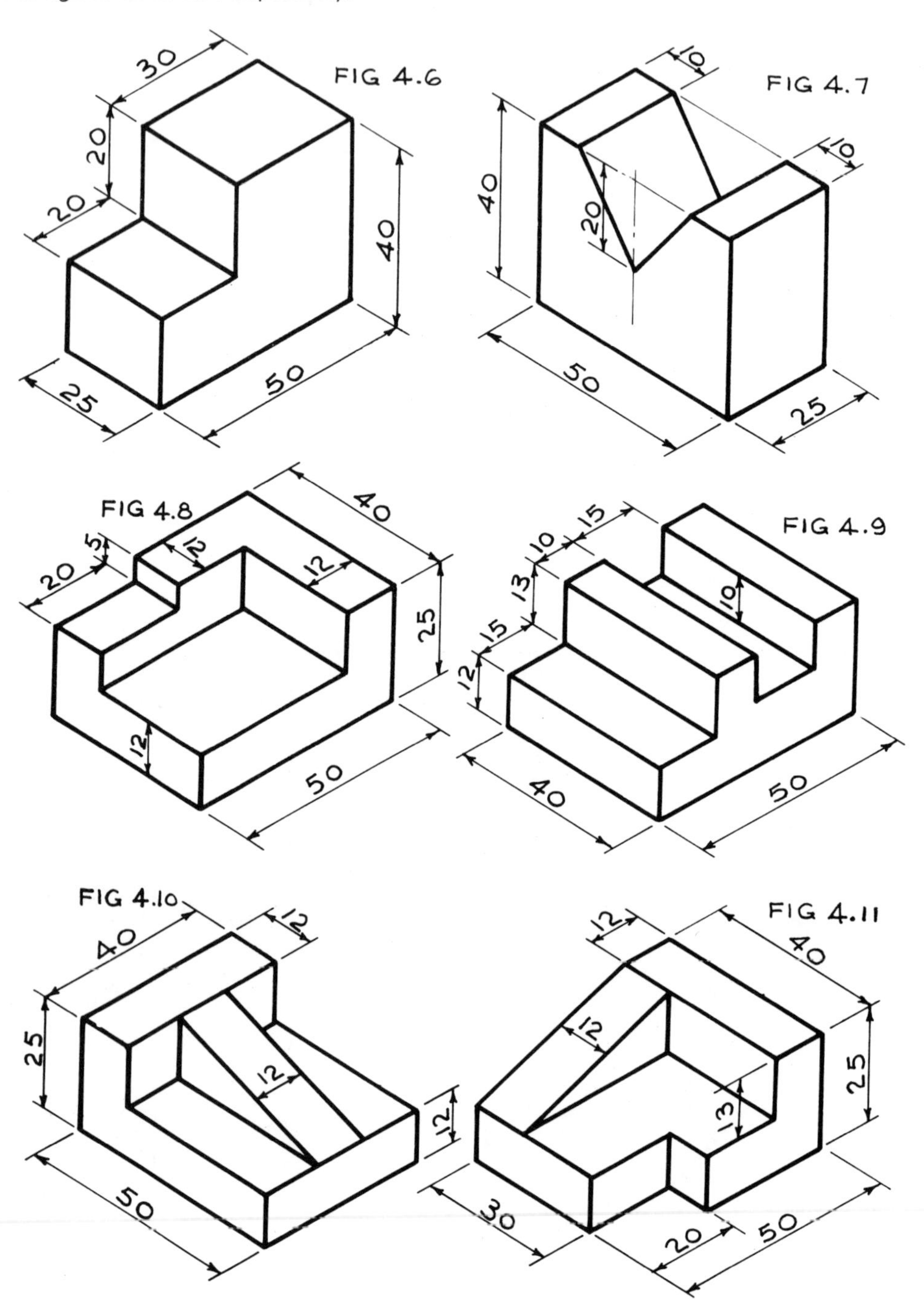

THE EFFECT OF TILTING THE GIVEN ELEVATION

If the given elevation is tilted at an angle to the initial horizontal datum line (XY line) the resulting First Auxiliary Elevations are as shown in figures 4.12 to 4.15. The student should notice that this results in a much clearer pictorial type view of the object. The exercises on this procedure, which follow, also serve to emphasise the two very simple principles for obtaining auxiliary elevations.

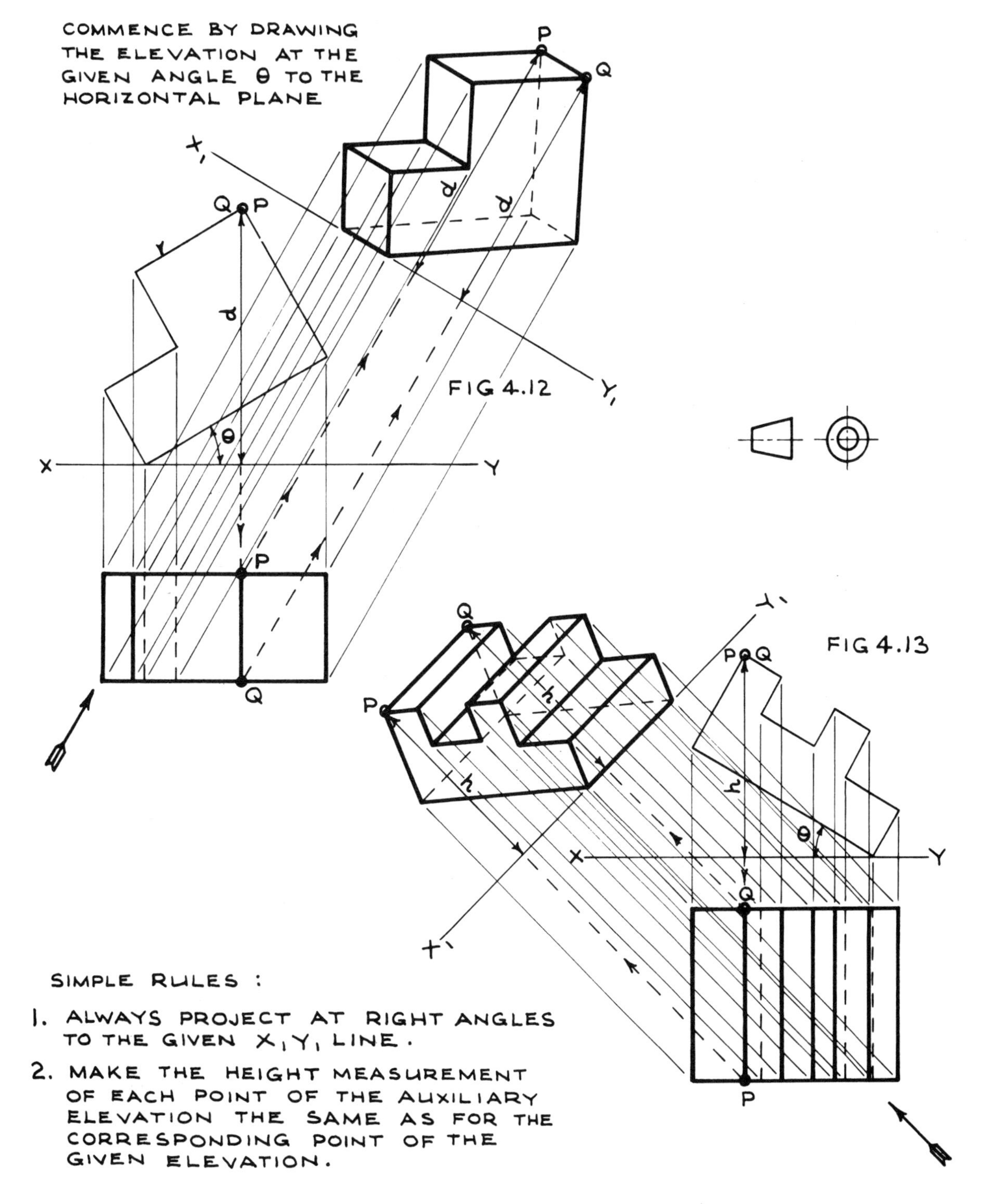

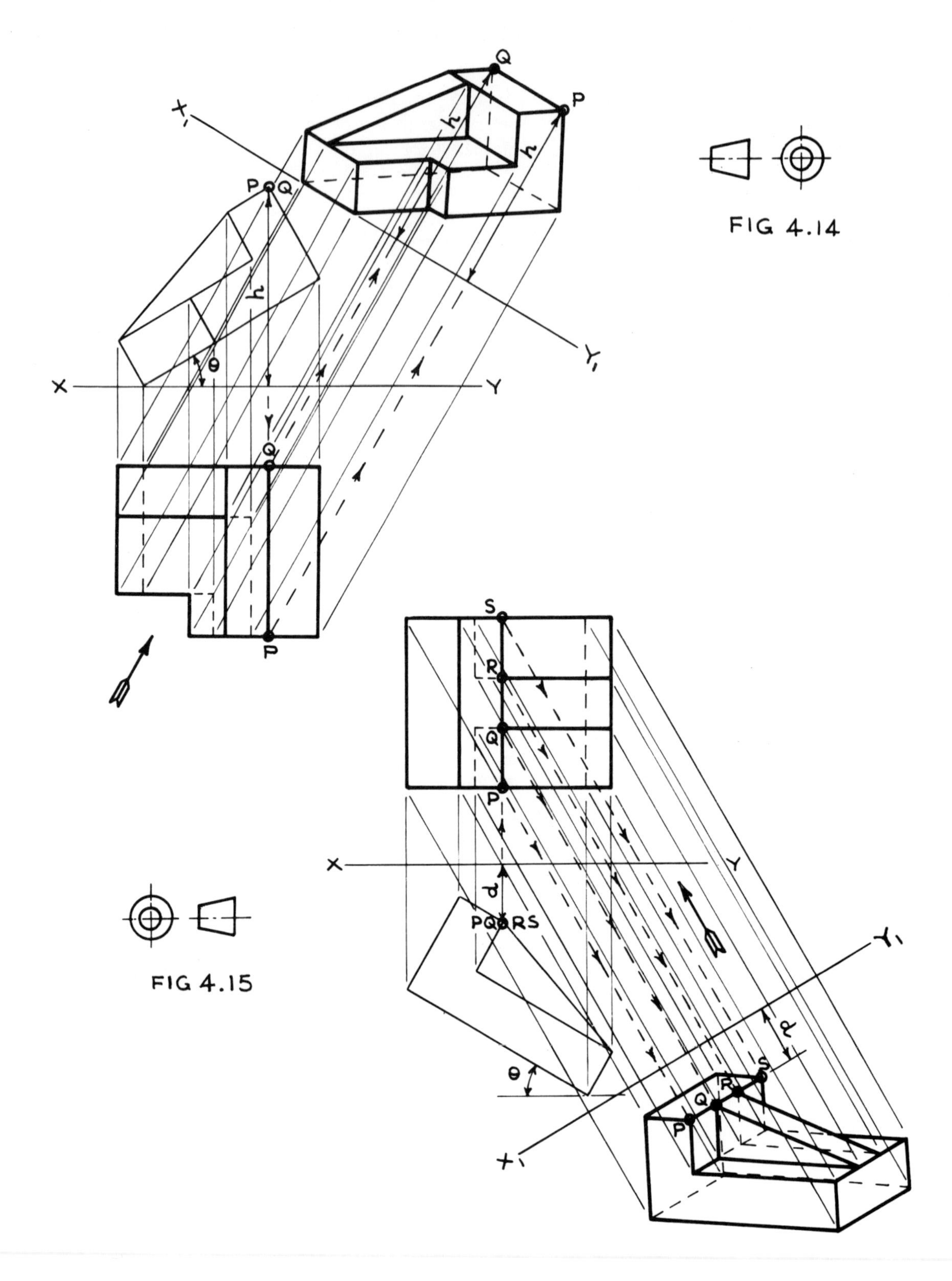

NOTE: θ HAS BEEN MADE 30° FOR CONVENIENCE

Exercise 13. Using First Angle orthographic projection draw a tilted Front Elevation and corresponding Plan of the object pictured in figure 4.6. Use any convenient angle of inclination to the horizontal plane (XY line). Add one First Auxiliary Elevation referred to a new datum line (X_1Y_1 line) which is inclined at any convenient angle to the XY line.
Exercises 14 to *16.* Repeat the instruction for *Exercise 13* but refer to figures 4.7 to 4.9 respectively.
Exercises 17 and *18.* Using Third Angle orthographic projection repeat the instructions as for *Exercise 13* but refer to figures 4.10 and 4.11 respectively.

FIRST AUXILIARY PLANS

By a similar process to that employed for finding First Auxiliary Elevations, First Auxiliary Plans may be constructed. However, in this case a simple inclined plane is employed as an auxiliary plane for projection of the new view. The vertical trace line of this auxiliary inclined plane becomes the new datum line (X_1Y_1 line).

Some examples of Auxiliary Plans are shown in figures 4.16 to 4.21, based on the information given by the pictorial drawings, figures 4.6 to 4.11. As well as the simple case, a more advanced example is shown in which the given Plan is set at an angle to the vertical plane (XY line).
Exercises 19 to *22.* Using First Angle projection draw a simple elevation and plan of each of the objects shown pictorially in figures 4.6 to 4.9. Add a First Auxiliary Plan in each case, using an X_1Y_1 line inclined at any convenient angle to the XY line.
Exercises 23 and *24.* Similar to the previous exercises but use Third Angle projection and refer to figures 4.10 and 4.11 respectively.
Exercises 25 to *27.* Starting with a simple plan rotated, as in figure 4.17, draw First Auxiliary Plans for the objects shown in figures 4.6, 4.9 and 4.10 respectively. Use First Angle projection.
Exercises 28 and *29.* Similar to the previous exercises but use Third Angle projection and refer to figures 4.7 and 4.9 respectively.
Exercise 30. A short cylinder 90 mm in diameter and 25 mm long is shown in First Angle projection in figure 4.22. Obtain an Auxiliary Elevation on X_1Y_1 which makes 50° with XY.
Exercise 31. A portion of a right cylinder, 90 mm in diameter, is shown in First Angle projection in figure 4.23. Obtain a First Auxiliary Elevation on X_1Y_1 which is inclined to 70° to XY.
Exercise 32. Two views of an object are given in Third Angle projection in figure 4.24. Draw a First Auxiliary Plan of the object when viewed in the direction of the arrow which is at 50° to XY.
Exercise 33. Two views of a cut cylinder are given in Third Angle projection in figure 4.25. Obtain a new elevation below X_1Y_1 which makes 40° with XY.
Exercise 34. The Elevation and Plan of a cube, 65 mm side, are given in First Angle projection in figure 4.26. Draw a First Auxiliary Plan to show the true section shape when cut by inclined plane V'T'H'.
Exercise 35. Project a First Auxiliary Plan of the portion of the cylinder, shown in Third Angle projection in figure 4.27, so as to show the true shape of the top face.
Exercise 36. A right circular cone 60 mm in diameter of base and altitude 70 mm is shown in Third Angle projection, Elevation and Plan, in figure 4.28. Project a First Auxiliary Elevation of the cone using X_1Y_1 which is inclined at 45° to XY.

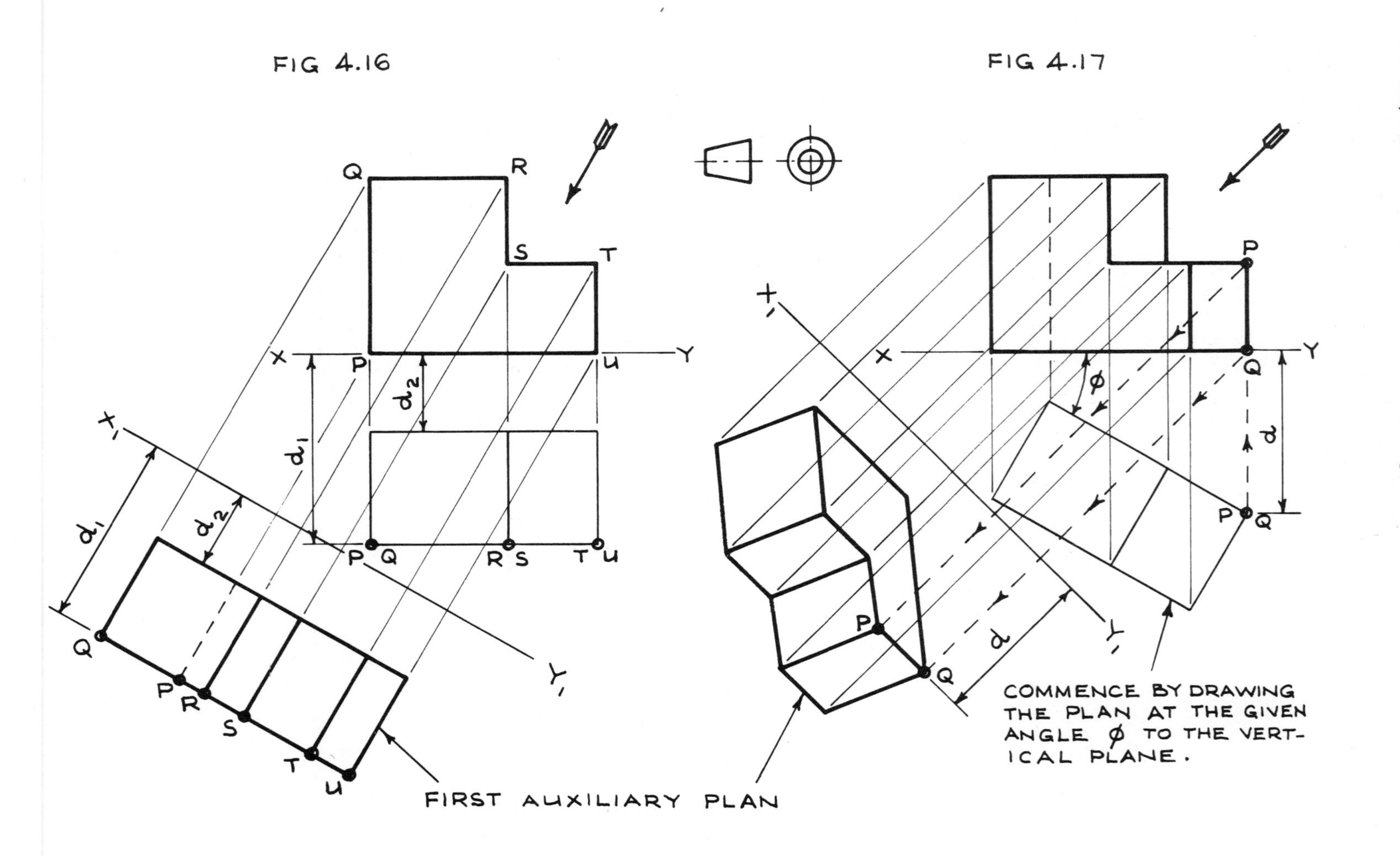

SIMPLE RULES:

1. ALWAYS PROJECT AT RIGHT ANGLES TO THE GIVEN X_1Y_1 LINE.
2. MAKE THE DISTANCE OF EACH POINT OF THE AUXILIARY PLAN FROM THE X_1Y_1 LINE THE SAME AS THE DISTANCE OF THE CORRESPONDING POINT OF THE GIVEN PLAN FROM THE XY LINE.

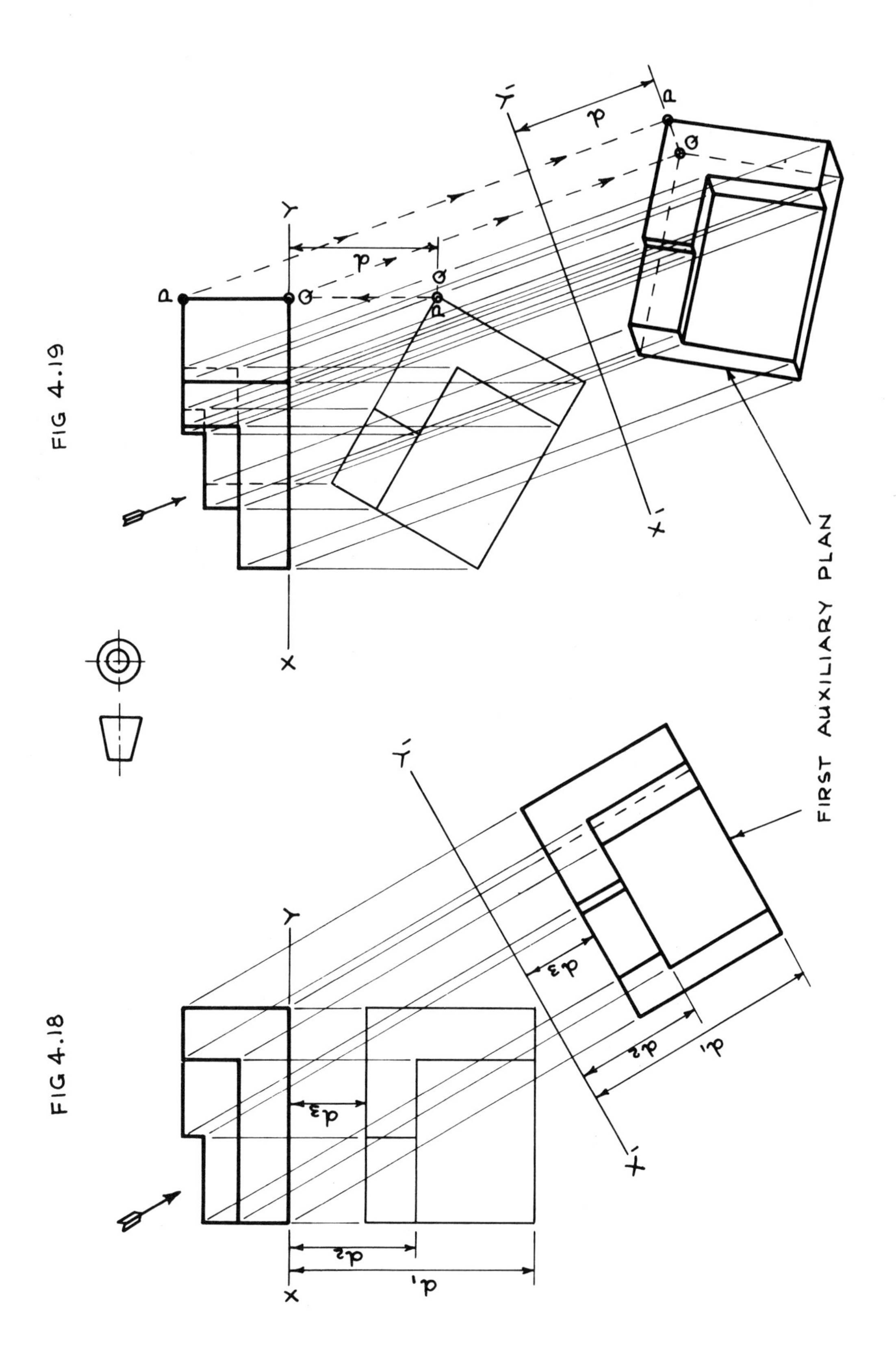
FIG 4.19
FIG 4.18
FIRST AUXILIARY PLAN
X
Y
X'
Y'
P
Q
p
d1
d2
d3

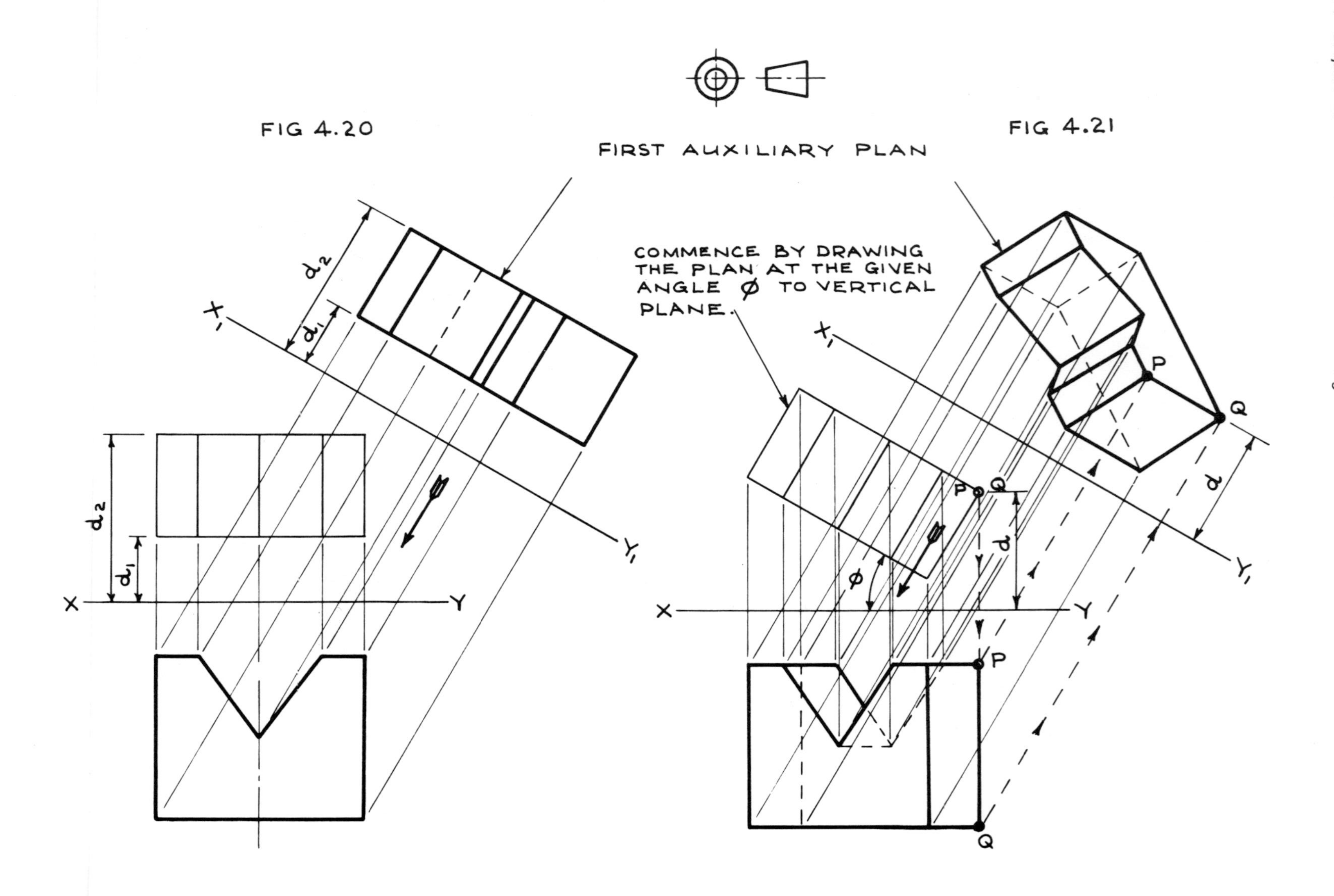

FIG 4.20

FIG 4.21

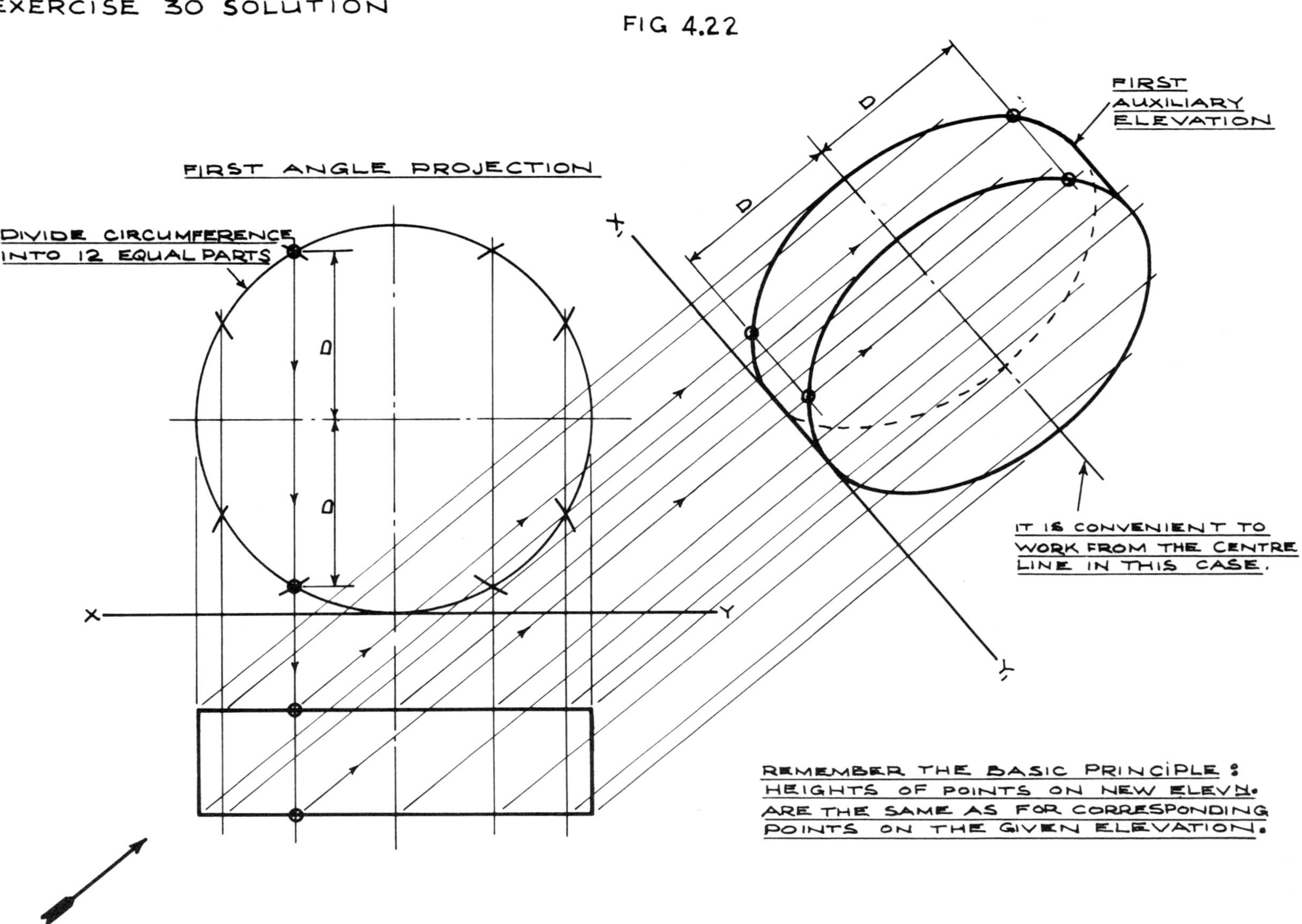
EXERCISE 30 SOLUTION
FIG 4.22
FIRST ANGLE PROJECTION
DIVIDE CIRCUMFERENCE INTO 12 EQUAL PARTS
D
D
D
D
X
Y
X1
Y1
FIRST AUXILIARY ELEVATION
IT IS CONVENIENT TO WORK FROM THE CENTRE LINE IN THIS CASE.
REMEMBER THE BASIC PRINCIPLE: HEIGHTS OF POINTS ON NEW ELEVN. ARE THE SAME AS FOR CORRESPONDING POINTS ON THE GIVEN ELEVATION.

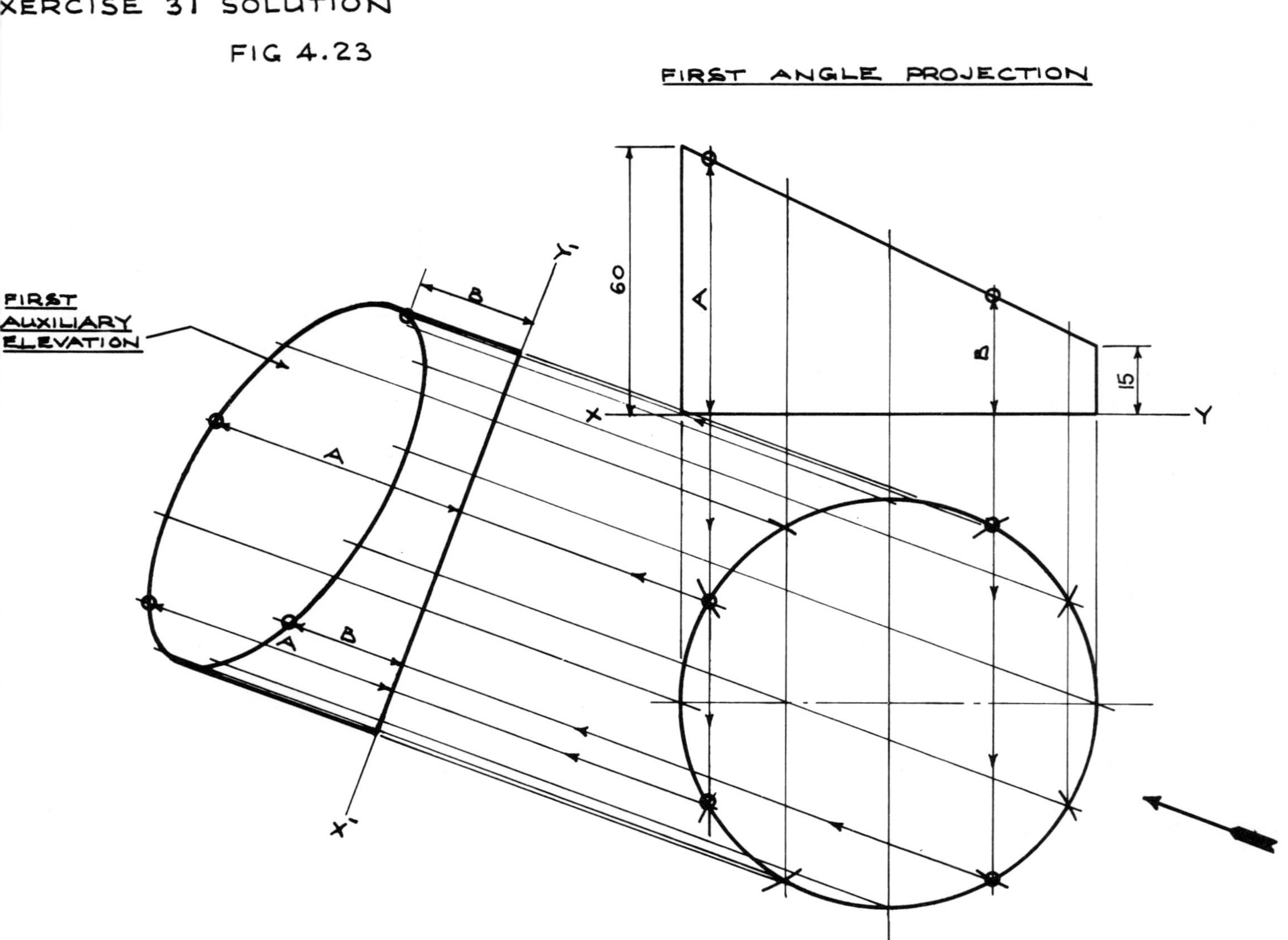
EXERCISE 31 SOLUTION
FIG 4.23
FIRST ANGLE PROJECTION
FIRST AUXILIARY ELEVATION
60
15
A
B
X
Y
X'
Y'

EXERCISES 32 AND 33 SOLUTIONS

FIG 4.24

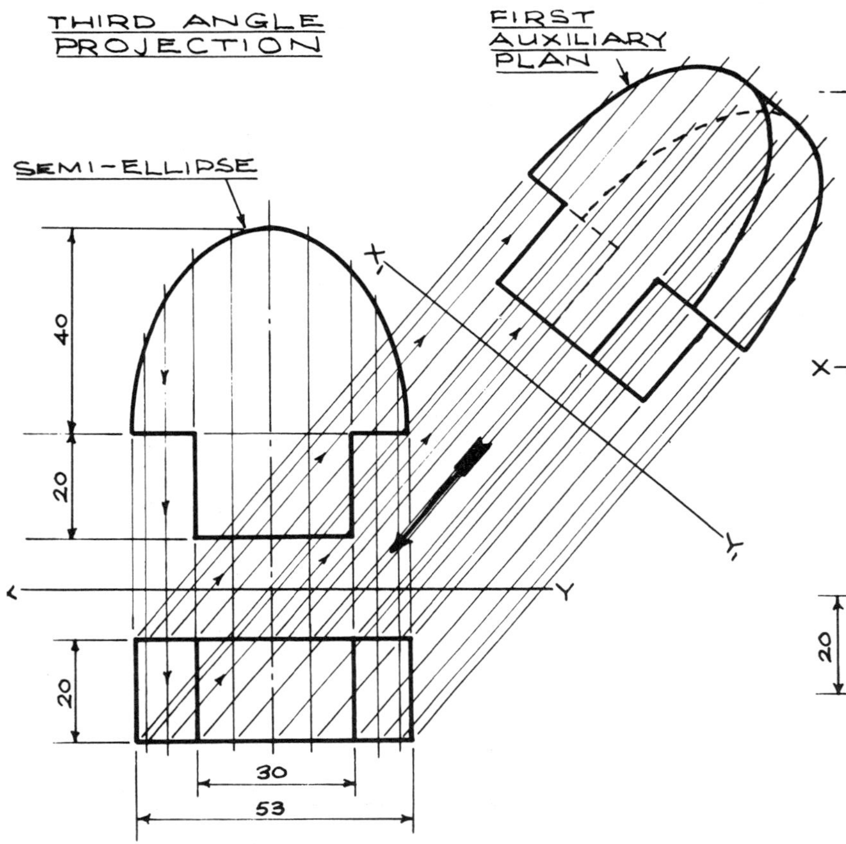

THIRD ANGLE PROJECTION

FIRST AUXILIARY PLAN

SEMI-ELLIPSE

40

20

20

30

53

X

Y

X_1

Y_1

FIG 4.25

THIRD ANGLE PROJECTION

P

Y_1

X

Y

X_1

P

P

35°

20

Ø 60

FIRST AUXILIARY ELEVATION

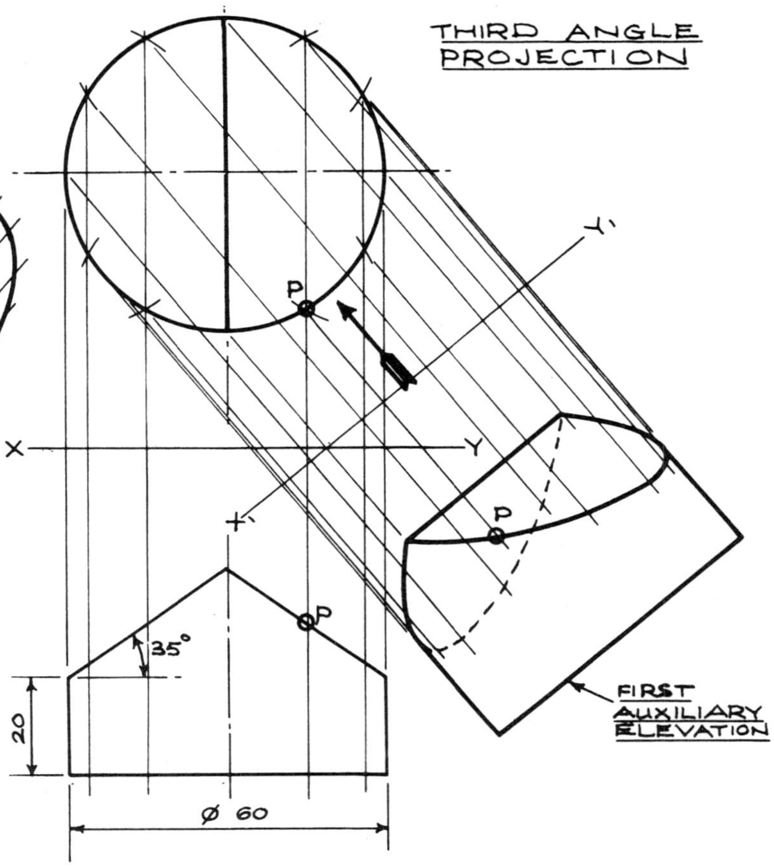

DISTANCE OF P BELOW XY IS THE SAME AS THE DISTANCE BELOW X_1Y_1

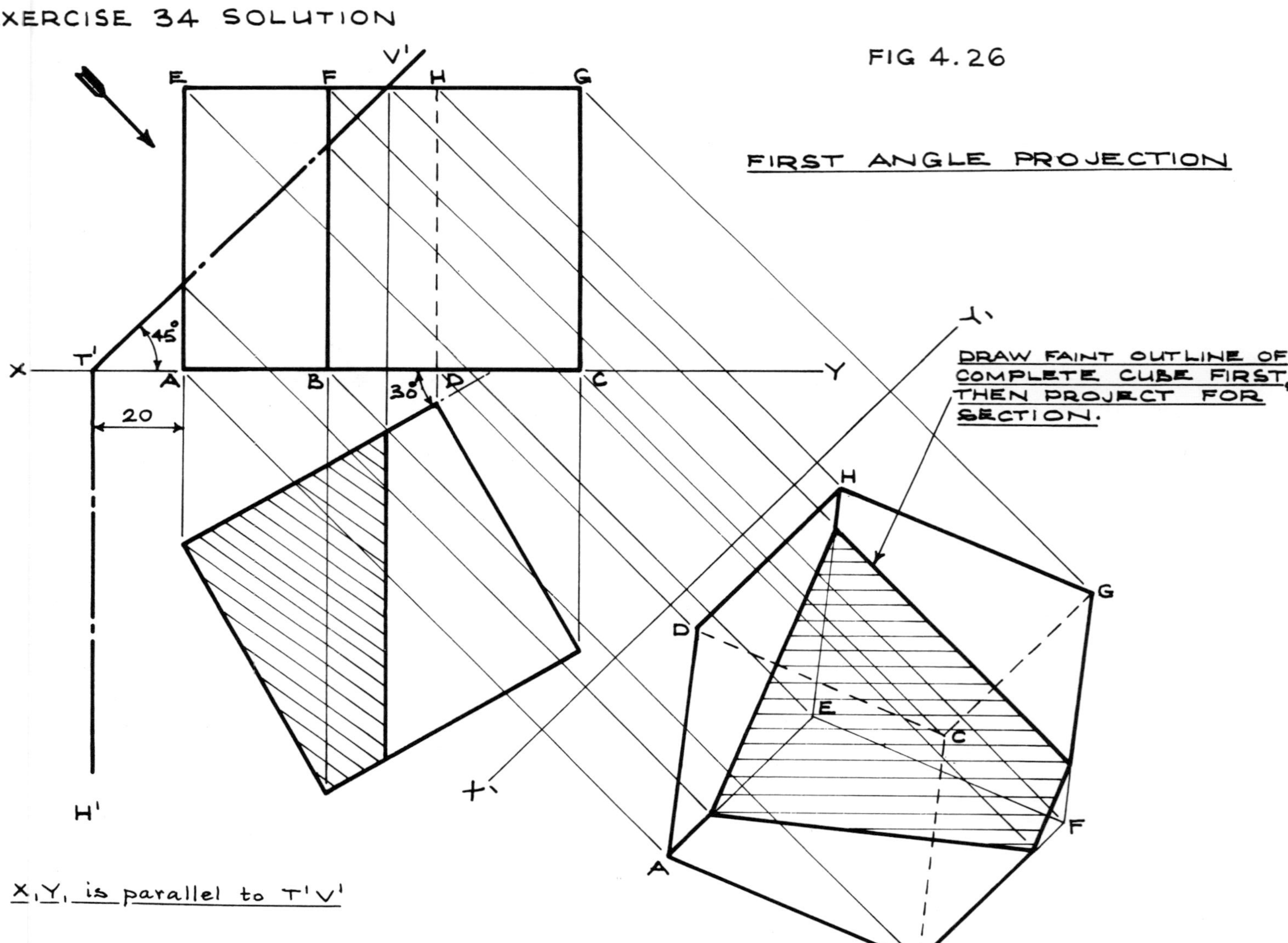

EXERCISE 34 SOLUTION
FIG 4.26
FIRST ANGLE PROJECTION
DRAW FAINT OUTLINE OF COMPLETE CUBE FIRST, THEN PROJECT FOR SECTION.
X₁Y₁ is parallel to T'V'
20
45°
30°

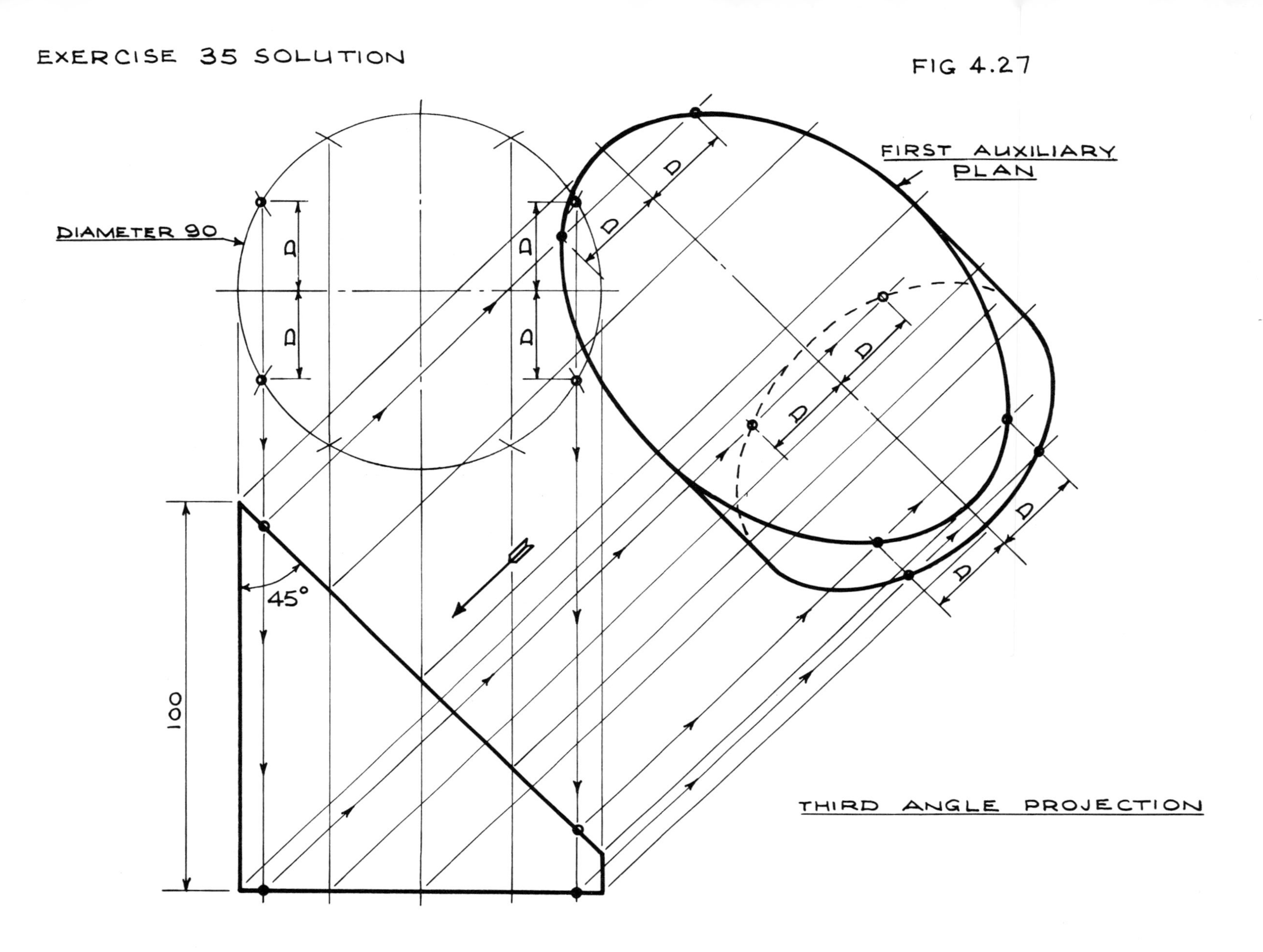

EXERCISE 35 SOLUTION
FIG 4.27
FIRST AUXILIARY PLAN
DIAMETER 90
D
45°
100
THIRD ANGLE PROJECTION

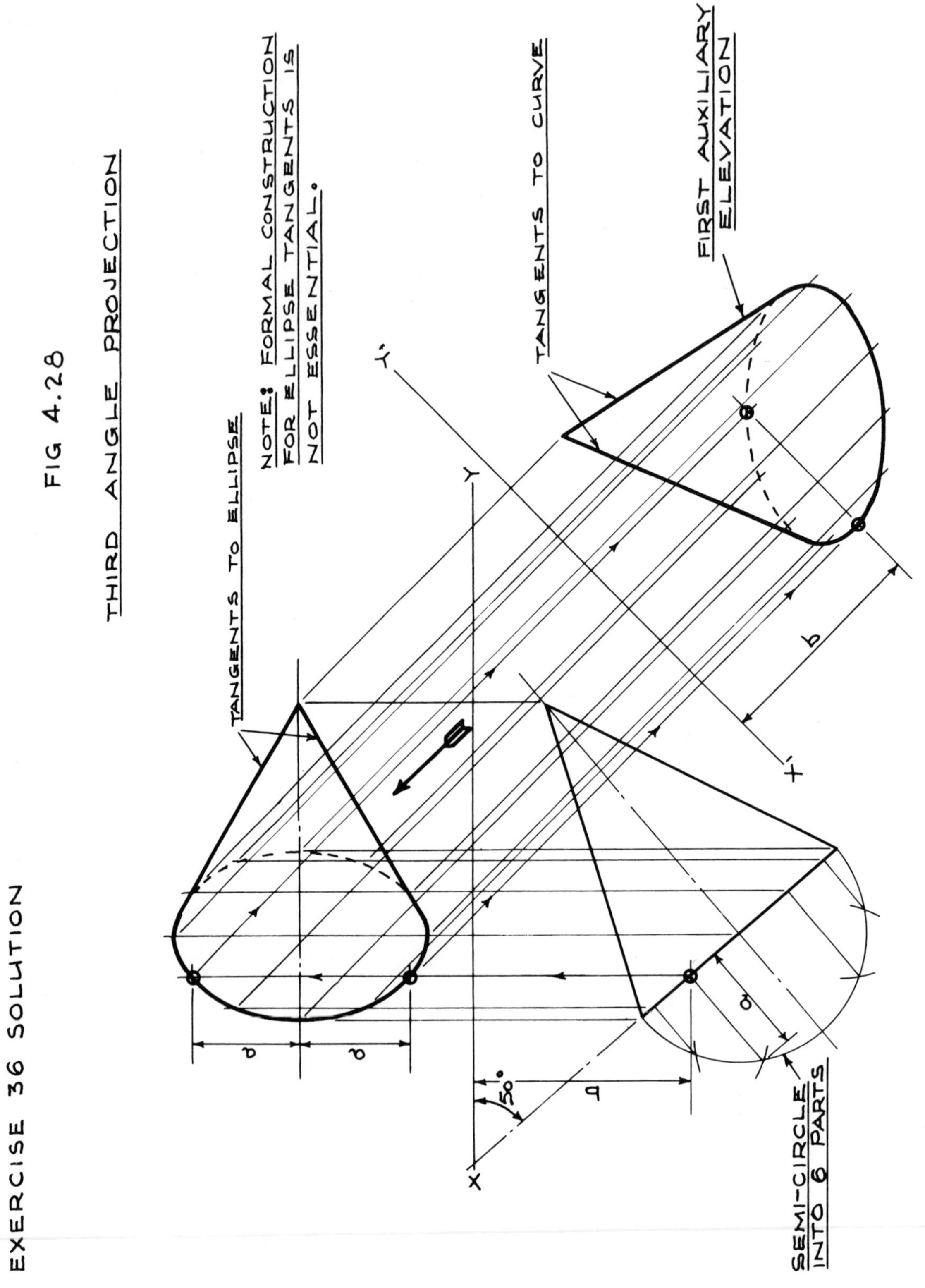

EXERCISE 36 SOLUTION

FIG 4.28

THIRD ANGLE PROJECTION

TANGENTS TO ELLIPSE

NOTE: FORMAL CONSTRUCTION FOR ELLIPSE TANGENTS IS NOT ESSENTIAL.

a

a

X

Y

50°

b

Y_1

TANGENTS TO CURVE

a

SEMI-CIRCLE INTO 6 PARTS

X_1

b

FIRST AUXILIARY ELEVATION

SECOND AUXILIARY VIEWS

A second auxiliary view is obtained by projecting from a first auxiliary view. For example, a second Auxiliary Elevation is projected from a First Auxiliary Plan. Figure 4.29 shows two examples of second Auxiliary Elevations, one projected in the direction of arrow 'A', at right angles to $X_A Y_A$, and the other in the direction of arrow 'B' which is perpendicular to $X_B Y_B$.

The principles for projection of Second Auxiliary Elevations are similar to those used for First Auxiliary Elevations and these may be re-stated as follows:

(i) Always project at right-angles to the second new datum line, from the First Auxiliary Plan;

(ii) Ensure that heights of all points of the given Elevation measured above the $X_1 Y_1$ line are the same as heights of corresponding points measured above the second new datum line.

The foregoing principles are clearly illustrated in figure 4.29 where a particular point P is tracted through the various projections.

Note. Second Auxiliary Plans may be obtained by a similar process. In figure 4.33 an example of a Second Auxiliary Plan is shown in First Angle projection.

Two quite common errors associated with auxiliary projection should be noted:

(i) The error of attempting to project a Second Auxiliary Elevation from the initial given Plan;

(ii) The error of attempting to project an elevation from an elevation of a plan from a plan. Elevations must always be derived from plans and plans from elevations.

THE EFFECTS PRODUCED BY AUXILIARY PROJECTIONS

At this stage it may be profitable to summarise the separate and combined effects of auxiliary projections.

(i) Auxiliary Elevations produce the effect of rotation of the object in a horizontal plane, about a vertical axis.

(ii) Auxiliary Plans produce the effect of rotation of the object in a vertical plane, about a horizontal axis.

(iii) By combination of effects (i) and (ii) a view of an object may be obtained, in a Second Auxiliary view, showing it inclined to both the principal planes.

A Second Auxiliary view is sometimes employed to obtain a drawing of an object as seen looking along a particular line of the object.

Most of the exercises to follow are of a more advanced nature and serve to illustrate the practical applications of auxiliary projection.

Exercise 37. Using your own dimensions, obtain two Second Auxiliary Elevations of the simple 'L' shaped object shown in figure 4.29.

Exercise 38. A rectangular prism measures 70 x 60 x 35 mm. Obtain a view of the prism along an internal diagonal AB. (Alternative solutions are given to this exercise.)

Exercise 39. The Elevation and Plan of a hexagonal pyramid are given in Third Angle projection in figure 4.32. Obtain an Elevation and Plan of the pyramid in the position such that face ABC is parallel to the HP and edge of base BC is at 45° to the VP. Base of hexagonal pyramid is 25 mm side and altitude is 60 mm. (N.I.G.C.E.)

Exercise 40. Two views of a Bracket are given in First Angle projection in figure 4.33. Project a First Auxiliary Elevation of the bracket when viewed in direction of arrow 'A'. Obtain a second Auxiliary Plan of the bracket when viewed in direction of arrow 'B' which is at 90° to XY. (N.I.G.C.E.)

Exercise 41. The Elevation and Plan of a Casting are given in Third Angle projection in figure 4.34. Using First or Third Angle projection draw the plan and elevation of the casting and hence project an Auxiliary Plan on the ground line $X_1 Y_1$. From this Auxiliary Plan project an Auxiliary Elevation on the original ground line. Hidden detail need not be shown in the final view. (N.I.G.C.E.)

Exercise 42. Given the Elevation and Plan of a triangle ABC, figure 4.35, to obtain its true shape by an

FIG 4.29

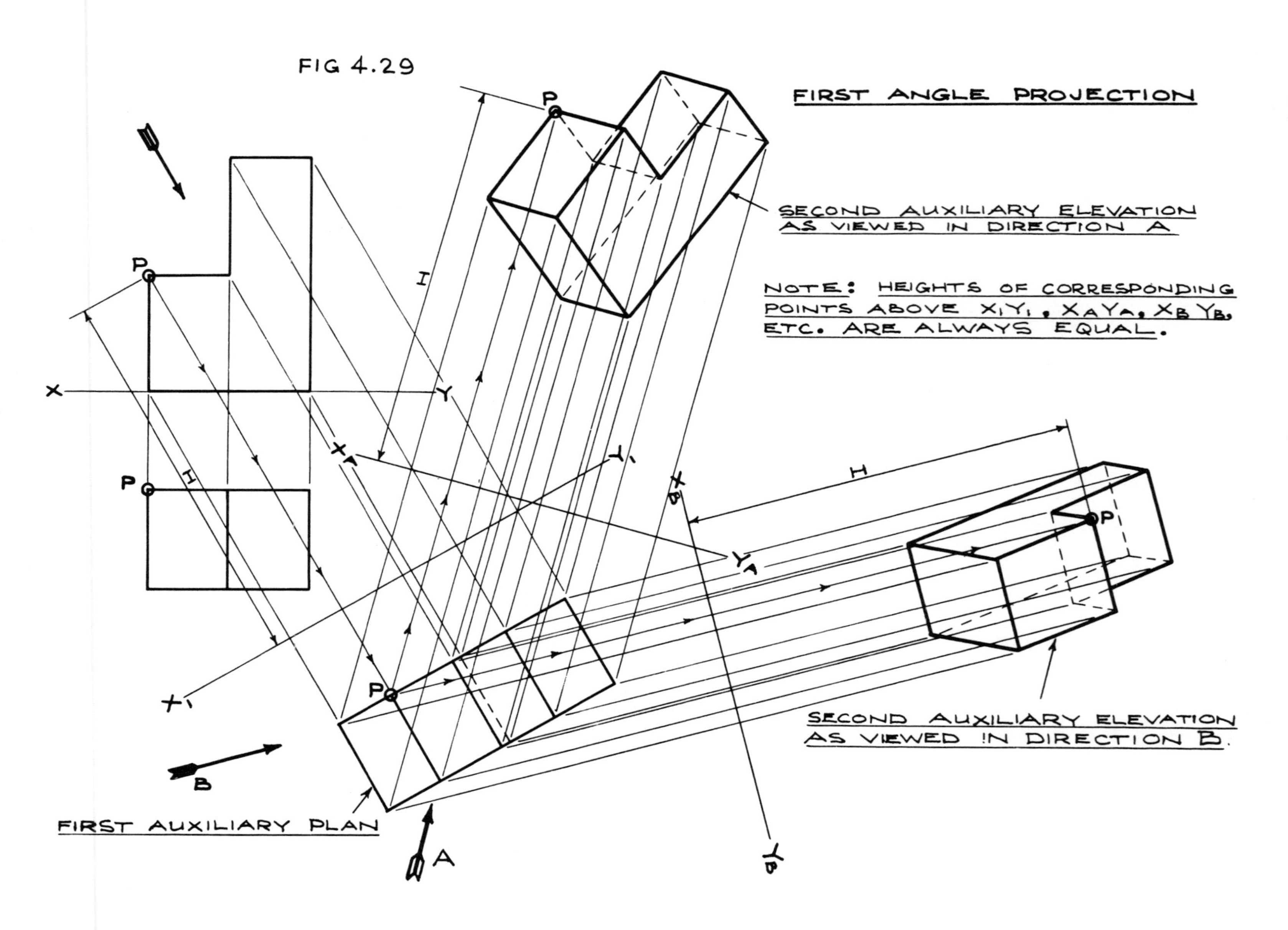
FIRST ANGLE PROJECTION
SECOND AUXILIARY ELEVATION AS VIEWED IN DIRECTION A
NOTE: HEIGHTS OF CORRESPONDING POINTS ABOVE X1Y1, XAYA, XBYB, ETC. ARE ALWAYS EQUAL.
SECOND AUXILIARY ELEVATION AS VIEWED IN DIRECTION B.
FIRST AUXILIARY PLAN
P
H
X
Y
X1
Y1
XB
YB
A
B

auxiliary projection method.

Exercise 43. A portion of a right square pyramid, side of base 50 and altitude 60, is given in Elevation and Plan in figure 4.36. Obtain the true angle between the planes A and B using an auxiliary projection method. (N.I.G.C.E.)

EXERCISE 38
A FIRST SOLUTION

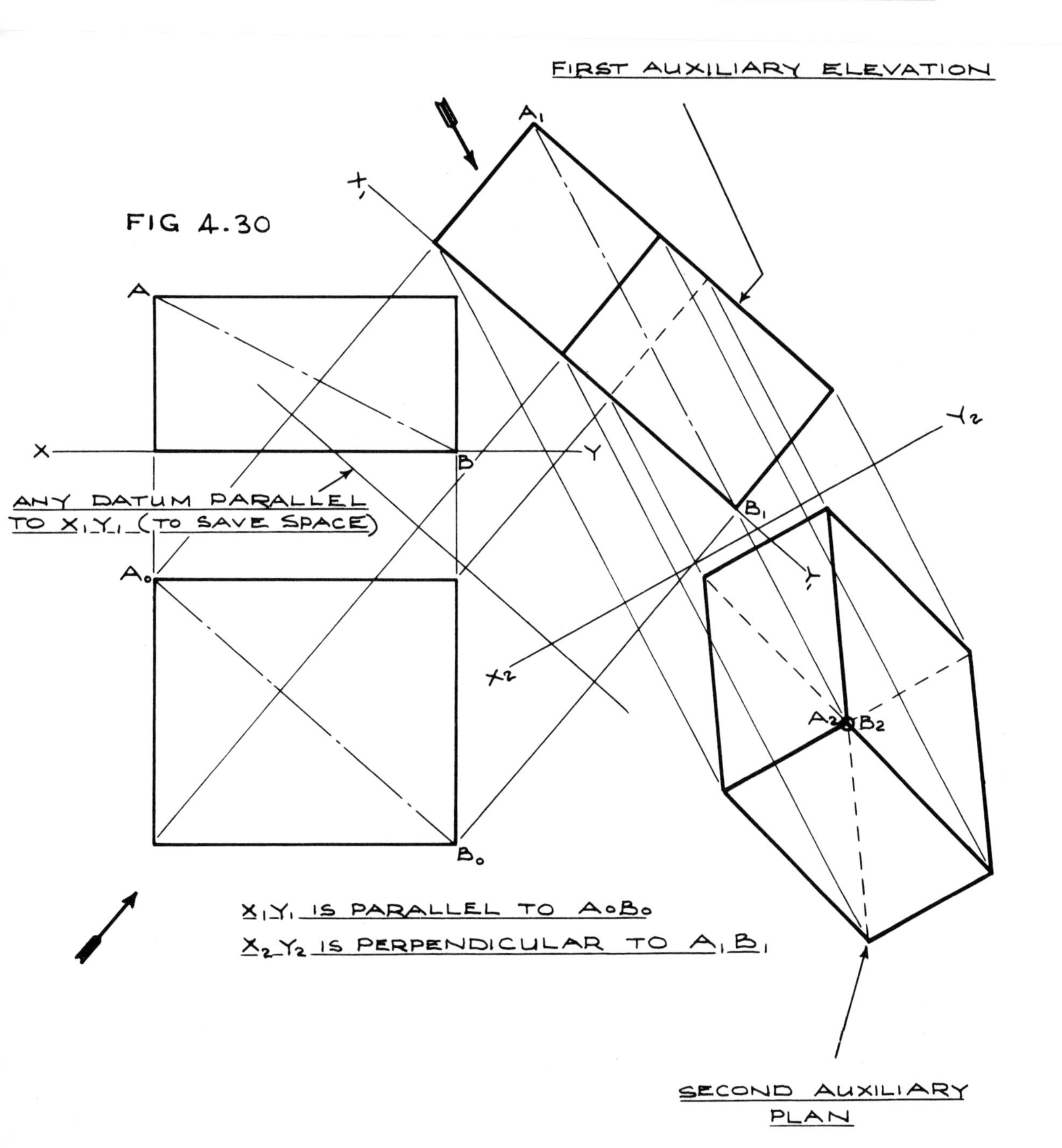

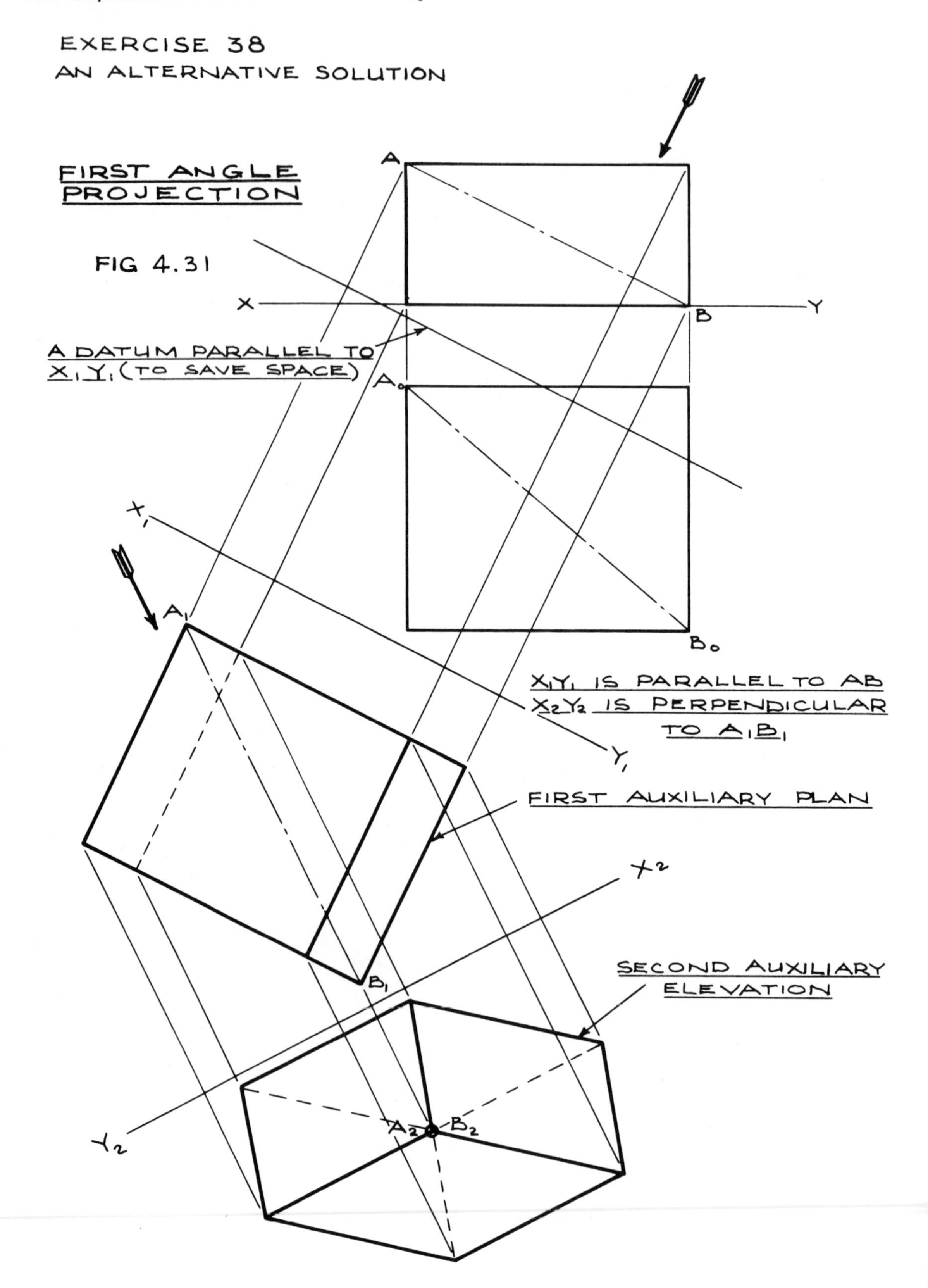
EXERCISE 38
AN ALTERNATIVE SOLUTION
FIRST ANGLE PROJECTION
FIG 4.31
A
X
Y
B
A DATUM PARALLEL TO X1Y1 (TO SAVE SPACE)
A0
X1
A1
B0
X1Y1 IS PARALLEL TO AB
X2Y2 IS PERPENDICULAR TO A1B1
Y1
FIRST AUXILIARY PLAN
X2
B1
SECOND AUXILIARY ELEVATION
Y2
A2
B2

EXERCISE 39 SOLUTION

THIRD ANGLE PROJECTION

FIG 4.32

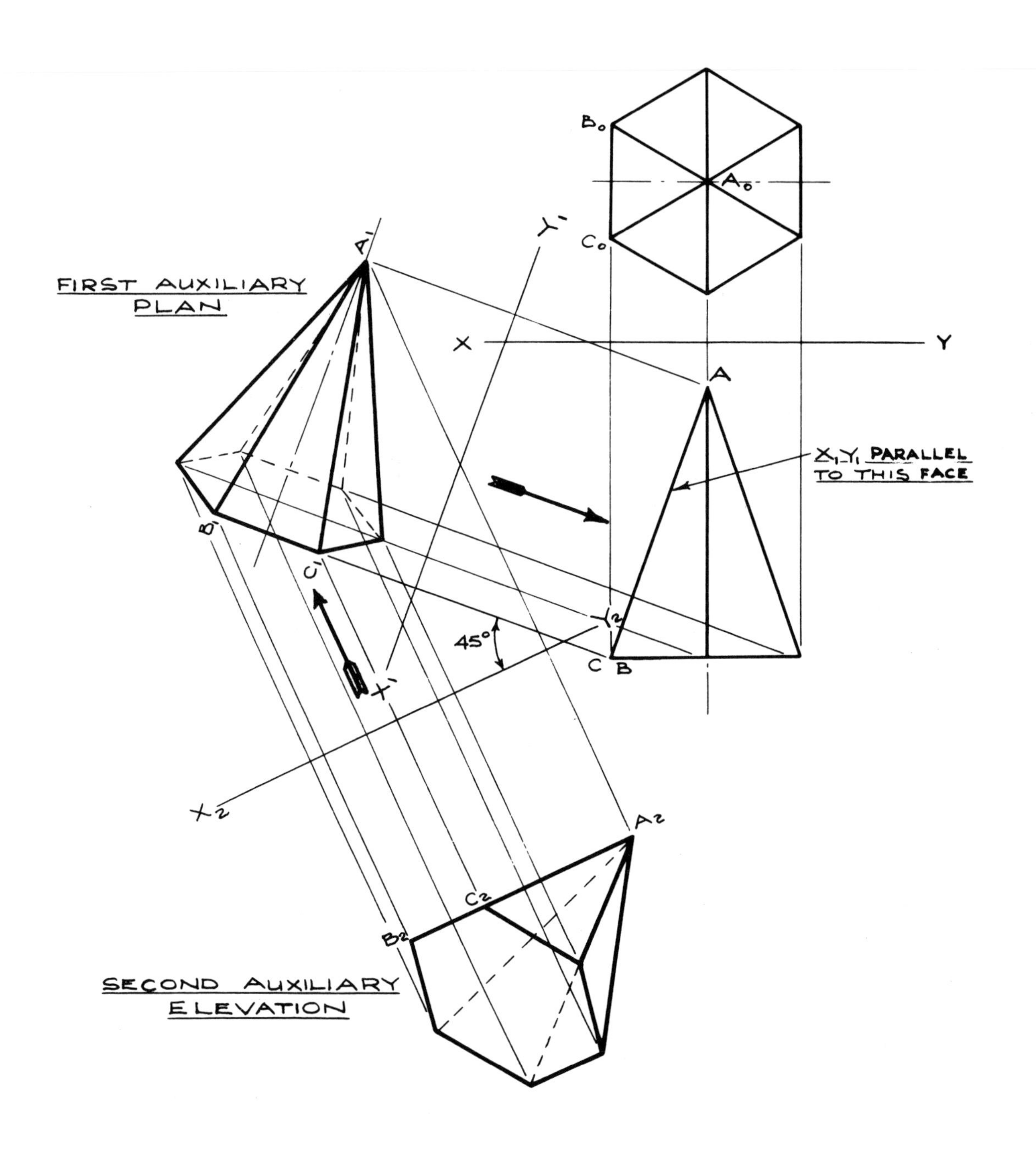

EXERCISE 40 SOLUTION

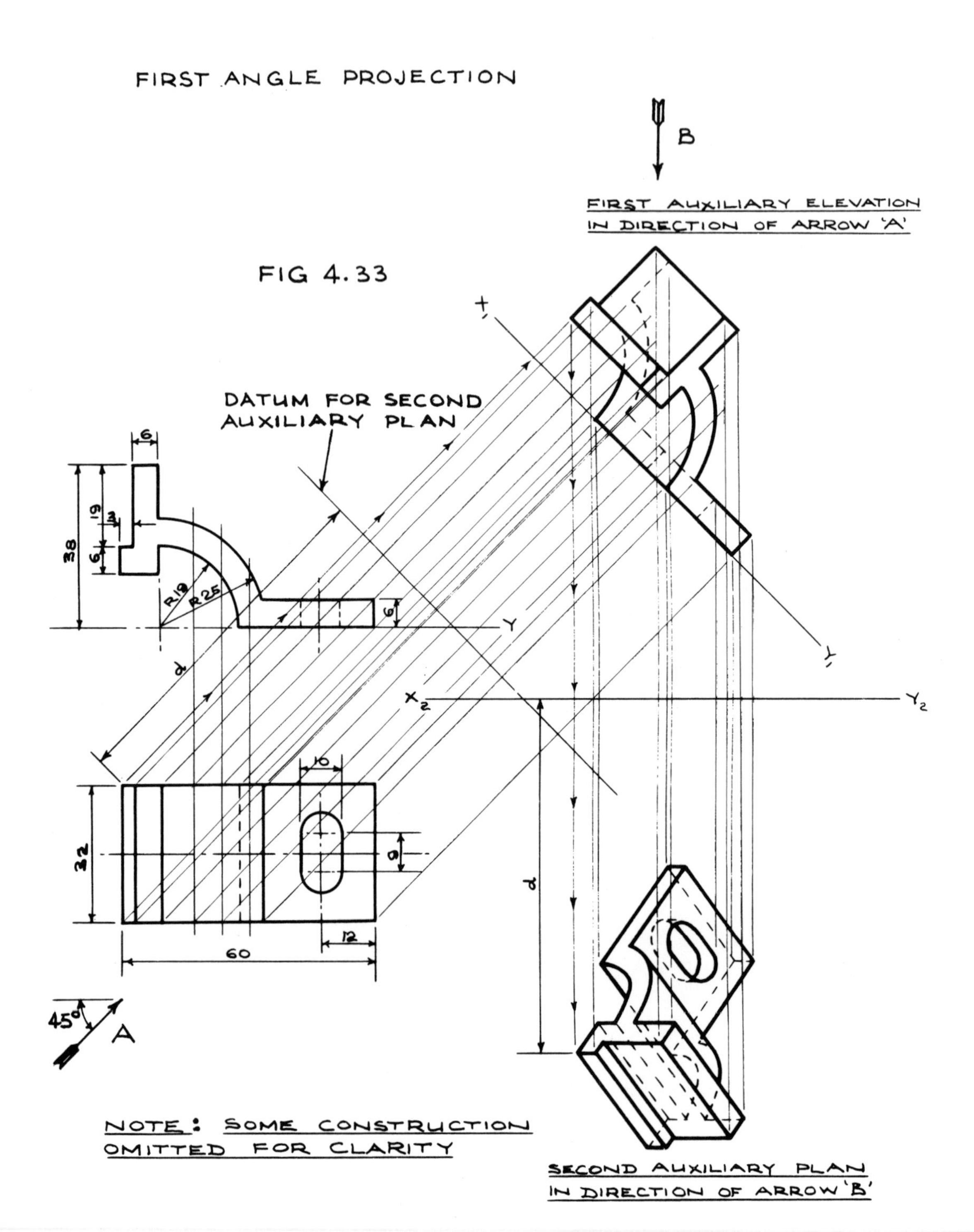

EXERCISE 41 SOLUTION

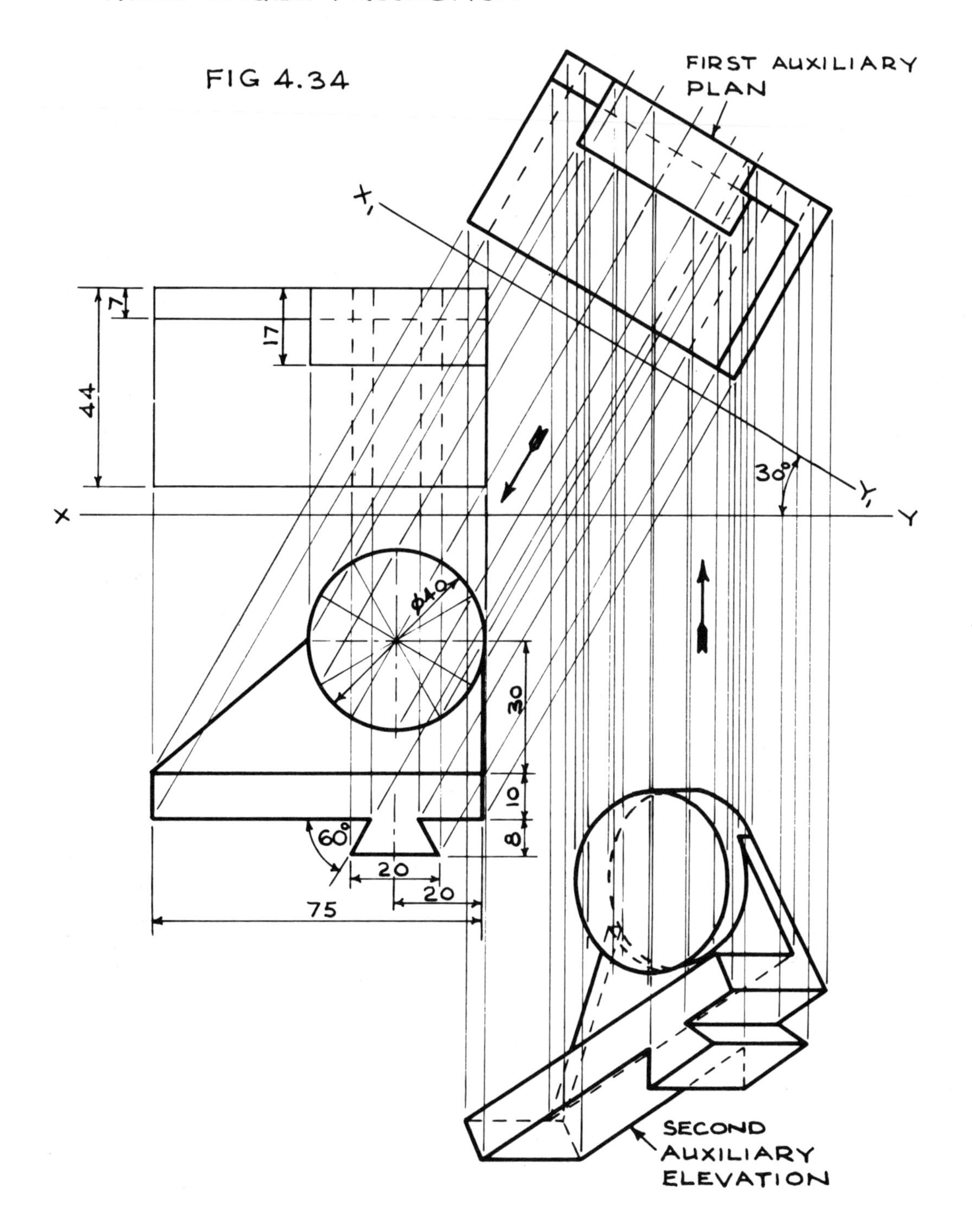

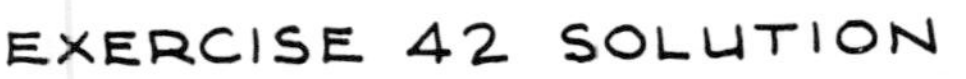
EXERCISE 42 SOLUTION

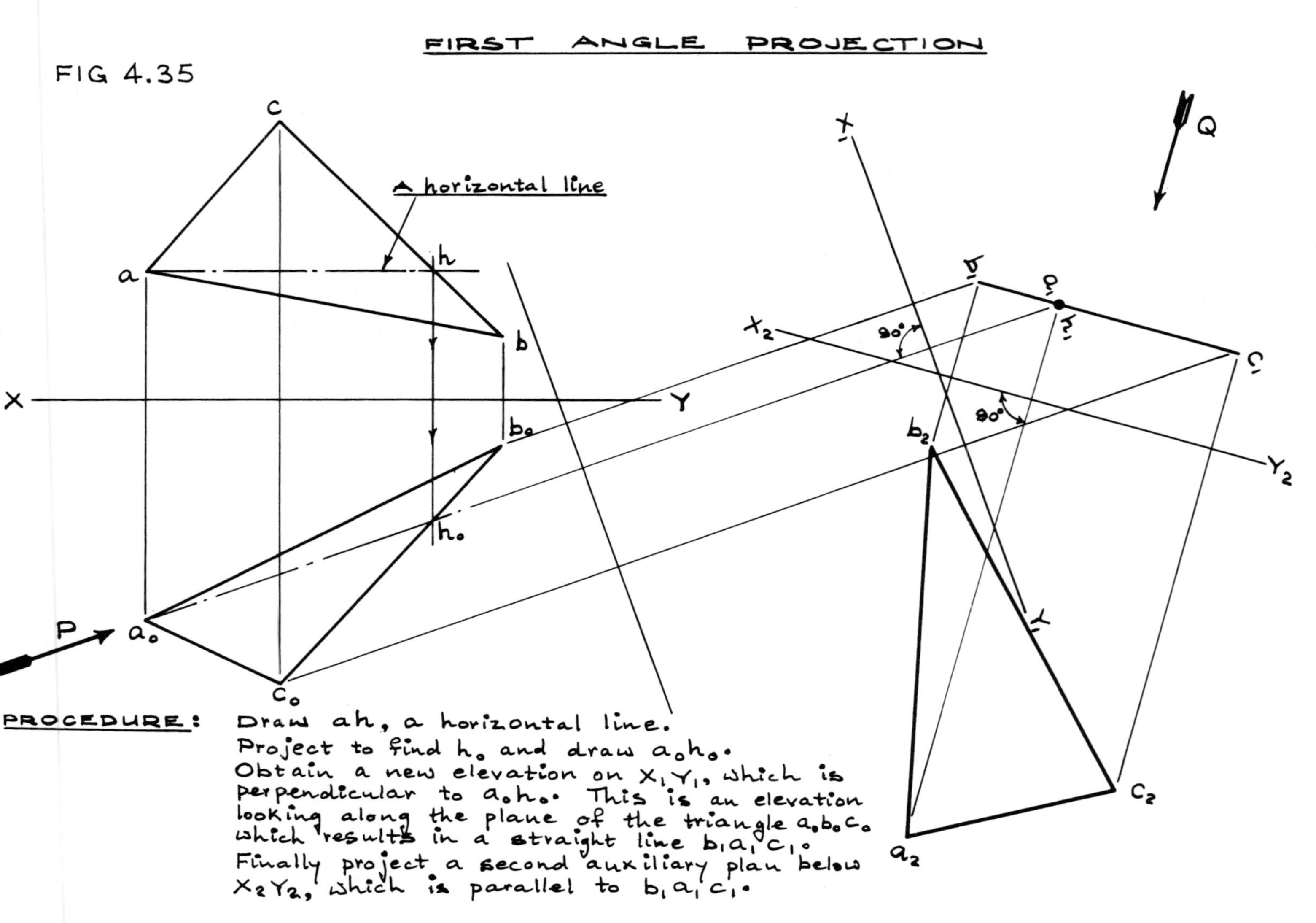
FIRST ANGLE PROJECTION
FIG 4.35
c
a
b
h
a horizontal line
X
Y
b₀
h₀
a₀
c₀
P
Q
X₁
Y₁
X₂
Y₂
b₁
a₁
h₁
c₁
b₂
a₂
c₂
90°
90°
PROCEDURE: Draw ah, a horizontal line.
Project to find h₀ and draw a₀h₀.
Obtain a new elevation on X₁Y₁, which is perpendicular to a₀h₀. This is an elevation looking along the plane of the triangle a₀b₀c₀ which results in a straight line b₁a₁c₁.
Finally project a second auxiliary plan below X₂Y₂, which is parallel to b₁a₁c₁.

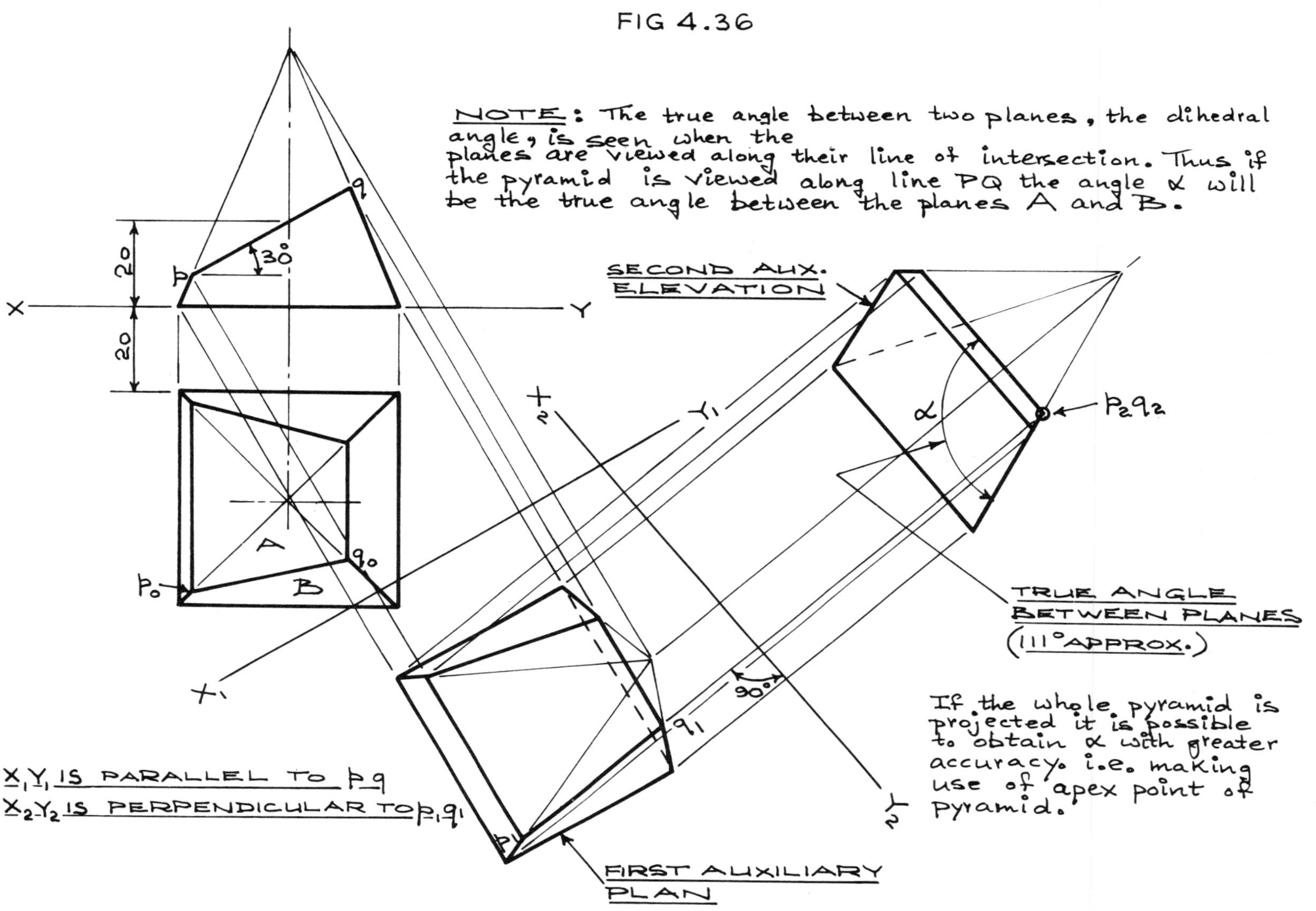
EXERCISE 43 SOLUTION
FIG 4.36
NOTE: The true angle between two planes, the dihedral angle, is seen when the planes are viewed along their line of intersection. Thus if the pyramid is viewed along line PQ the angle α will be the true angle between the planes A and B.
SECOND AUX. ELEVATION
TRUE ANGLE BETWEEN PLANES (111° APPROX.)
If the whole pyramid is projected it is possible to obtain α with greater accuracy. i.e. making use of apex point of pyramid.
FIRST AUXILIARY PLAN
X
Y
X1
Y1
X2
Y2
p
q
p0
q0
p1
q1
p2q2
30°
20
20
90°
A
B
α

X1Y1 IS PARALLEL TO pq
X2-Y2 IS PERPENDICULAR TO p1q1

5
SECTIONS, INTERSECTIONS AND DEVELOPMENTS

TRUE LENGTH OF A LINE

The basic method for finding the true length of a straight line is best illustrated by reference to the projections of a square pyramid. Figure 5.1 shows two views of a square pyramid with one sloping edge emphasised by a heavy line. If the pyramid is rotated through 45° about its vertical axis, then the elevation shown in figure 5.2 gives the true length of the edge in question.

Figures 5.3 and 5.4 show the 'rotated line' method for finding the true length, in first angle and third angle projection respectively. The true angle between the line and the horizontal plane is also obtained in this construction.

Alternatively the elevation of the line may be rotated about a horizontal axis and the true length obtained as in figure 5.5. By this method the true angle which the line makes with the vertical plane is obtained.

The most commonly applied construction for true length of a line, which is the one shown in figure 5.3, may be simplified to a right-angled triangle construction which is shown in figure 5.6. This construction will at once determine the true length of any line, the plan length and elevation height of which are known. This particular construction is extremely useful for development work where many true lengths may be required. It is not necessary to draw the hypotenuse of the right-angled triangle of figure 5.6 since compasses or dividers may be used to lift the length directly.

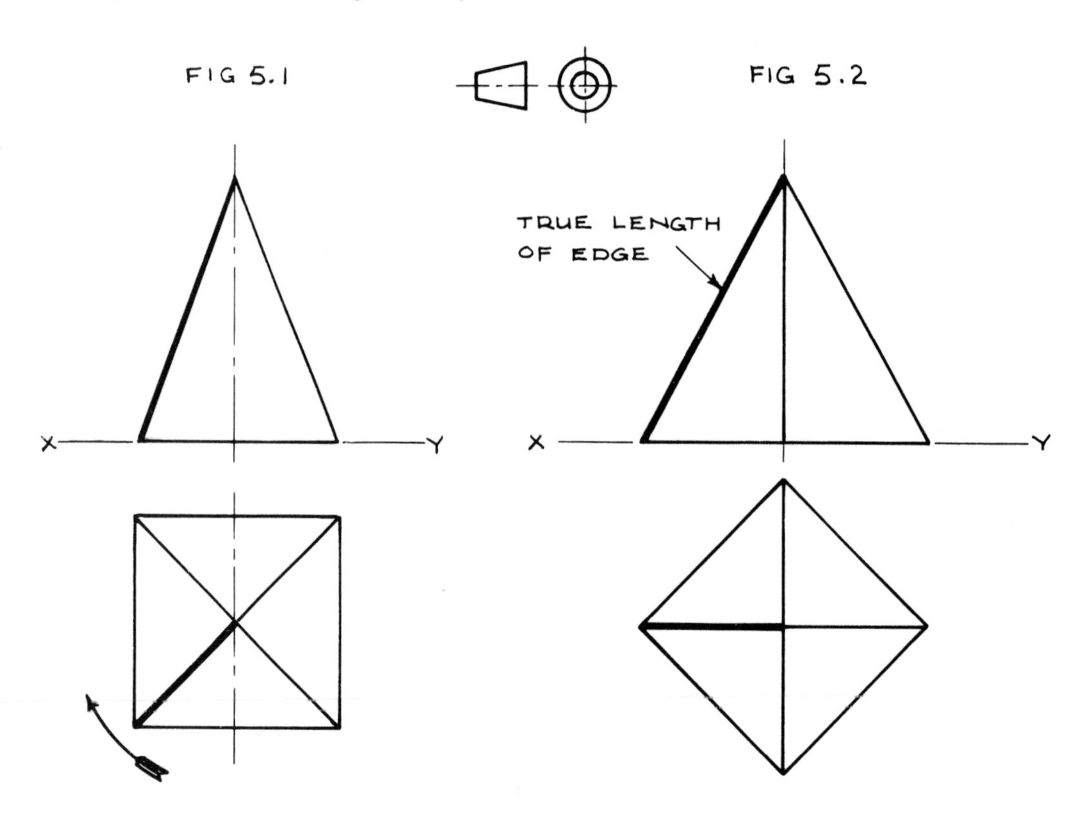

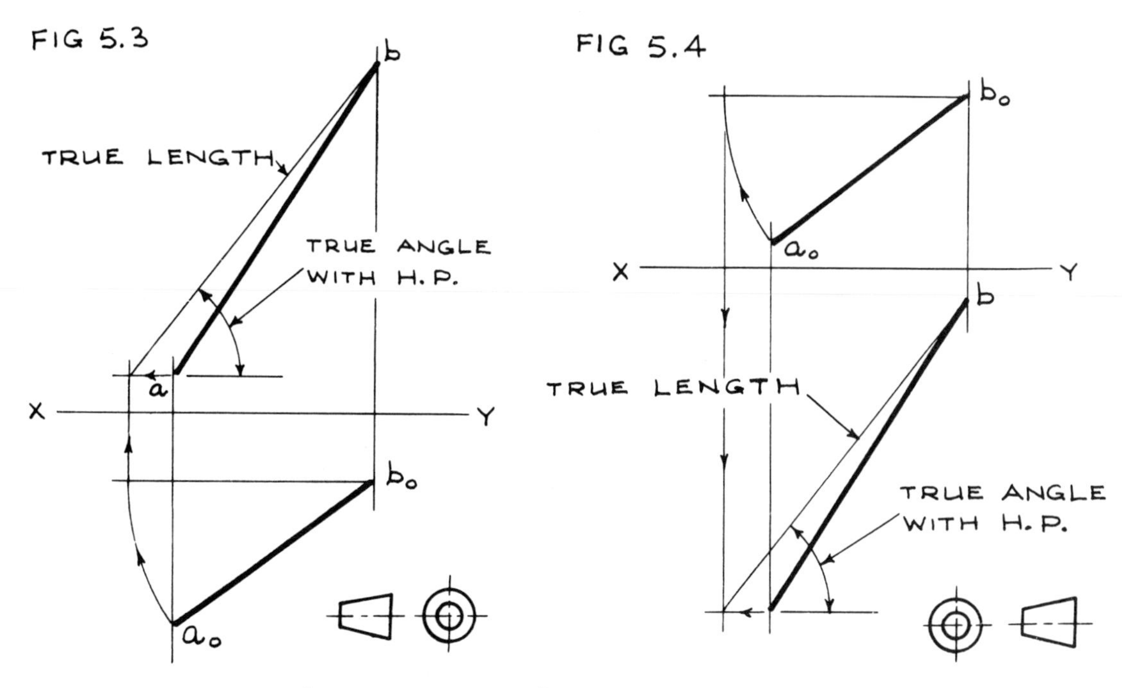

The elevation ab and the plan $a_o b_o$ of a line are given. To find the true length swing plan $a_o b_o$ about b_o using compasses and project as shown in Figs 5.3 and 5.4, in First Angle and Third Angle projection respectively.

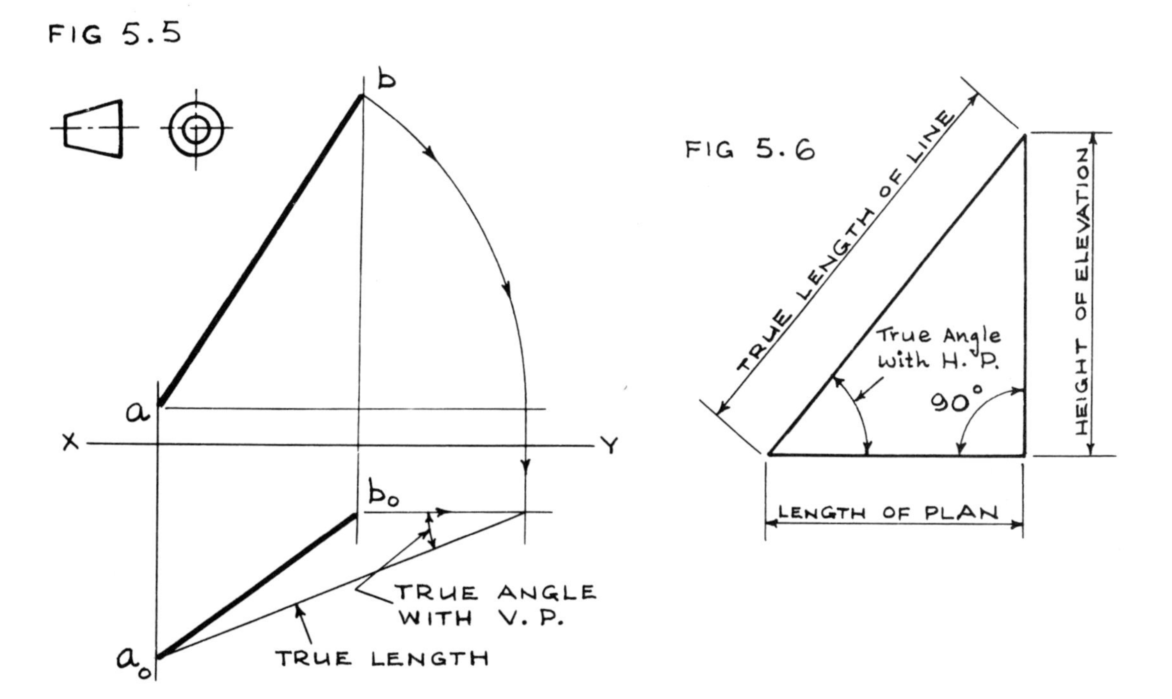

TRUE LENGTH BY AUXILIARY PROJECTION

This method is illustrated by figures 5.17 and 5.18. True length is obtained as in figure 5.17 by projecting a first auxiliary elevation of the line on X_1Y_1. The true angle between the line and the HP is also obtained.

Figure 5.18 shows true length being obtained by projection of a first auxiliary plan using X_1Y_1. By this method the true angle between the line and the VP is obtained.

TRACE POINTS OF A LINE

The trace points of a line are the points where the line, produced if necessary, meets the principal planes. Trace points of a line are illustrated in figure 5.7. The orthographic representation of the construction for determining trace points of a line is shown in figure 5.8.

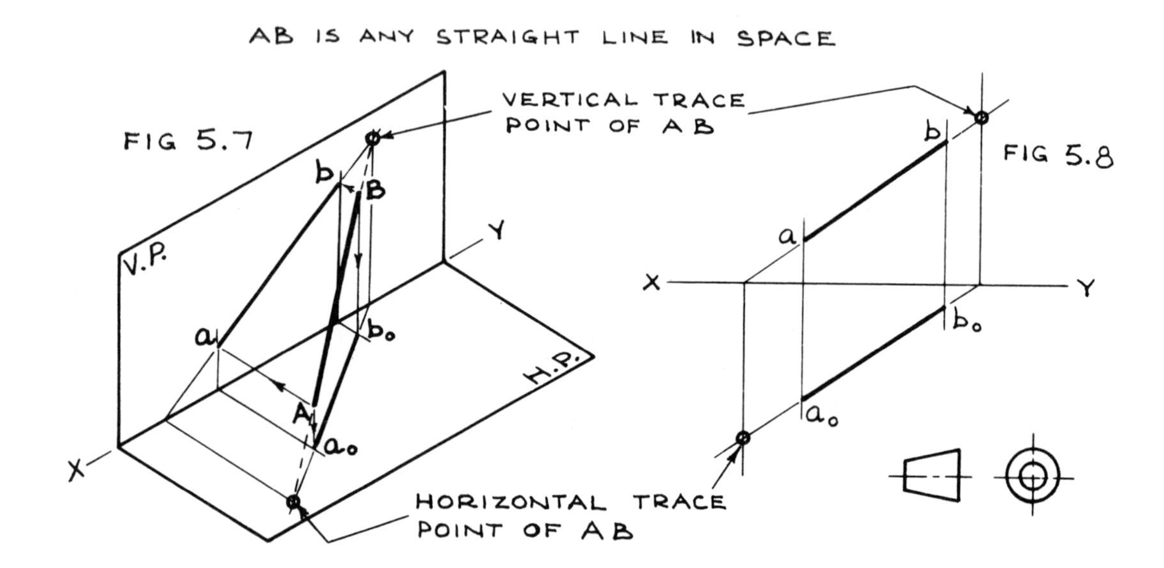

The point where a line, or the line produced, meets the vertical plane is called the VERTICAL TRACE point of the line.
Similarly, the point where the line, produced if necessary, meets the horizontal plane is known as the HORIZONTAL TRACE point of the line.

TRACE LINES OF A PLANE

The straight lines in which any plane intersects the principal planes are known as the traces of the plane. The horizontal and vertical trace lines of an inclined plane, a vertical plane and a general oblique plane are shown pictorially in figures 5.9, 5.11 and 5.13 respectively.

The orthographic representation of a plane by its trace lines is shown in figures 5.10, 5.12 and 5.14.

An inclined plane is inclined to the horizontal plane (HP) and makes an angle of 90° with the vertical plane (VP) as shown in figures 5.9 and 5.10.

A vertical plane is at right-angles to the HP and is inclined to the VP as shown in figures 5.11 and 5.12.

An oblique plane is inclined to both principal planes as shown in figures 5.13 and 5.14. Special types of oblique plane having their traces coincident with the XY line or parallel to the XY line are shown in figures 5.15 and 5.16 respectively.

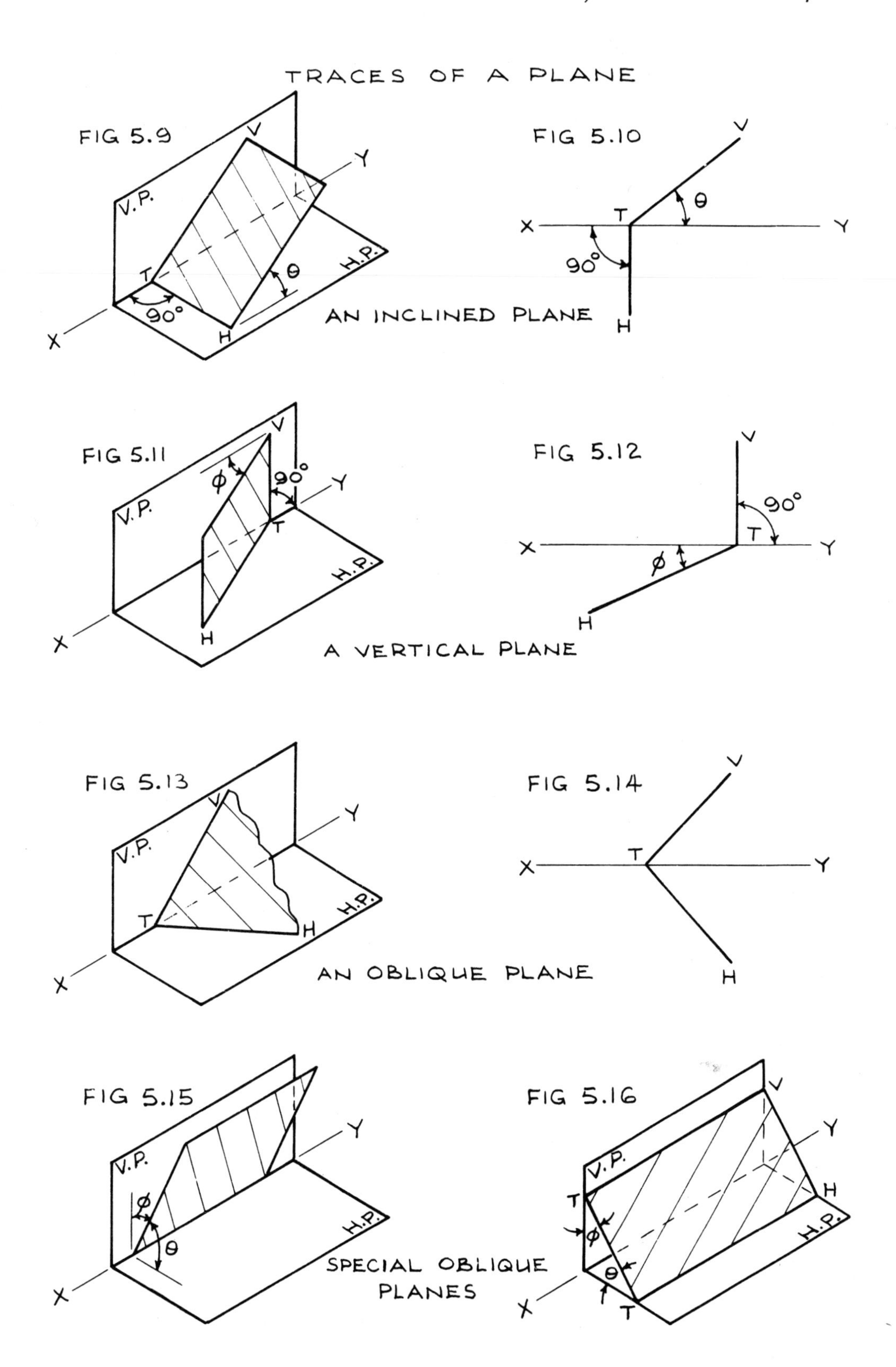
TRACES OF A PLANE
FIG 5.9
V.P.
H.P.
V
Y
T
X
H
90°
θ
FIG 5.10
V
θ
T
X
Y
90°
H
AN INCLINED PLANE
FIG 5.11
V.P.
H.P.
V
ϕ
90°
Y
T
X
H
FIG 5.12
V
90°
T
X
Y
ϕ
H
A VERTICAL PLANE
FIG 5.13
V.P.
H.P.
V
Y
T
H
X
FIG 5.14
V
T
X
Y
H
AN OBLIQUE PLANE
FIG 5.15
V.P.
H.P.
Y
ϕ
θ
X
FIG 5.16
V.P.
H.P.
V
Y
T
H
ϕ
θ
X
T
SPECIAL OBLIQUE PLANES

TRUE LENGTH OF A LINE

Another useful method of dealing with a line problem is by AUXILIARY PROJECTION.

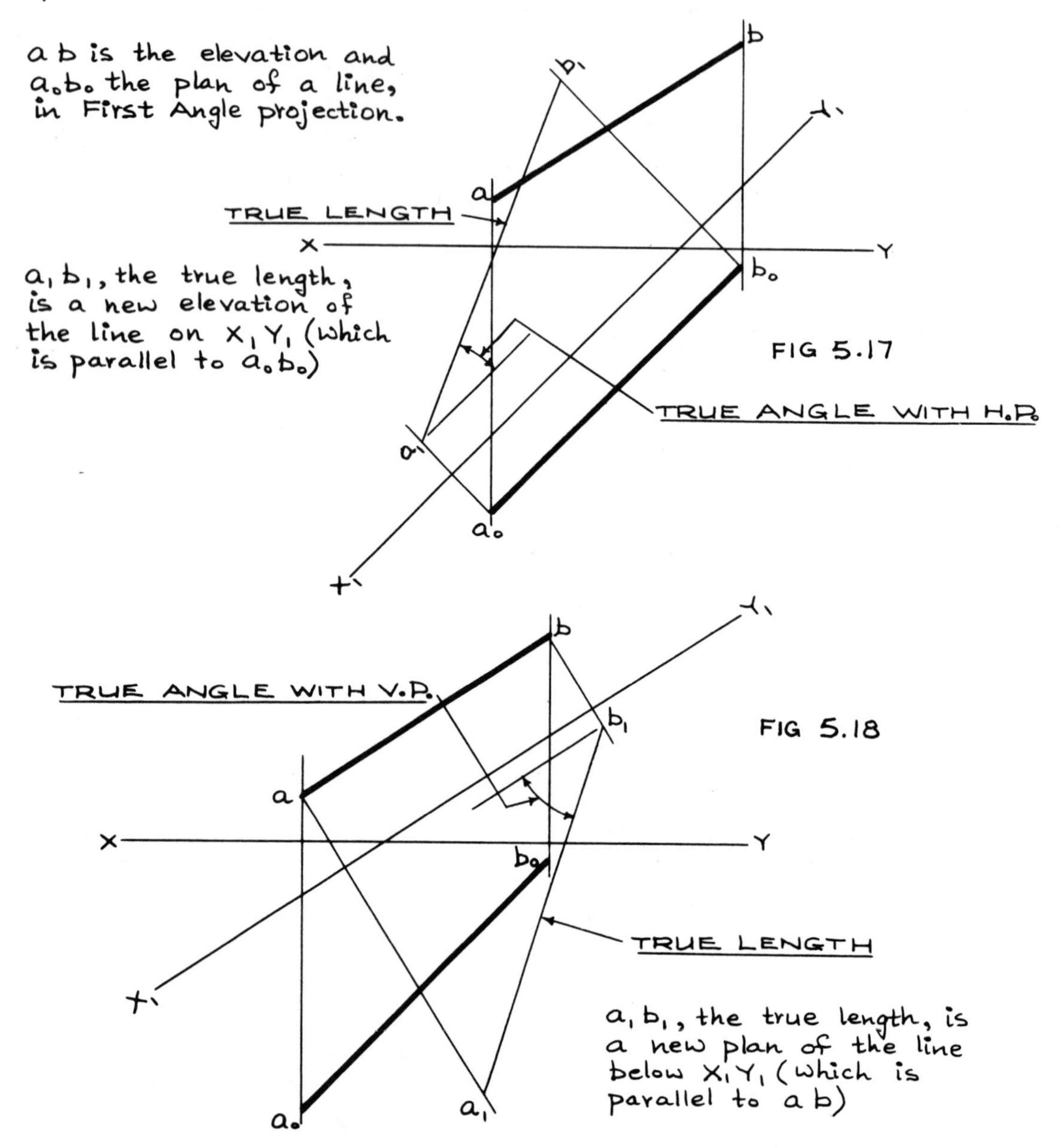

FIG 5.17

FIG 5.18

Remember the basic principles of Auxiliary Projection:

(i) For a new elevation the heights of a and b above XY are the same for a_1 and b_1 above X_1Y_1;

(ii) For a new plan the distances of a_0 and b_0 below XY are the same for a_1 and b_1 below X_1Y_1.

Exercise 1. A line AB in space is shown in the pictorial diagram, figure 5.19. 'A' is 15 above the HP and 10 from the VP, B is 35 above the HP and 40 from the VP. The distance between the projectors of A and B is 60. Determine the true length of AB, the true angles made with the HP and VP and the trace points.
Exercise 2. Given the elevation a b and the plan $a_o b_o$ of a straight line AB, determine its true length, trace points and true angle with the HP and the VP. 'A' is 20 above the HP and 60 from the VP, B is 50 above the HP and 20 from the VP. The distance between the projectors of the ends of AB is 65.
Exercise 3. The elevation and plan of a square pyramid, 50 mm side of base and altitude 60 mm, are given in First Angle projection in figure 5.23. The pyramid is intersected by an inclined plane making 30° with the HP, the vertical trace (vt) of which cuts the axis of the pyramid at the mid-point. Obtain a sectioned plan, a sectioned side view, the true shape of the section and develop the sloping surfaces of the pyramid below the cutting plane.
Exercise 4. Similar to previous exercise but in this case the pyramid is rotated about its axis, in the HP, through an angle of 45°. This is shown in Third Angle projection in figure 5.24.
Exercise 5. A regular hexagonal prism, 20 mm side of hexagon and 70 mm long, is resting with a rectangular face in the HP as shown in figure 5.25 plan view. The prism is intersected by a vertical plane vth. Obtain a sectioned front view, a sectioned side view, the true shape of the section and the development of the lateral surface of the larger portion of the sectioned solid.
Exercise 6. The front view and plan, in Third Angle projection, of a hexagonal pyramid of 25 mm side of hexagon and 60 mm altitude are given in figure 5.26. The pyramid is cut by an inclined plane VTH making 30° with XY and intersecting the pyramid axis at its mid-point. Draw a sectioned plan, a sectioned side view, the true shape of the section and the development of the sloping surfaces below VT.
Exercise 7. Two hollow square prisms, sides 35 mm and 25 mm respectively, have their axes intersecting at 65°. Obtain a view showing the lines of intersection and develop the surfaces of each prism (figure 5.27).
Exercise 8. A square prism of side 20 mm intersects a square pyramid, side of base 40 mm and altitude 65 mm, the axes intersecting at right-angles. Show the lines of intersection in front view and plan and also the development of both solids, excluding the base of the pyramid (figure 5.28).
Exercise 9. The front view and plan of a hexagonal pyramid, 30 mm side of base and 80 mm altitude, are given in First Angle projection in figure 5.29. Obtain a sectioned plan, sectioned side view and an auxiliary plan showing the true shape of the section when the pyramid is cut by inclined plane VTH, which intersects the mid-point of the axis at 45° to the HP.
Exercise 10. The front view and plan of a sheet metal hopper are given in Third Angle projection in figure 5.30. Draw these given views to a suitable scale and draw the development of the hopper having the joint along edge AB.
Exercise 11. A right cylinder, 40 mm diameter and 50 mm high has the top portion removed by inclined plane VTH as shown in figure 5.31. Draw a side view, true shape of sectioned surface and the development of the curved surface of the remaining portion of the cylinder.

EXERCISE I SOLUTION

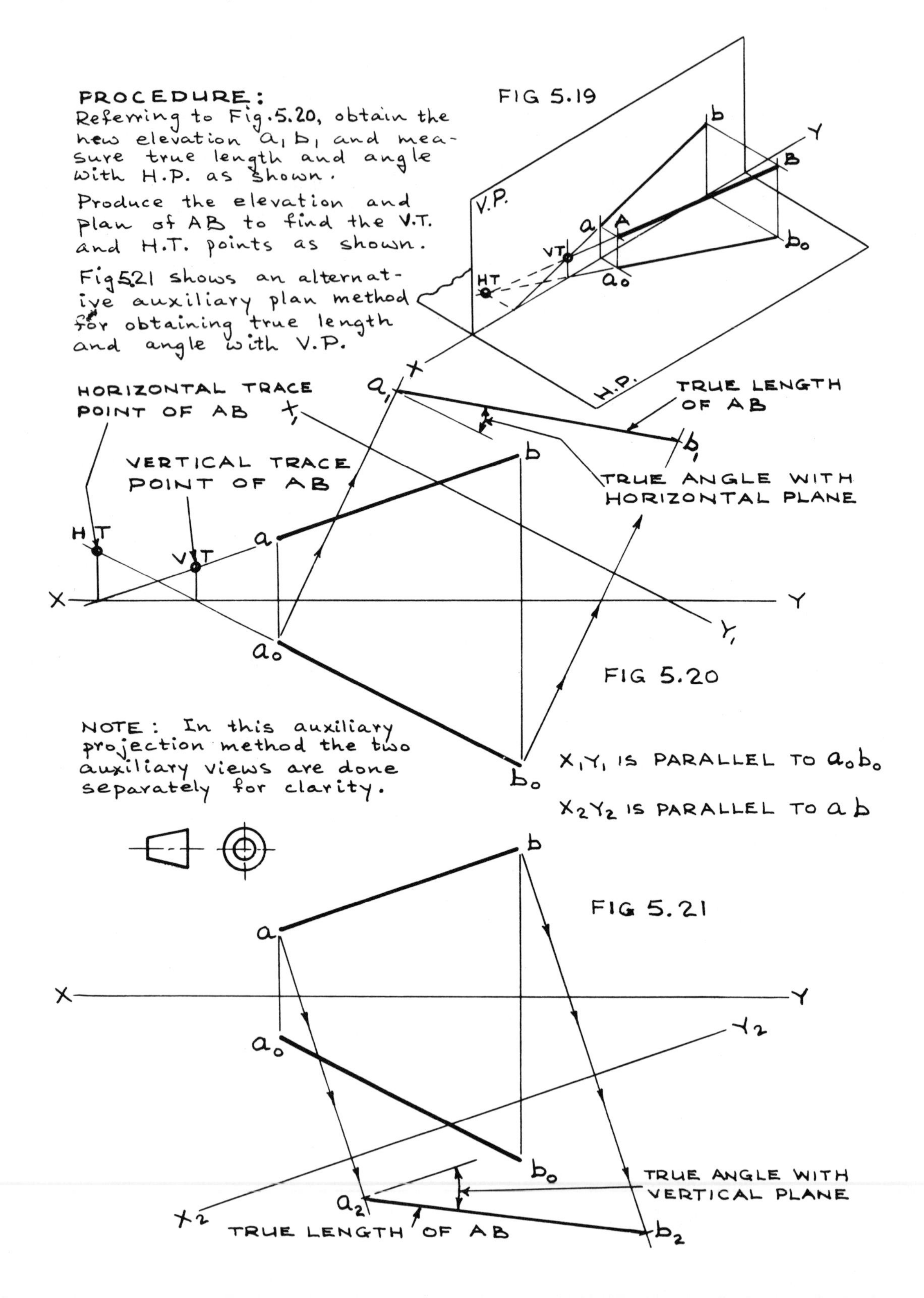

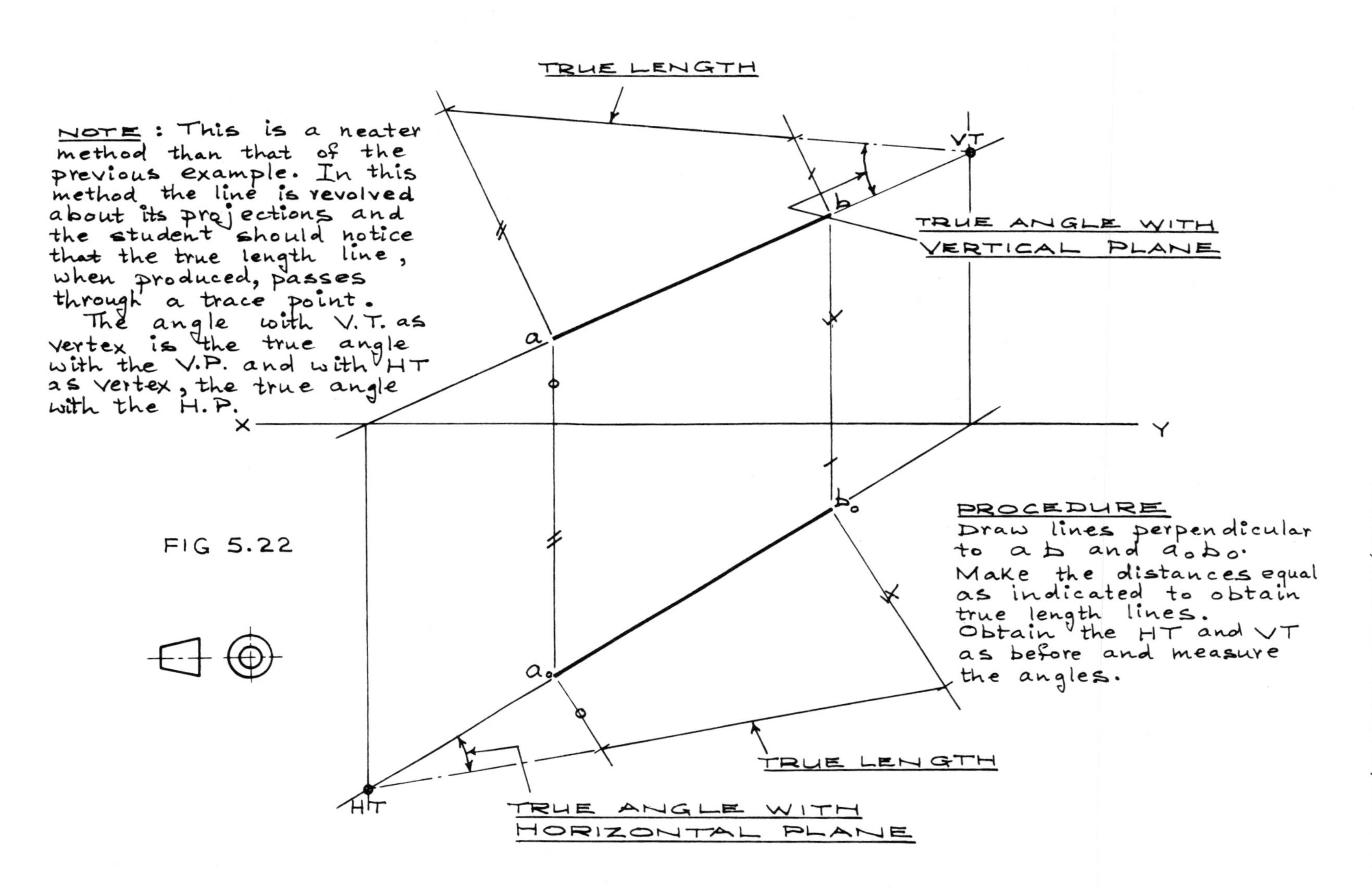
EXERCISE 2 SOLUTION
NOTE : This is a neater method than that of the previous example. In this method the line is revolved about its projections and the student should notice that the true length line, when produced, passes through a trace point.
The angle with V.T. as vertex is the true angle with the V.P. and with HT as vertex, the true angle with the H.P.
TRUE LENGTH
VT
b
TRUE ANGLE WITH VERTICAL PLANE
a
X
Y
b_o
a_o
PROCEDURE
Draw lines perpendicular to a b and $a_o b_o$.
Make the distances equal as indicated to obtain true length lines.
Obtain the HT and VT as before and measure the angles.
TRUE LENGTH
HT
TRUE ANGLE WITH HORIZONTAL PLANE

FIG 5.22

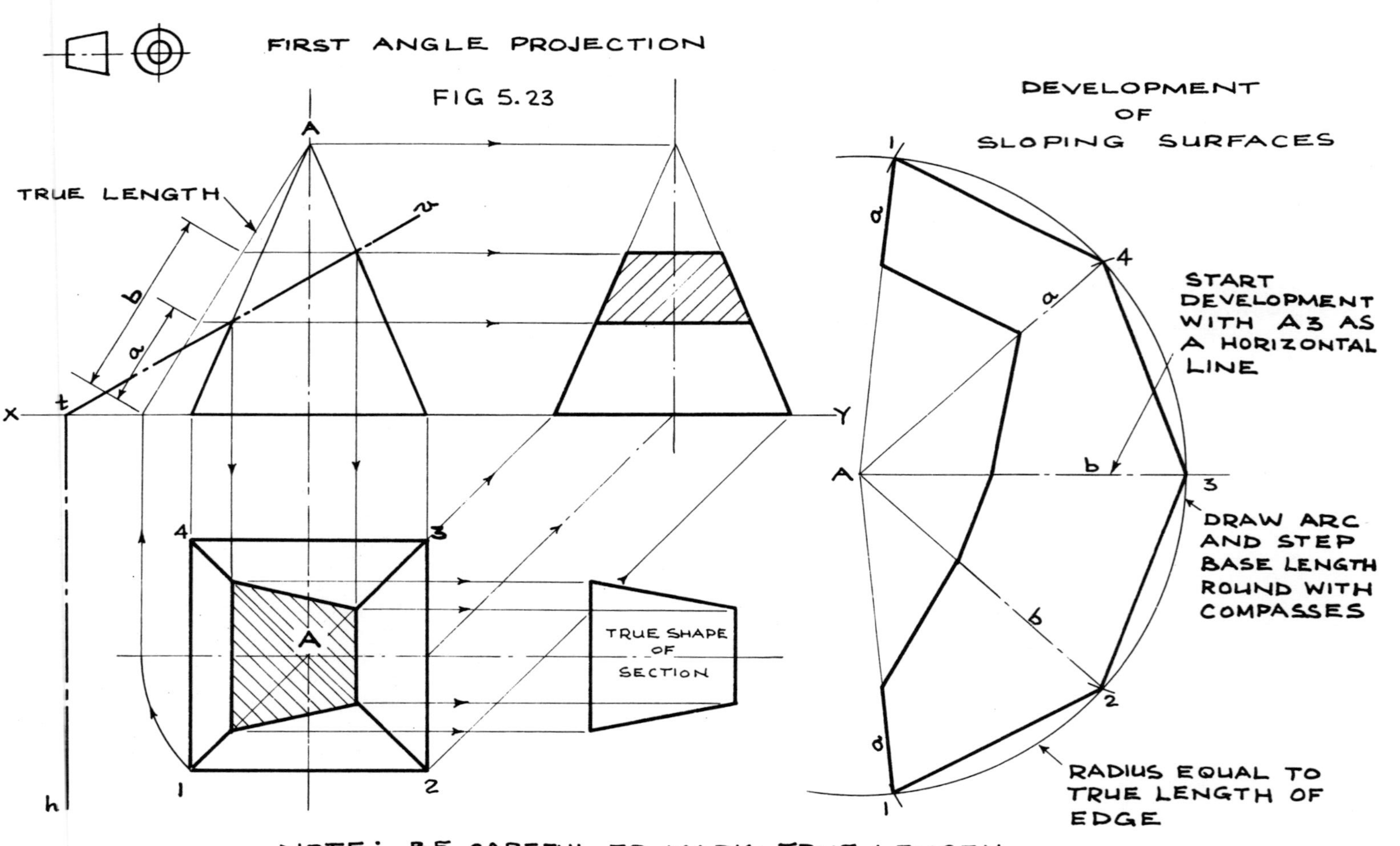
EXERCISE 3 SOLUTION
FIRST ANGLE PROJECTION
FIG 5.23
TRUE LENGTH
DEVELOPMENT OF SLOPING SURFACES
START DEVELOPMENT WITH A3 AS A HORIZONTAL LINE
DRAW ARC AND STEP BASE LENGTH ROUND WITH COMPASSES
RADIUS EQUAL TO TRUE LENGTH OF EDGE
TRUE SHAPE OF SECTION
NOTE: BE CAREFUL TO MARK TRUE LENGTH DISTANCES a AND b ON THE DEVELOPMENT
X
Y
A
1
2
3
4
a
b

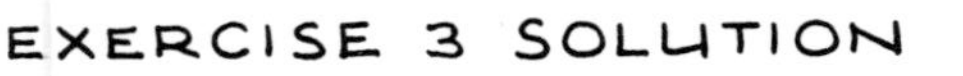

EXERCISE 4 SOLUTION

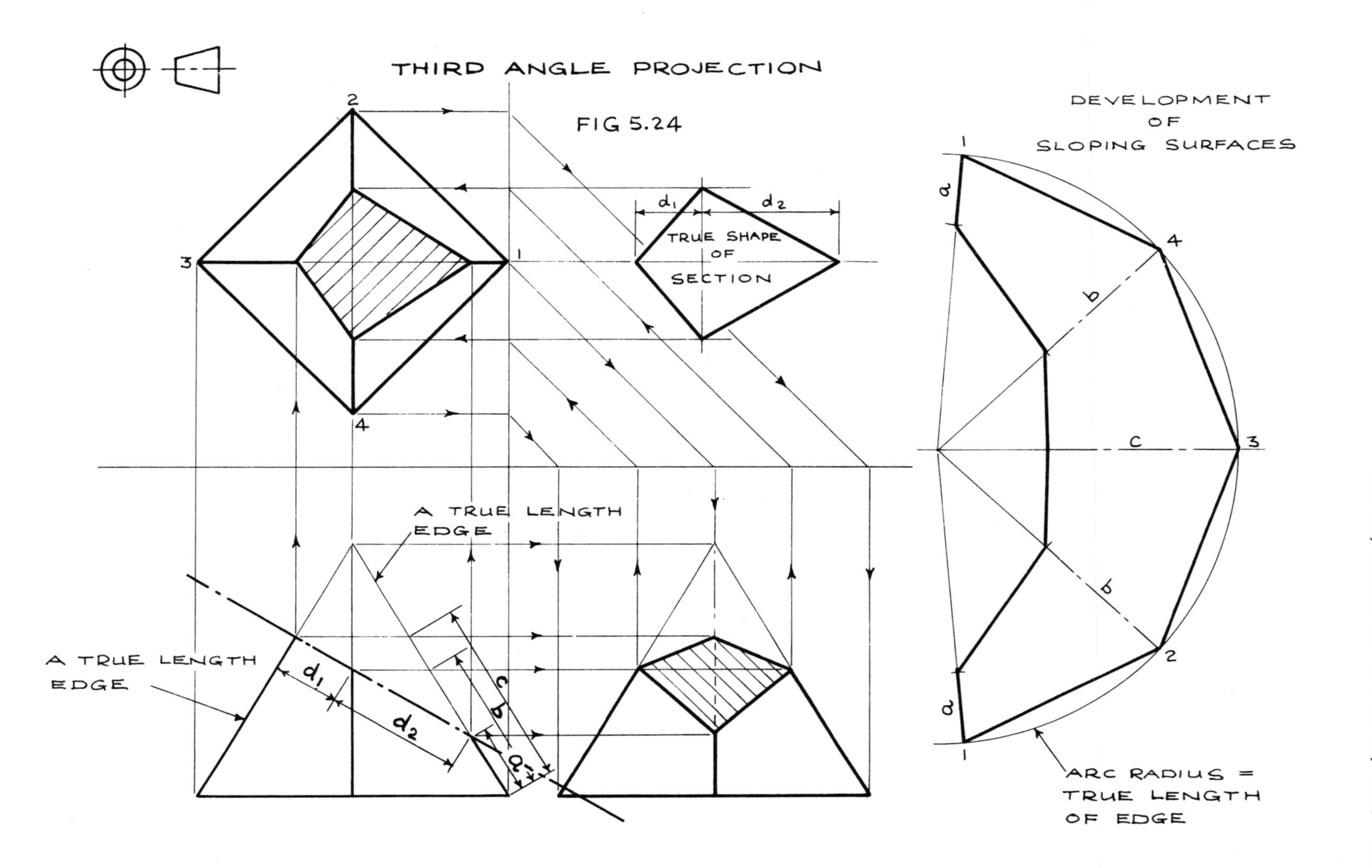
THIRD ANGLE PROJECTION
FIG 5.24
TRUE SHAPE OF SECTION
DEVELOPMENT OF SLOPING SURFACES
A TRUE LENGTH EDGE
A TRUE LENGTH EDGE
ARC RADIUS = TRUE LENGTH OF EDGE
1
2
3
4
a
b
c
d1
d2

EXERCISE 5 SOLUTION

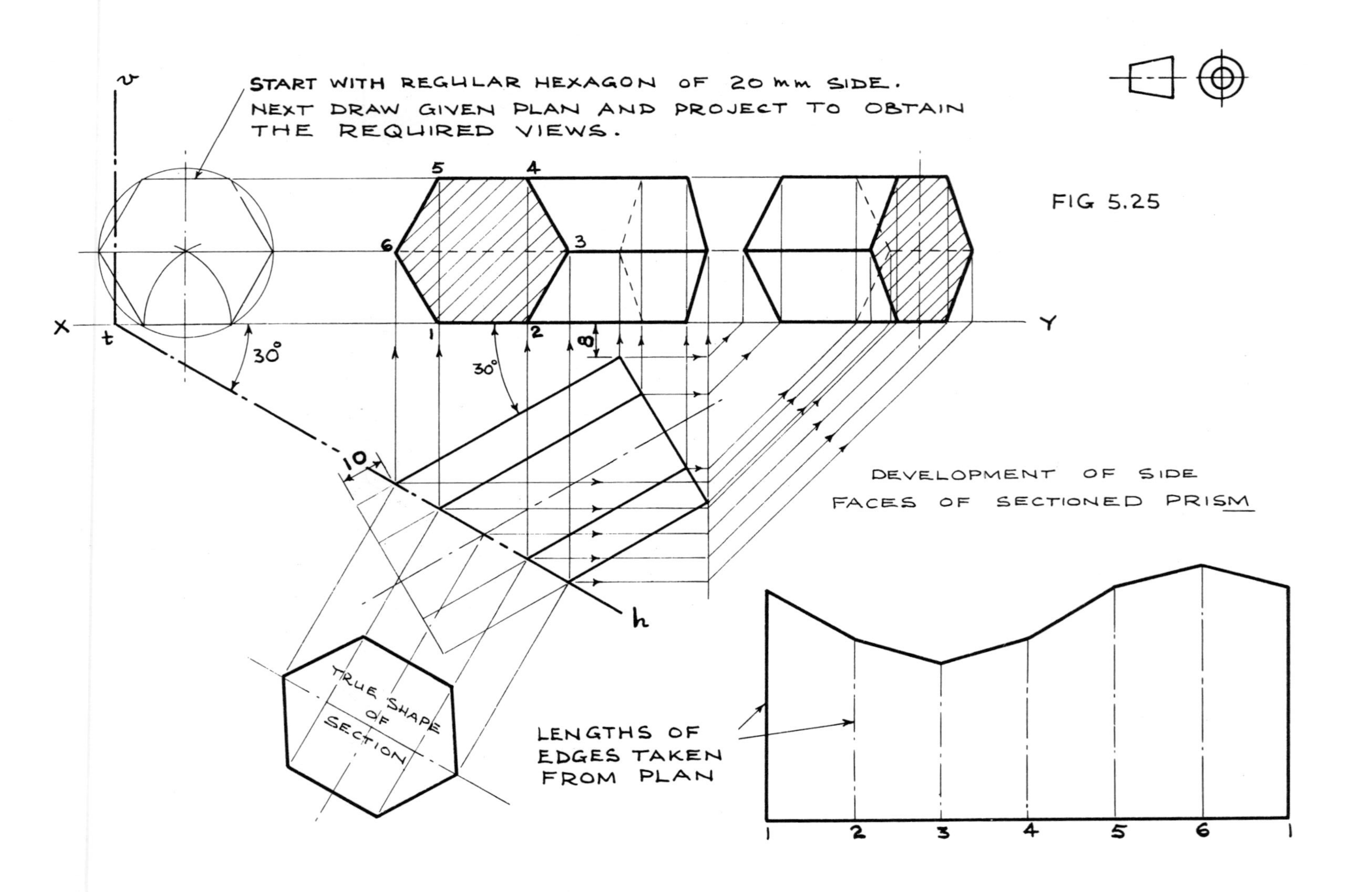
START WITH REGULAR HEXAGON OF 20 mm SIDE.
NEXT DRAW GIVEN PLAN AND PROJECT TO OBTAIN
THE REQUIRED VIEWS.
FIG 5.25
DEVELOPMENT OF SIDE
FACES OF SECTIONED PRISM
LENGTHS OF
EDGES TAKEN
FROM PLAN
TRUE SHAPE OF SECTION
30°
30°
10
8
X
Y
h
t
v
1
2
3
4
5
6
1
2
3
4
5
6
1

EXERCISE 6 SOLUTION

PROCEDURE:

Start with plan view — a regular hexagon.
Next project to obtain front view and side view.

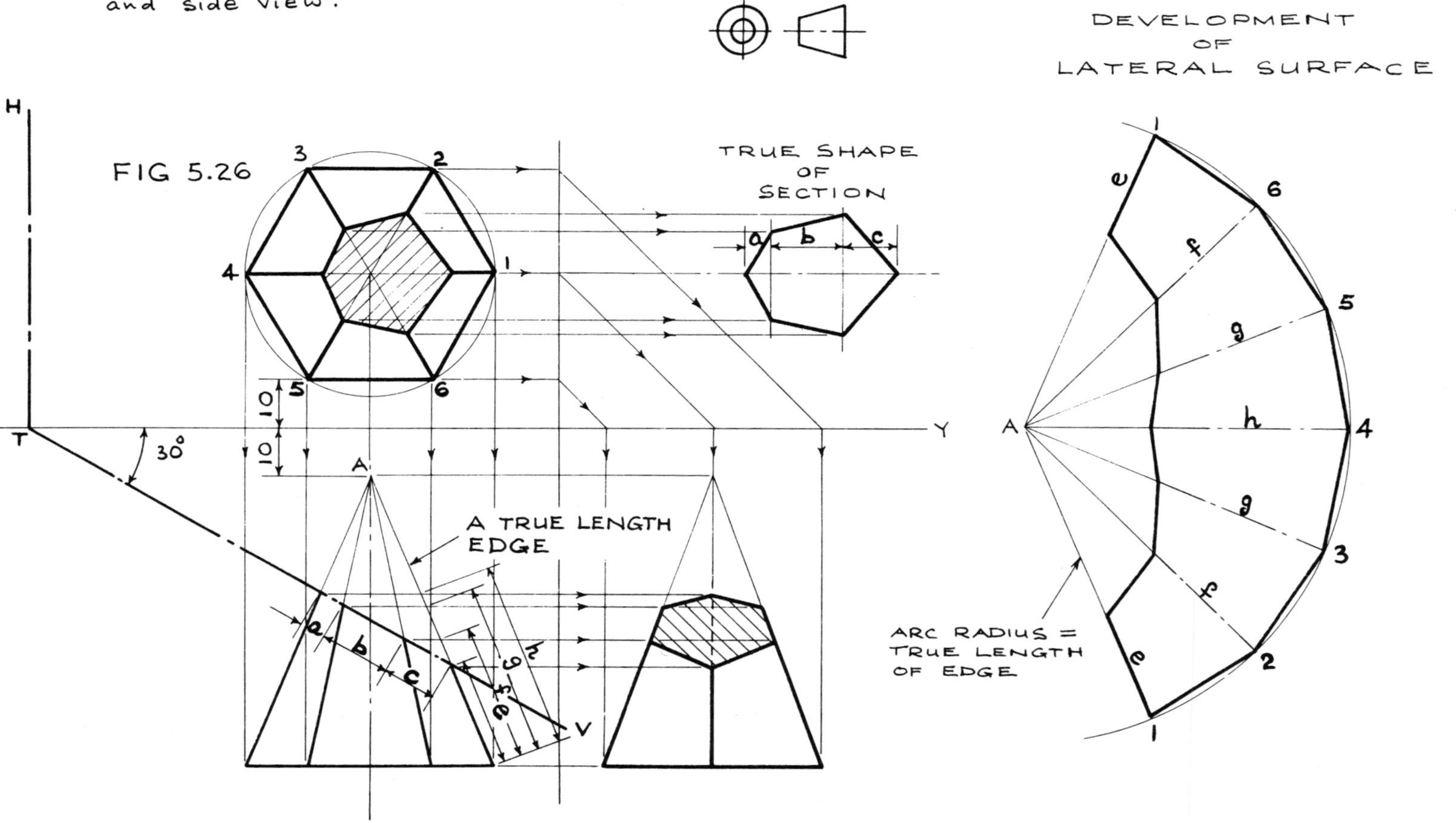

FIG 5.26

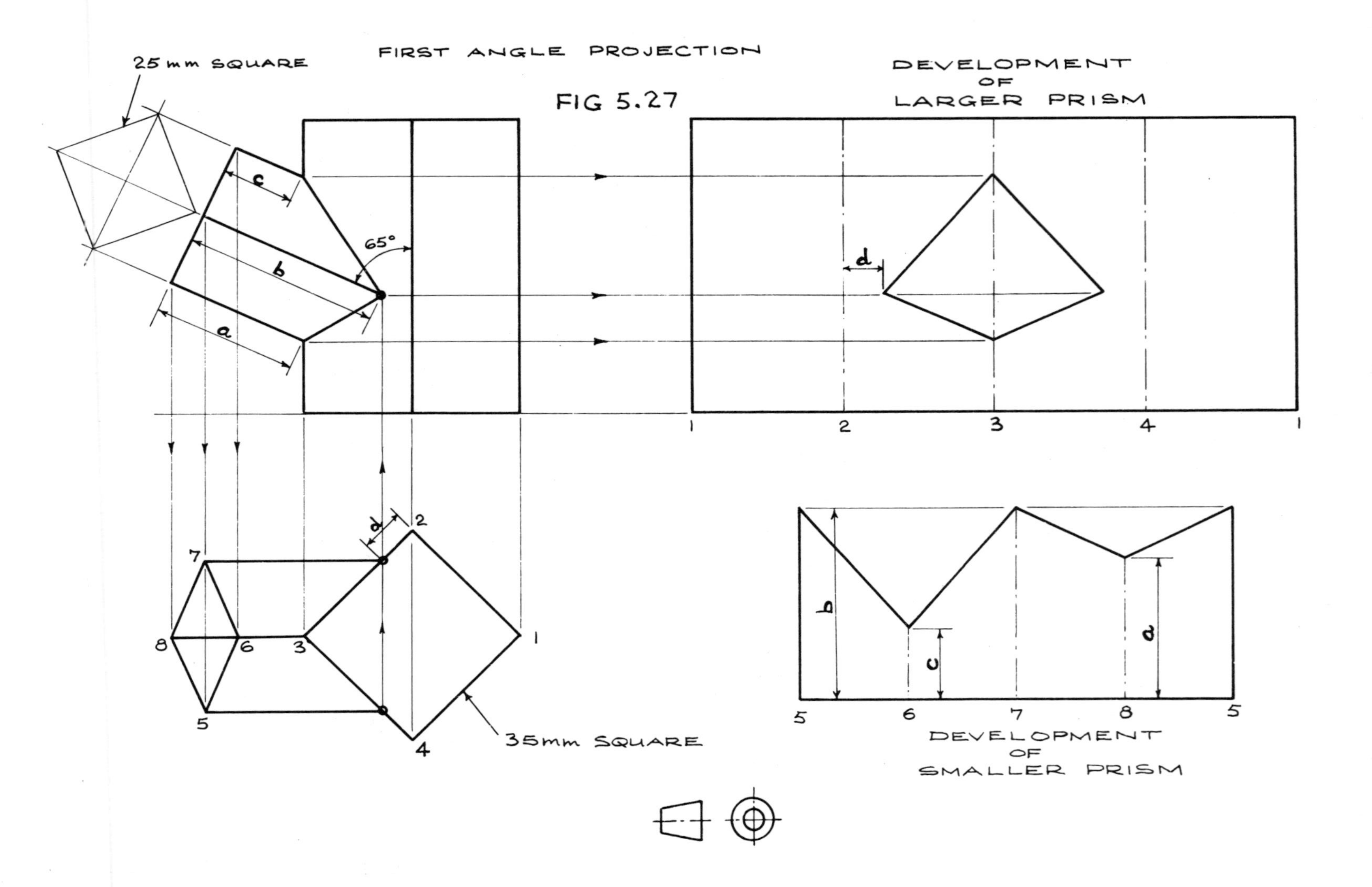

EXERCISE 7 SOLUTION
FIRST ANGLE PROJECTION
FIG 5.27
25 mm SQUARE
DEVELOPMENT OF LARGER PRISM
65°
a
b
c
d
1
2
3
4
5
6
7
8
35mm SQUARE
DEVELOPMENT OF SMALLER PRISM

EXERCISE 8 SOLUTION

FIG 5.28

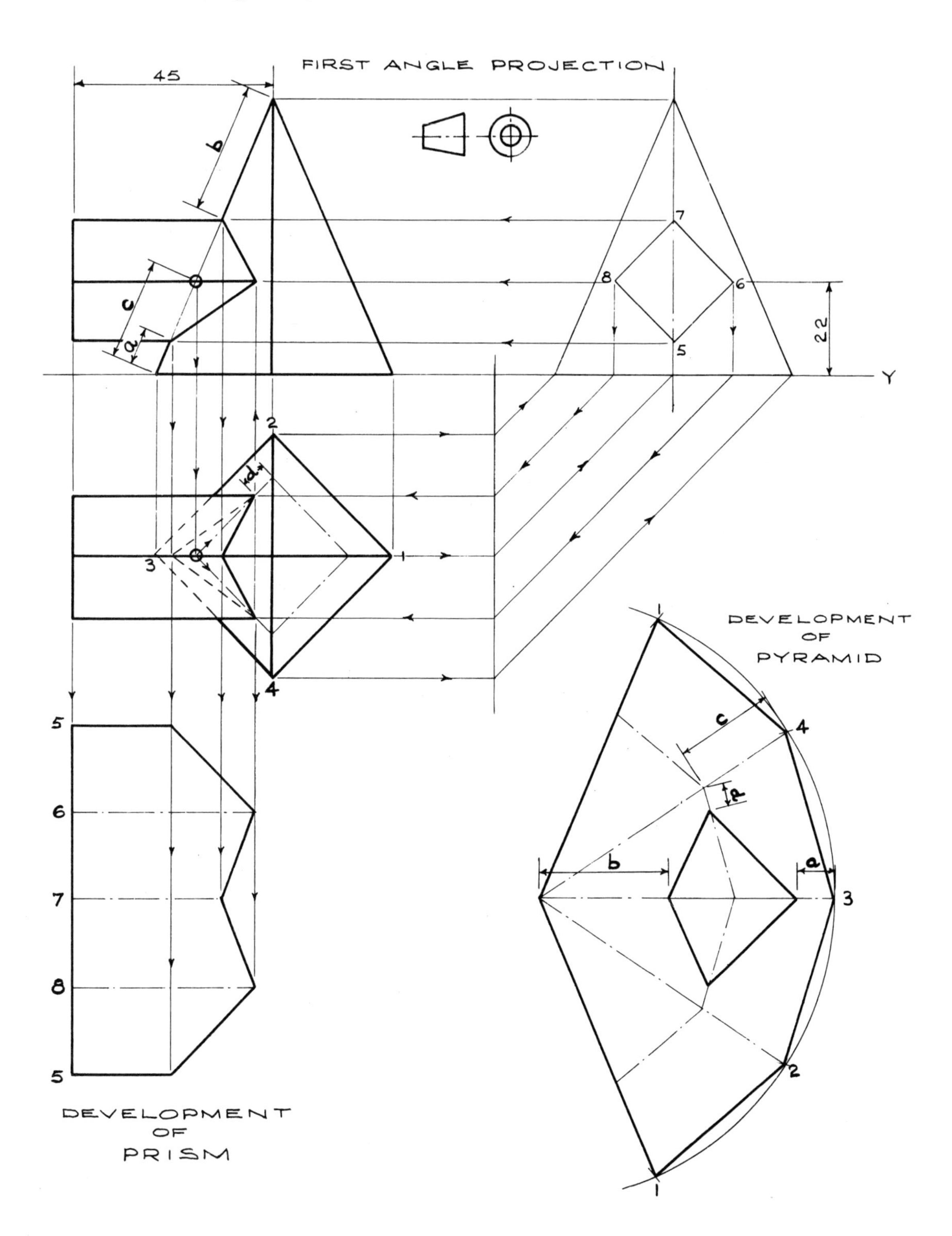

EXERCISE 9 SOLUTION

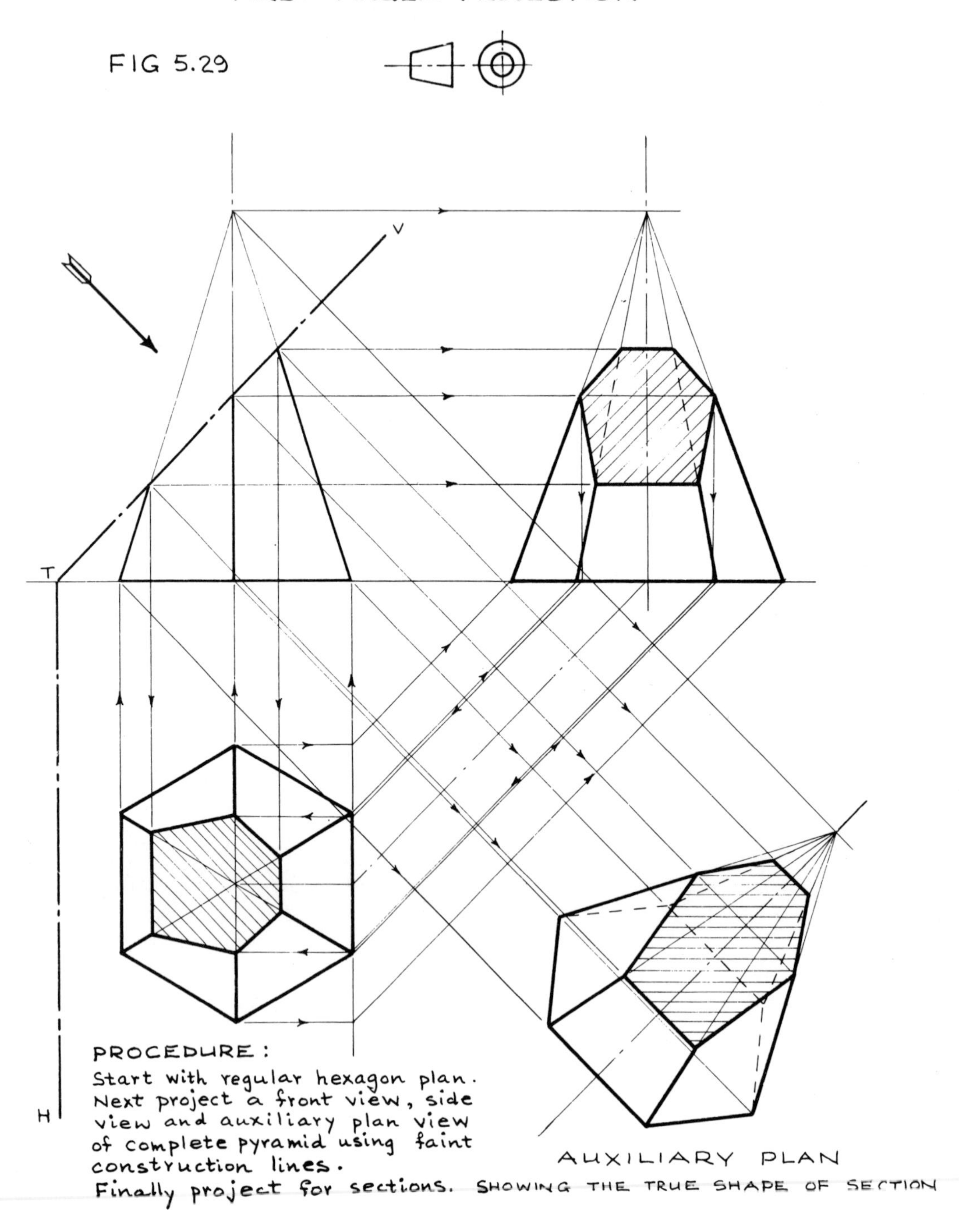

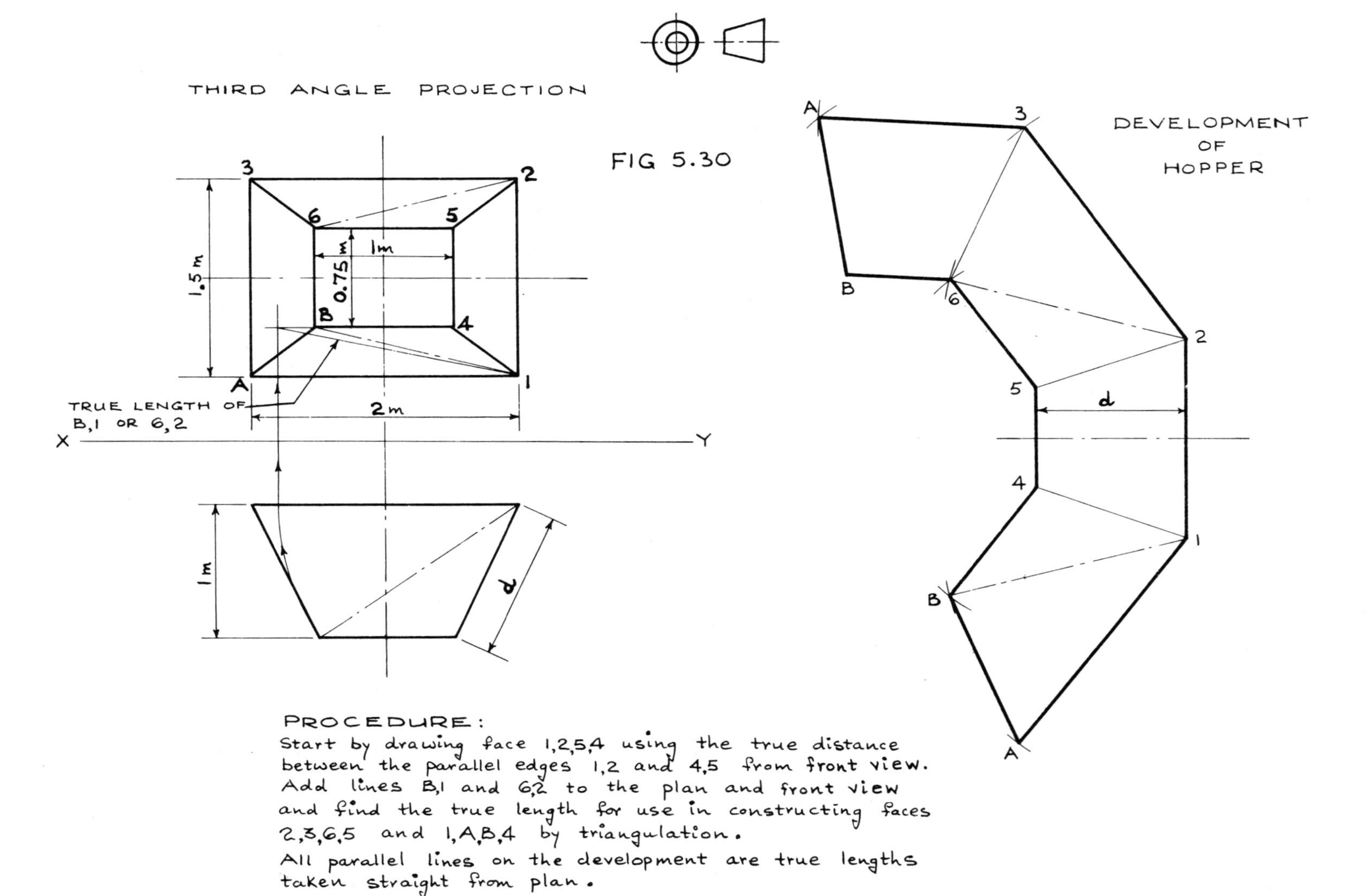

EXERCISE 10 SOLUTION
THIRD ANGLE PROJECTION
FIG 5.30
DEVELOPMENT OF HOPPER
1m
0.75 m
1.5 m
2m
1m
d
TRUE LENGTH OF B,1 OR 6,2
X
Y
PROCEDURE:
Start by drawing face 1,2,5,4 using the true distance between the parallel edges 1,2 and 4,5 from front view.
Add lines B,1 and 6,2 to the plan and front view and find the true length for use in constructing faces 2,3,6,5 and 1,A,B,4 by triangulation.
All parallel lines on the development are true lengths taken straight from plan.

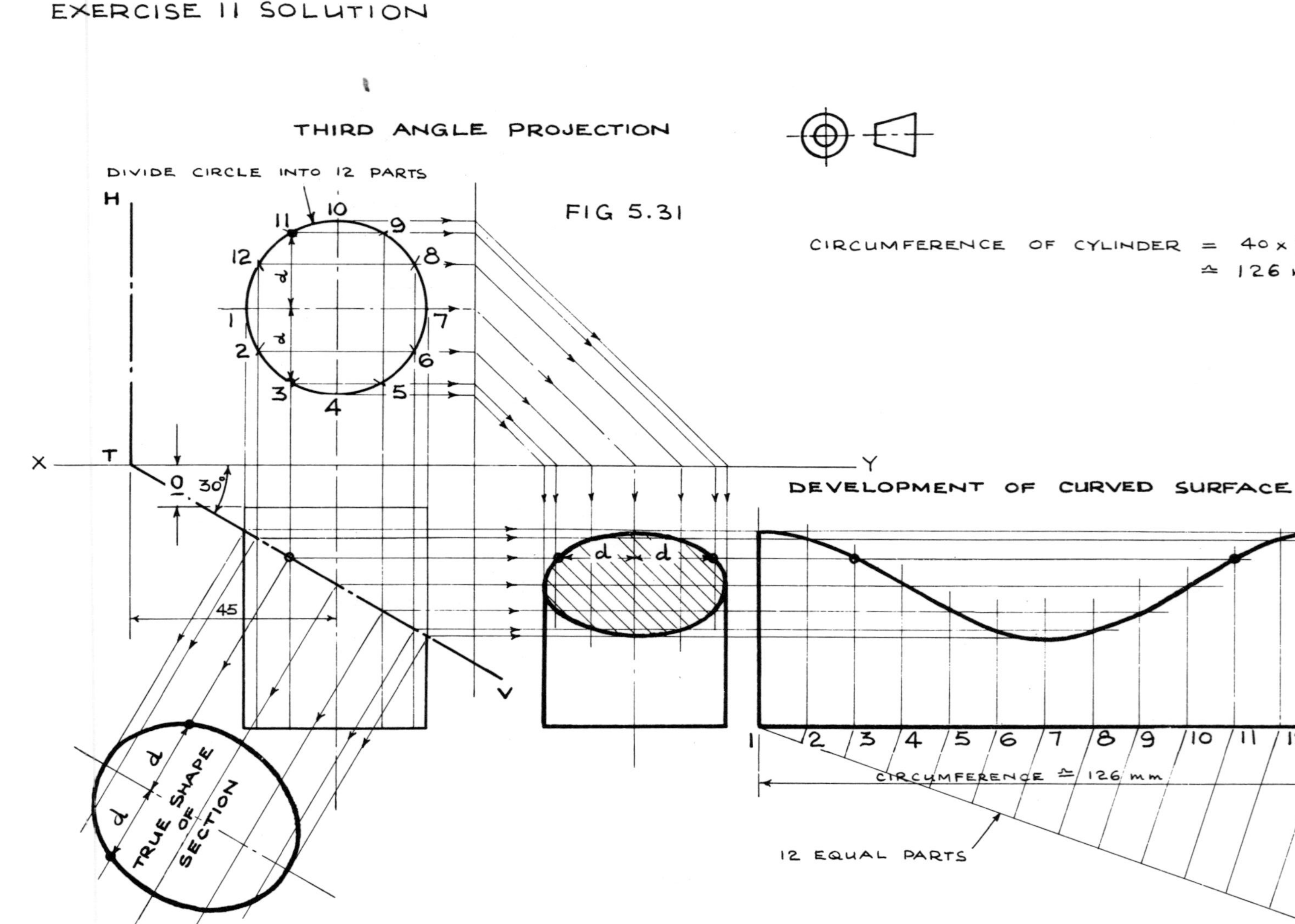
EXERCISE 11 SOLUTION
THIRD ANGLE PROJECTION
DIVIDE CIRCLE INTO 12 PARTS
FIG 5.31
CIRCUMFERENCE OF CYLINDER = 40 × 22/7
≏ 126 mm
DEVELOPMENT OF CURVED SURFACE
CIRCUMFERENCE ≏ 126 mm
12 EQUAL PARTS
TRUE SHAPE OF SECTION
45
30°
X
Y
H
T
V

INTERSECTION OF CYLINDERS

The simplest case of equal cylinders intersecting at right-angles is shown in figure 5.32. The front view showing the lines of intersection is simply obtained by drawing straight lines. Similarly for two equal cylinders with axes intersecting at any angle, straight lines are obtained as shown in figure 5.33.

In the cases of unequal cylinders intersecting, curved lines are obtained in front view for the lines of intersection as shown in figures 5.34 and 5.35. The method of obtaining these curves is the same in each case and is a very general method which may be employed to many other cases of intersecting solids. Briefly, horizontal planes through 1, 11 and 5, 7 give rise to points 5, 7 in front view, figure 5.34, and to points 5, 7, 11, 1 in figure 5.35. Similarly further front view points 4, 8, in figure 5.34 and 4, 8, 10, 2 in figure 5.35 are obtained by using horizontal planes through 2, 10 and 4, 8. Special points such as 6 in figure 5.34 and 6, 0 in figure 5.35 are found where projectors from plan views intersect the centre line of each front view.

Exercise 12. Two hollow right cylinders, 35 mm diameter, meet at right-angles as shown by the given front view shown in figure 5.36. Using Third Angle projection, add a plan view, side view and also show the development of one of the cylinders.

Exercise 13. Two cylindrical pipes, 40 mm and 37 mm diameter respectively, have axes intersecting at 45° as shown in the front view of figure 5.37. Obtain the curve of intersection in front view and develop each pipe.

Exercise 14. Develop the curved surface of the oblique cylinder shown in figure 5.38 front view and plan.

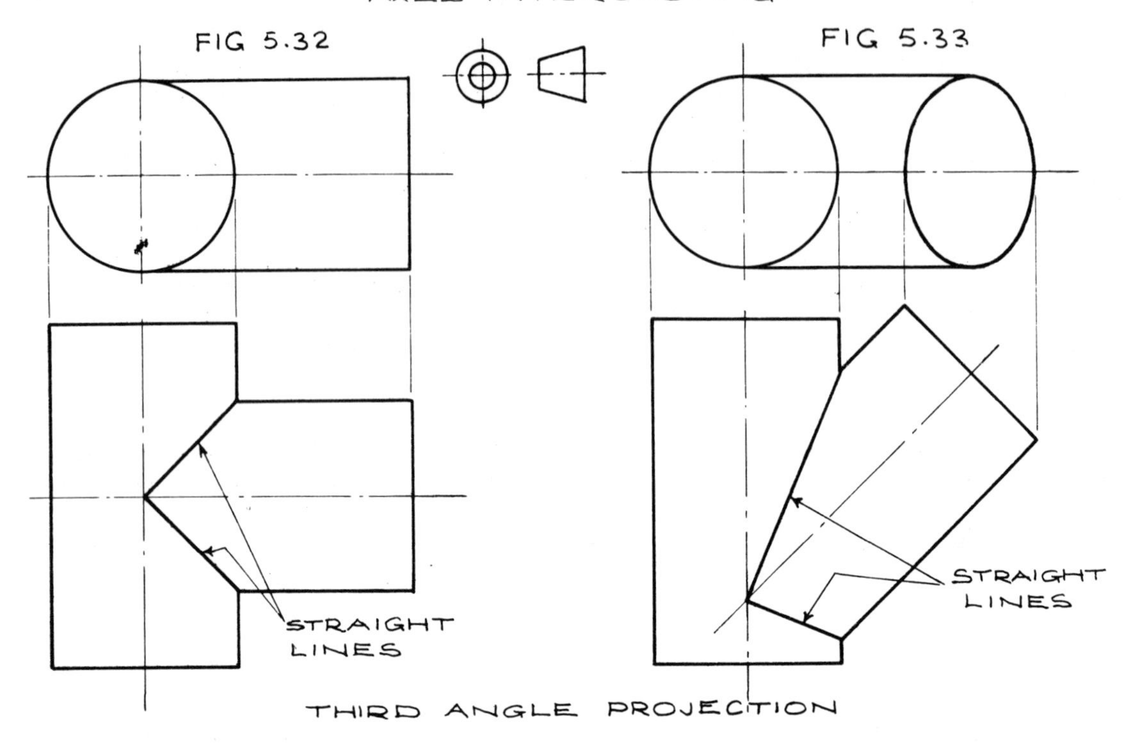

THIRD ANGLE PROJECTION

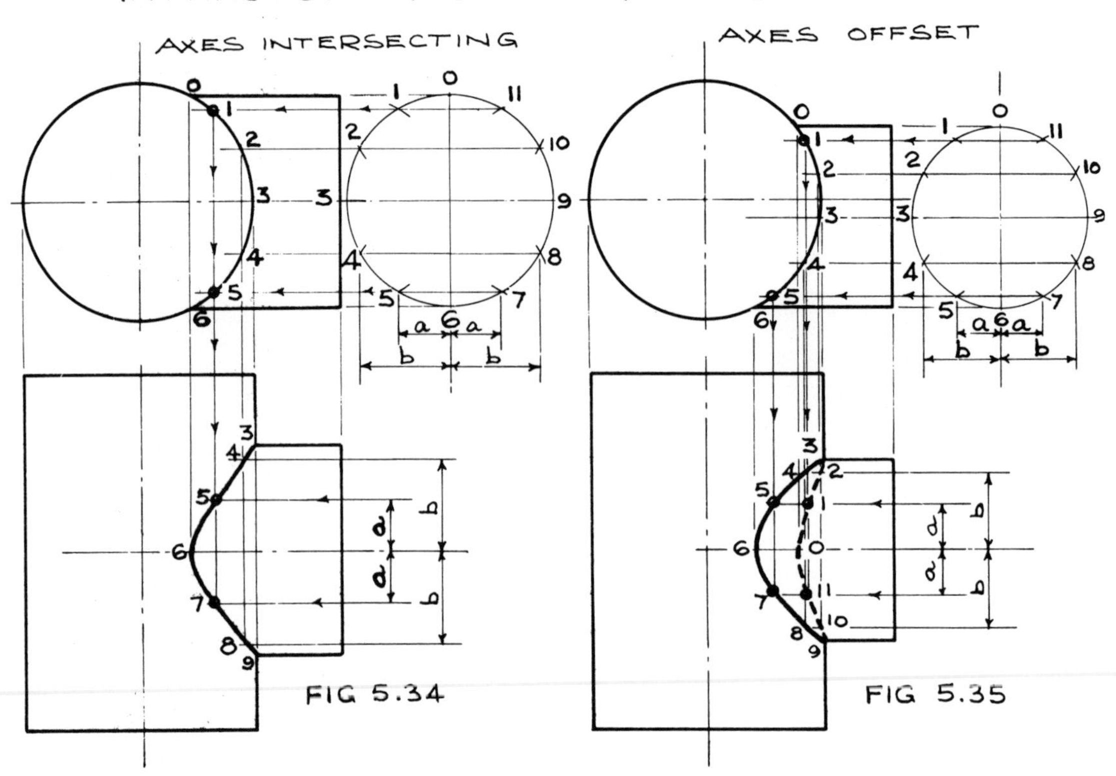

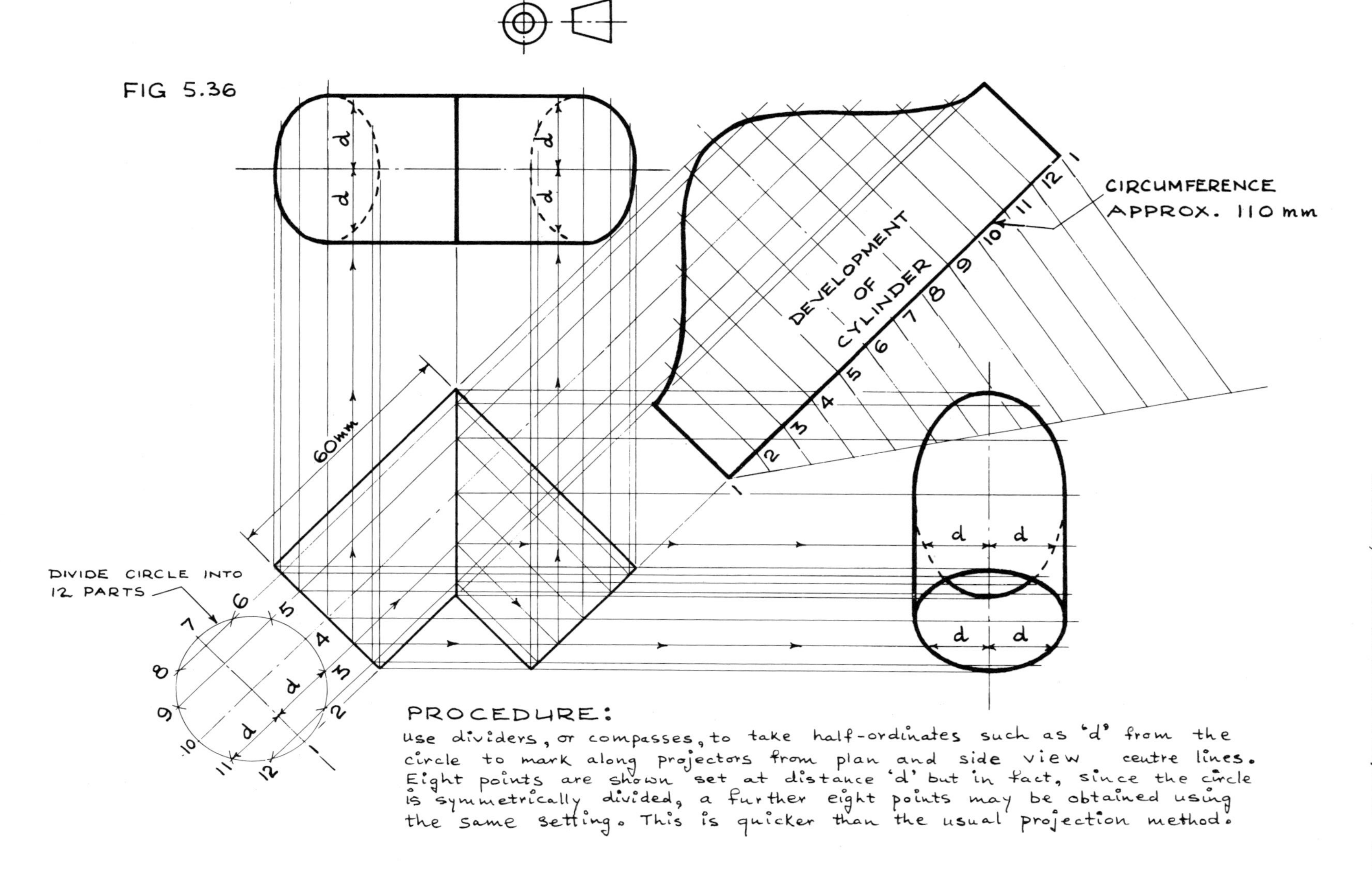
EXERCISE 12 SOLUTION
FIG 5.36
DIVIDE CIRCLE INTO 12 PARTS
60mm
DEVELOPMENT OF CYLINDER
CIRCUMFERENCE APPROX. 110 mm
PROCEDURE:
Use dividers, or compasses, to take half-ordinates such as 'd' from the circle to mark along projectors from plan and side view centre lines. Eight points are shown set at distance 'd' but in fact, since the circle is symmetrically divided, a further eight points may be obtained using the same setting. This is quicker than the usual projection method.

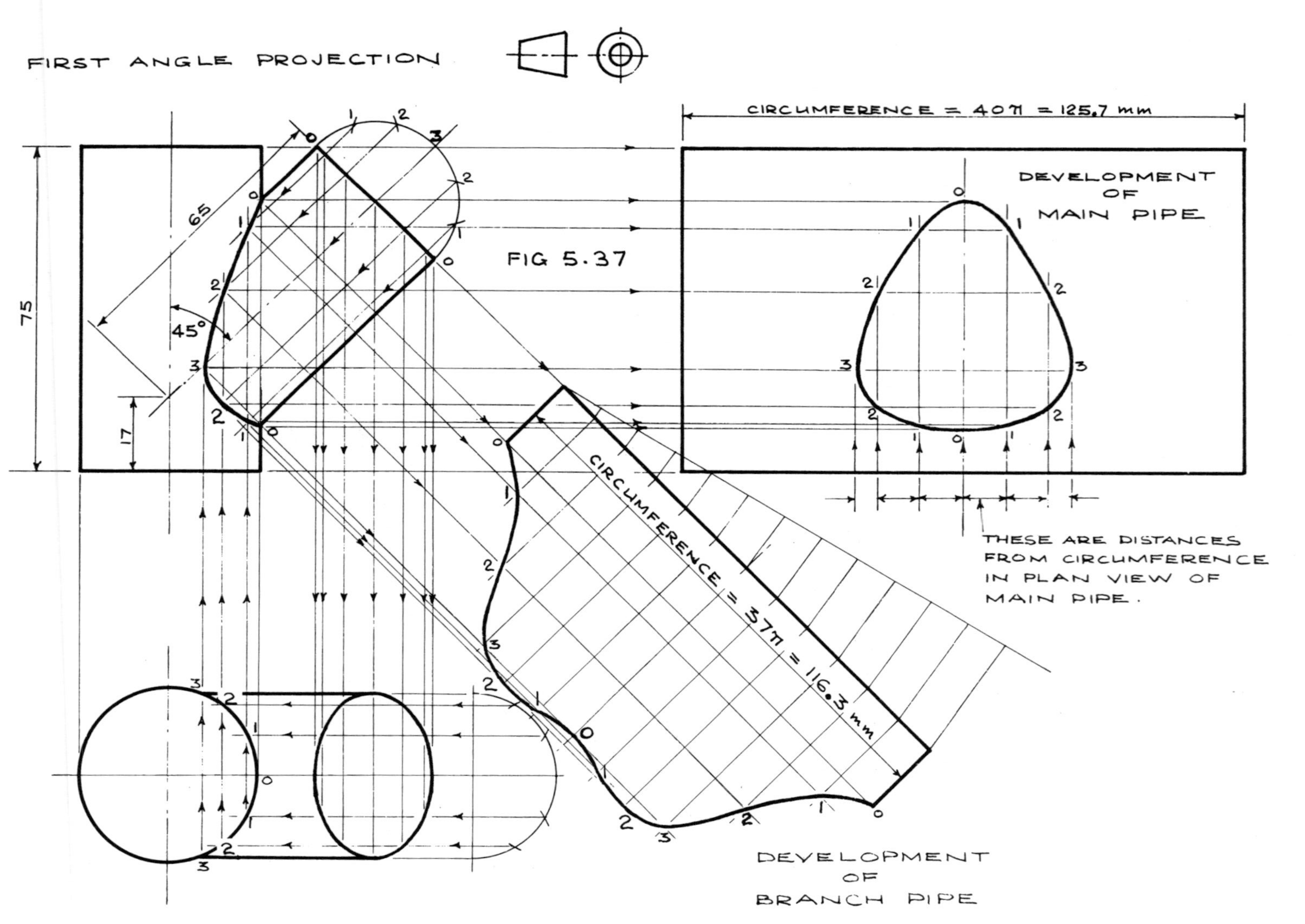
EXERCISE 13 SOLUTION
FIRST ANGLE PROJECTION
CIRCUMFERENCE = 40π = 125.7 mm
DEVELOPMENT OF MAIN PIPE
THESE ARE DISTANCES FROM CIRCUMFERENCE IN PLAN VIEW OF MAIN PIPE.
CIRCUMFERENCE = 37π = 116.3 mm
DEVELOPMENT OF BRANCH PIPE
65
45°
17
75

FIG 5.37

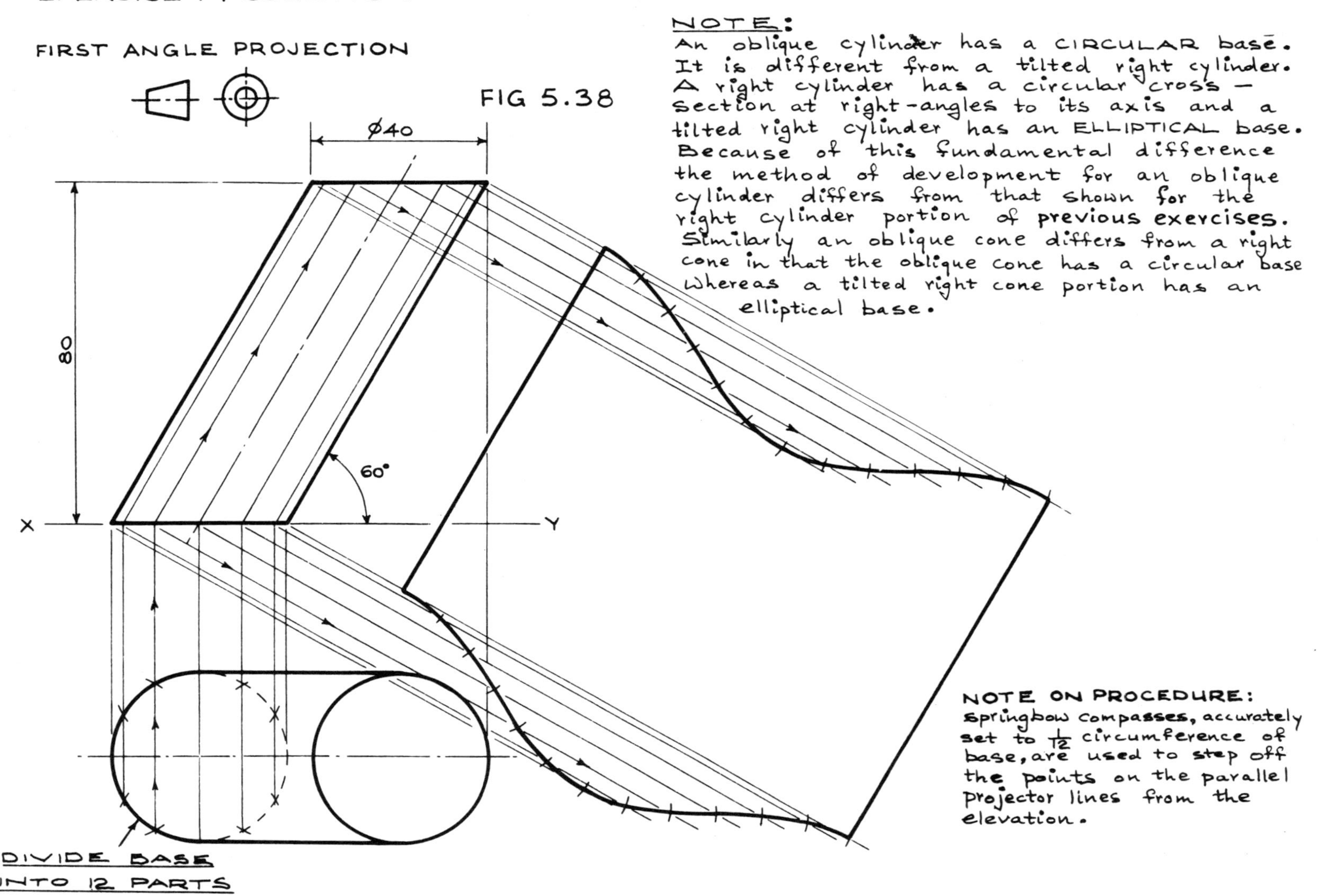
EXERCISE 14 SOLUTION
FIRST ANGLE PROJECTION
FIG 5.38
Ø40
80
60°
X
Y
NOTE:
An oblique cylinder has a CIRCULAR base. It is different from a tilted right cylinder. A right cylinder has a circular cross-section at right-angles to its axis and a tilted right cylinder has an ELLIPTICAL base. Because of this fundamental difference the method of development for an oblique cylinder differs from that shown for the right cylinder portion of previous exercises. Similarly an oblique cone differs from a right cone in that the oblique cone has a circular base whereas a tilted right cone portion has an elliptical base.
NOTE ON PROCEDURE:
Springbow compasses, accurately set to 1/12 circumference of base, are used to step off the points on the parallel projector lines from the elevation.
DIVIDE BASE
INTO 12 PARTS

USE OF GENERATOR LINES

Some geometrical solids may be said to be 'generated' by the revolution of a line about an axis of the solid, e.g. a right cylinder may be said to be generated by the revolution, about its axis, of a straight line parallel to the axis. Again, a right circular cone may be said to be generated by a straight line, one end of which is fixed and the other end is moved in a circle. Thus any straight line drawn on the curved surface of a cylinder or cone may be referred to as a generator line.

Generator lines are useful in development of surface problems as is demonstrated in solutions to exercises.

In the three views of a right cone in First Angle projection, figure 5.39, twelve equally spaced generators are shown. Two generators, No.'s 2 and 12, are emphasised and the student should carefully check the actual generators which are visible in each view.

In the Third Angle projection views, figure 5.40, generators numbered 5 and 9 are emphasised.

It may be worth noticing that, in this particular example, there is less work in projecting the three views in First Angle projection, figure 5.39, than in obtaining the Third Angle views, figure 5.40.

Exercise 15. The front view and plan of a right circular cone, 60 mm diameter of base and altitude 70 mm, is given in First Angle projection in figure 5.42. The cone is cut by inclined plane VTH which intersects the axis mid-point and makes 45° with the HP. Obtain a sectioned side view, a sectioned plan, the true shape of the section and the development of the curved surface below VTH.

Exercise 16. Given the front view of a right circular cone, 70 mm diameter of the base and 75 mm altitude, obtain the sectioned plan, sectioned side view and the true shape of the section when cut by inclined plane VTH which intersects axis 30 mm above the HP and is at 30° to HP (figure 5.43).

Exercise 17. Similar to the previous exercise but in this case VTH is an inclined plane parallel to the sloping surface of the cone. Figure 5.44 is in Third Angle projection.

Exercise 18. Draw sectioned views of a sphere cut by inclined plane VTH as shown in Third Angle projection in figure 5.45.

Exercise 19. The front view of a right cylinder with portions removed is given in figure 5.46. Obtain a plan view of the object.

Exercise 20. The front view and side view of an object are given in figure 5.47. Project a plan view.

Exercise 21. An oblique pentagonal pyramid is given in Third Angle projection front view and plan in figures 5.48. Draw the development of the sloping surfaces.

Exercise 22. An oblique cone is given in First Angle projection front view and plan in figure 5.49. Obtain the development of the curved surface of the cone. Base diameter is 60 mm and vertical height of apex A is 80 mm.

Exercise 23. The front view and plan of a sheet metal transition piece, used to connect a 0.4 m diameter cylindrical pipe to a 0.6 m square section pipe, are given in Third Angle projection in figure 5.50. Using a scale of 1:10 obtain the development of this transition piece, making the joint along E1.

Exercise 24. Two views of a sheet metal transition piece to connect a square pipe with a rectangular pipe are given in Third Angle projection in figure 5.51. Draw the development of the transition piece.

Exercise 25. Obtain the curves of intersection in front view when a right cylinder penetrates a hexagonal prism as shown in figure 5.52.

Exercise 26. A cylinder penetrates an equilateral triangular prism as shown in figure 5.53. Show the curve of intersection in front view.

Exercise 27. A right cone, 80 mm diameter and 80 mm altitude, is intersected by a right cylinder, 40 mm diameter, with axis horizontal as shown in Third Angle projection in figure 5.54. The axis of the cylinder is 30 mm above the cone base. Obtain the intersection line of these solids, in plan and elevation.

Exercise 28. A right cone, 80 mm diameter of base and 70 mm altitude, is intersected by a right cylinder, 50 mm diameter. The axes of the solids are parallel and the cylinder axis is off-set 12.5 mm from the cone axis as shown in figure 5.56. Obtain the intersection line of the solids using two different methods.

Exercise 29. Determine the curves of intersection when a cylinder 60 mm diameter penetrates a sphere of 100 mm diameter, the cylinder being off-set from the sphere centre as indicated in plan in figure 5.58.

Exercise 30. Obtain the curves of intersection in plan and front view when a sphere of 80 mm diameter penetrates a right cone, 100 mm diameter of base and 100 mm altitude, as shown in figure 5.59. The axis of the cone and the sphere centre are in a plane parallel to the VP.

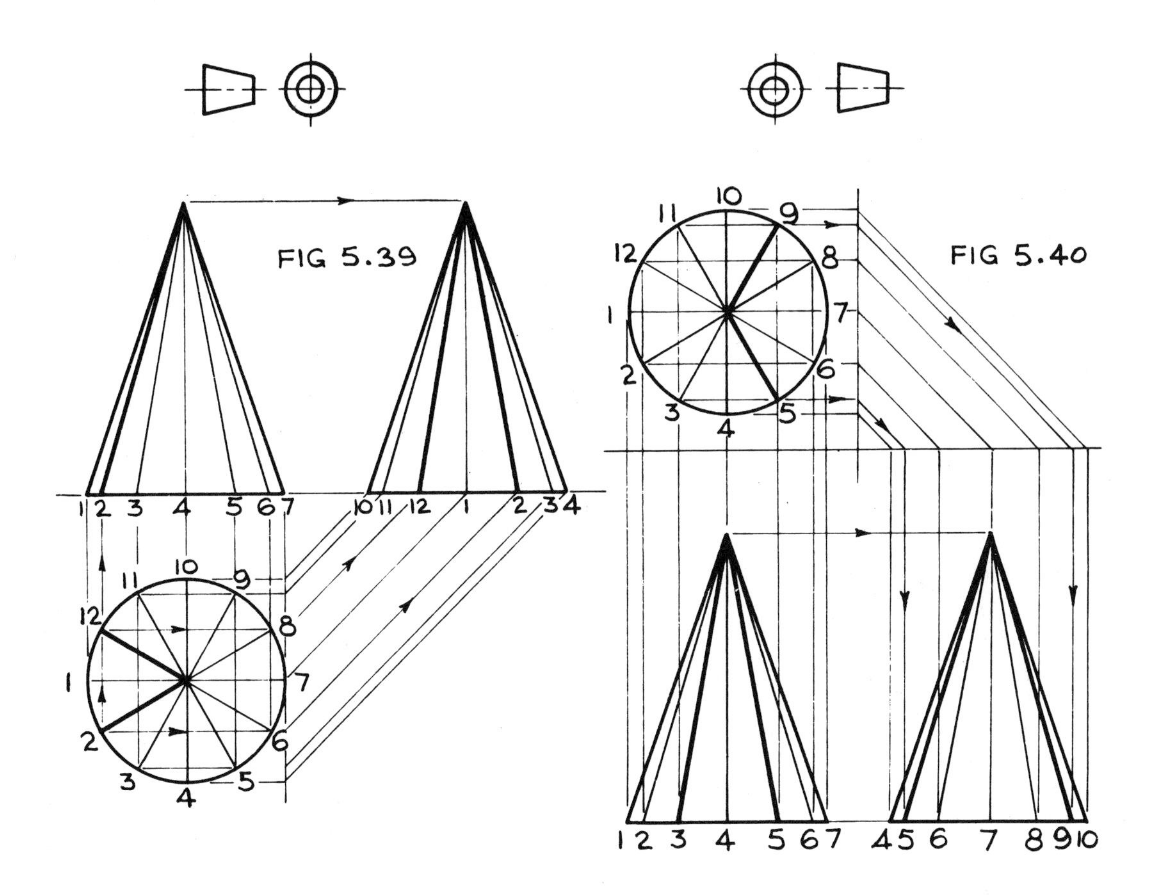

SECTIONS OF A RIGHT CIRCULAR CONE

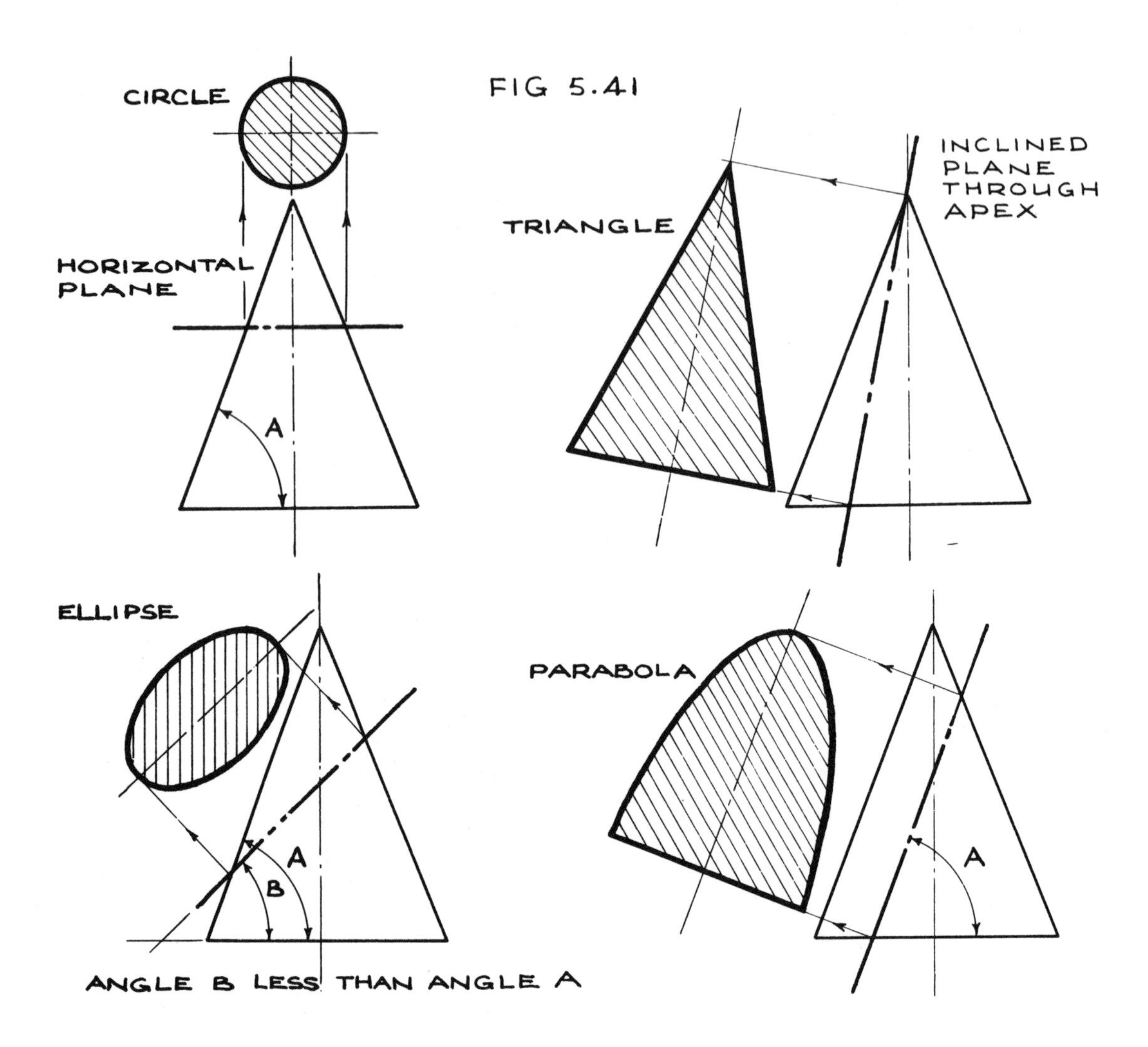

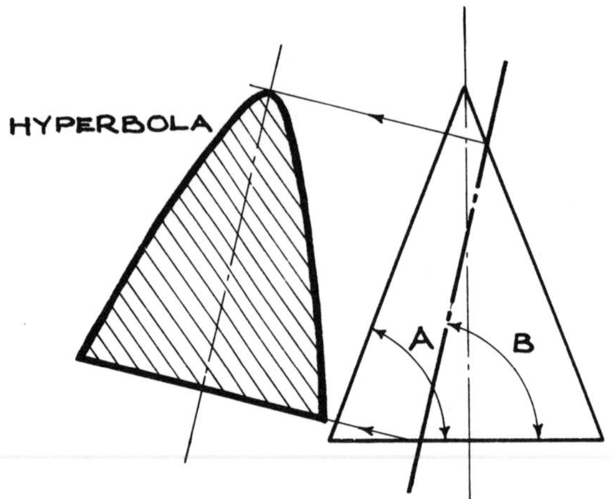

NOTE: In Chapter 6 the ellipse, parabola and hyperbola are treated as LOCI and alternative methods are shown for their construction.

In the following three exercises these curves are treated as sections of a right circular cone and are dealt with by the general geometrical method for any shape of section.

EXERCISE 15 SOLUTION

FIRST ANGLE PROJECTION

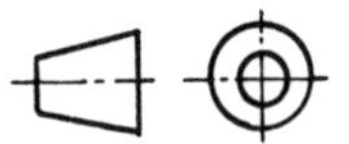

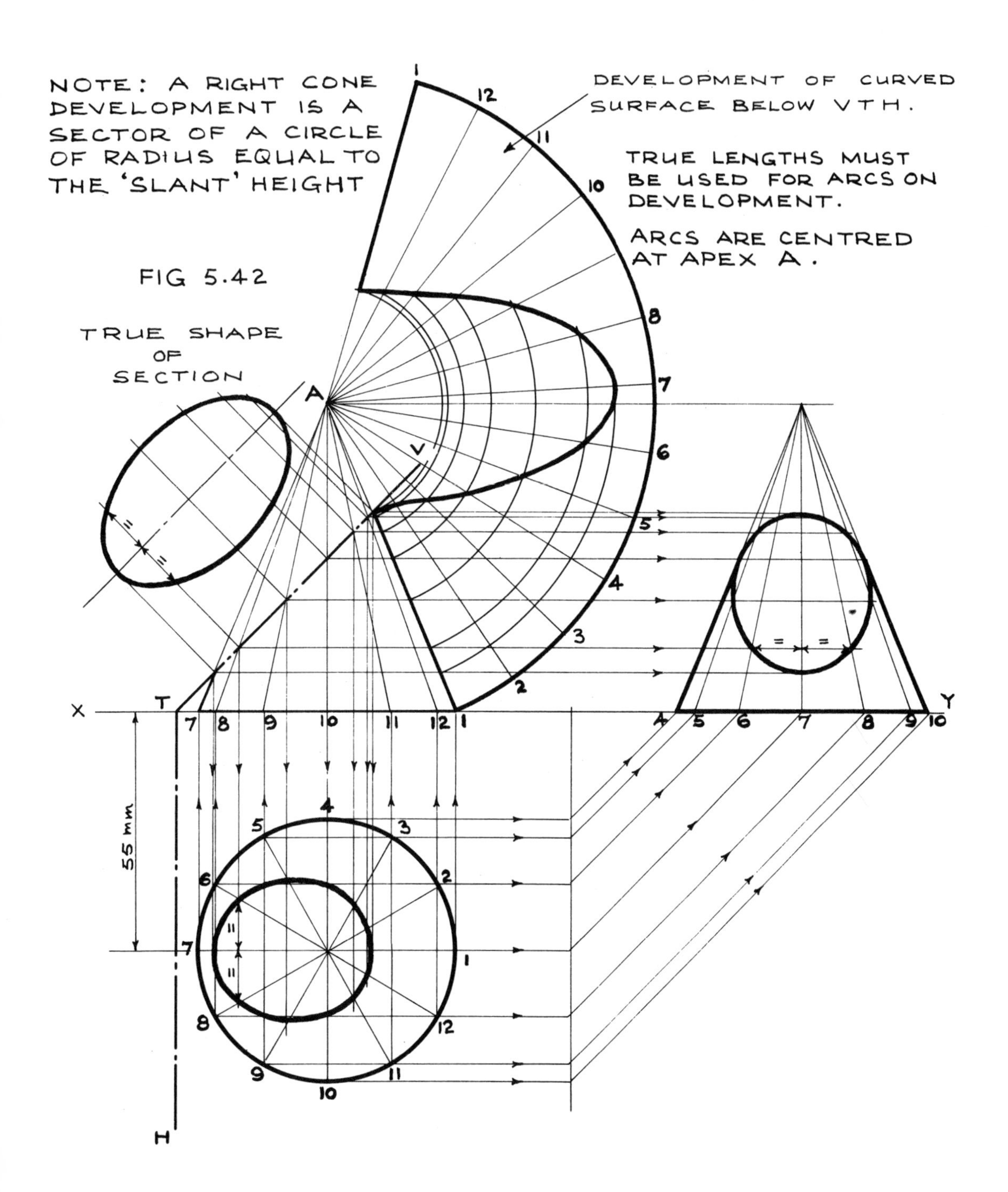

FIG 5.42

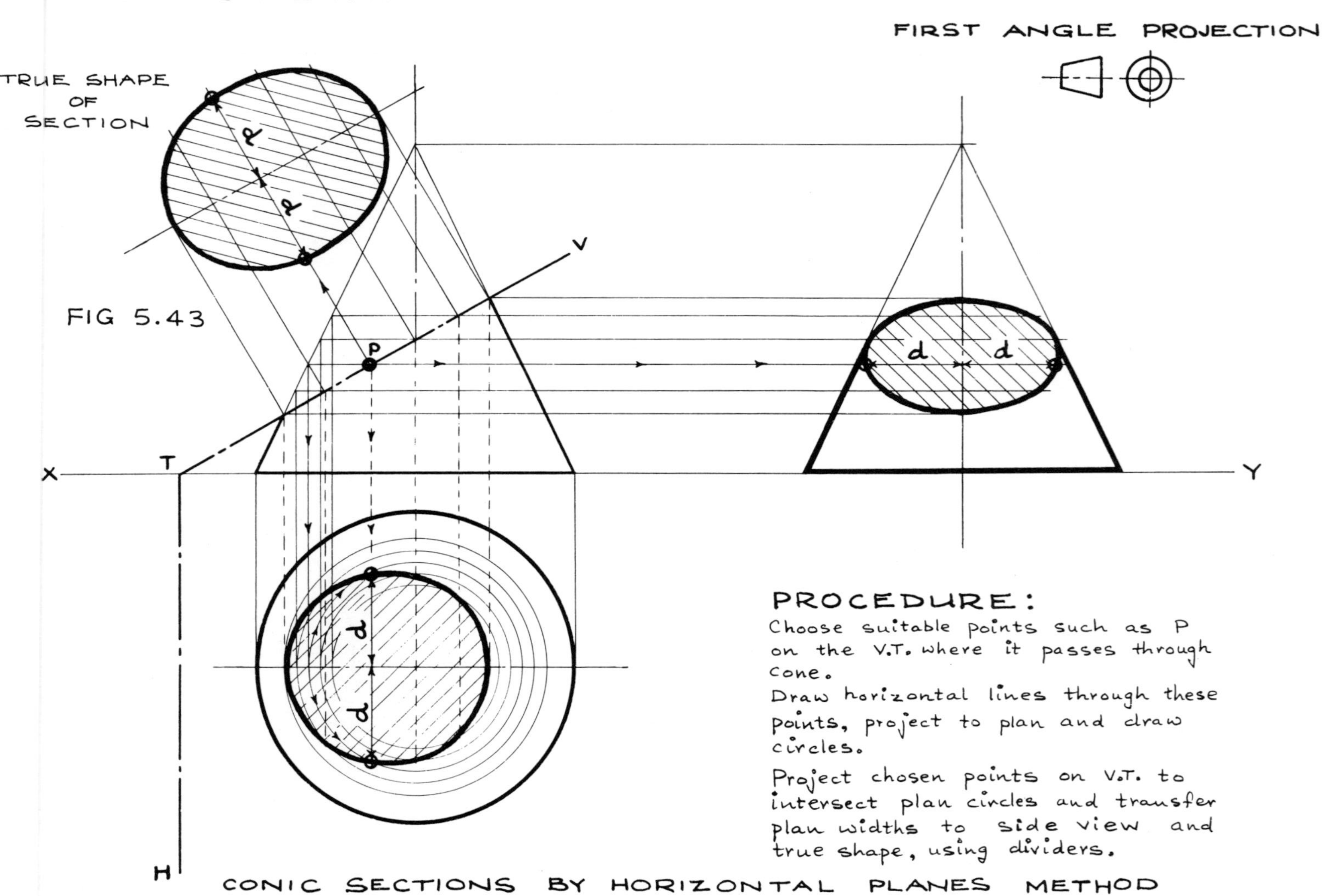
EXERCISE 16 SOLUTION
FIRST ANGLE PROJECTION
TRUE SHAPE OF SECTION
FIG 5.43
V
P
d
d
T
X
Y
H
PROCEDURE:
Choose suitable points such as P on the V.T. where it passes through cone.
Draw horizontal lines through these points, project to plan and draw circles.
Project chosen points on V.T. to intersect plan circles and transfer plan widths to side view and true shape, using dividers.
CONIC SECTIONS BY HORIZONTAL PLANES METHOD

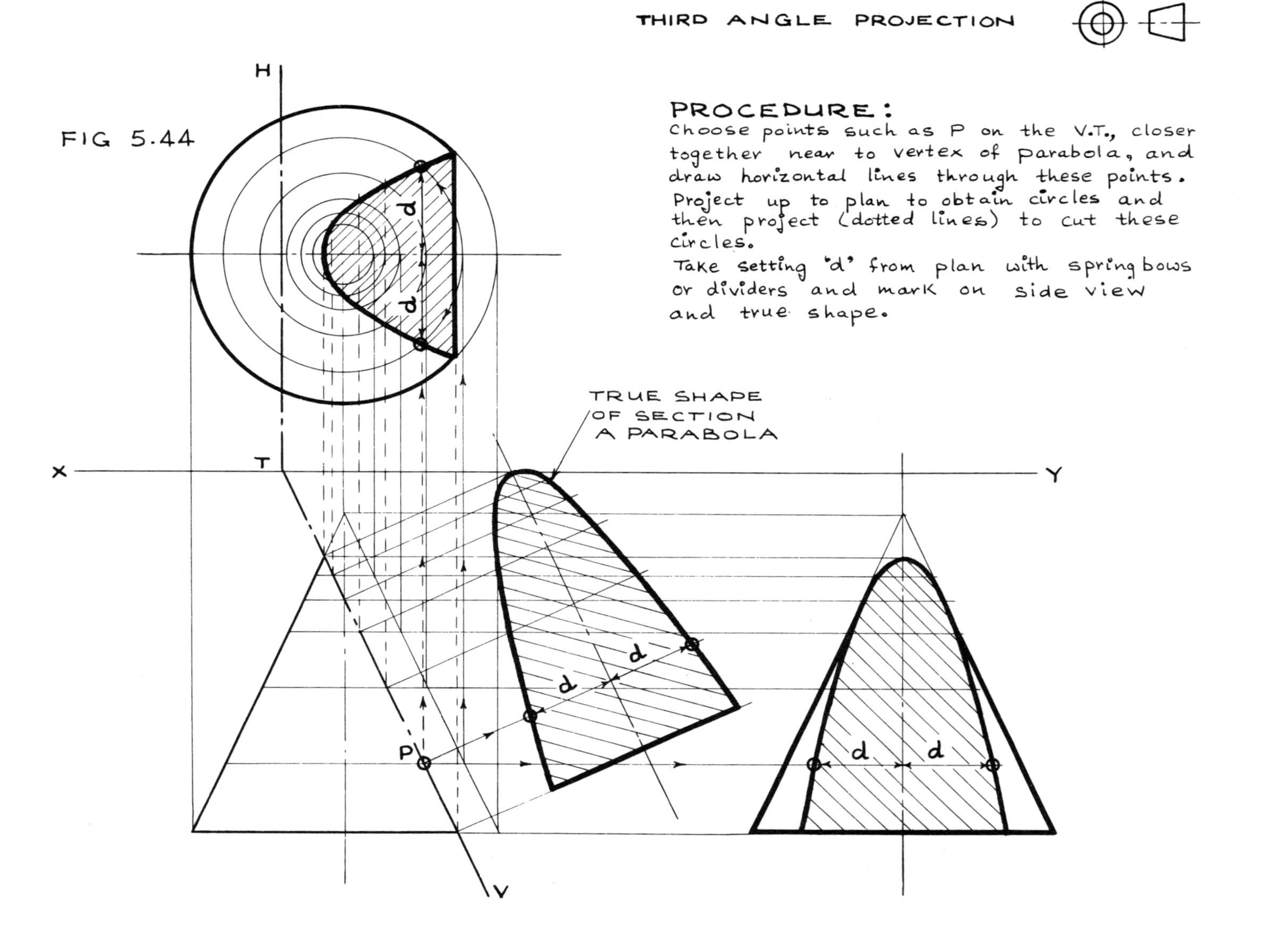
EXERCISE 17 SOLUTION
THIRD ANGLE PROJECTION
FIG 5.44
PROCEDURE:
Choose points such as P on the V.T., closer together near to vertex of parabola, and draw horizontal lines through these points.
Project up to plan to obtain circles and then project (dotted lines) to cut these circles.
Take setting 'd' from plan with springbows or dividers and mark on side view and true shape.
TRUE SHAPE OF SECTION A PARABOLA
H
T
X
Y
P
V
d
d
d
d
d
d

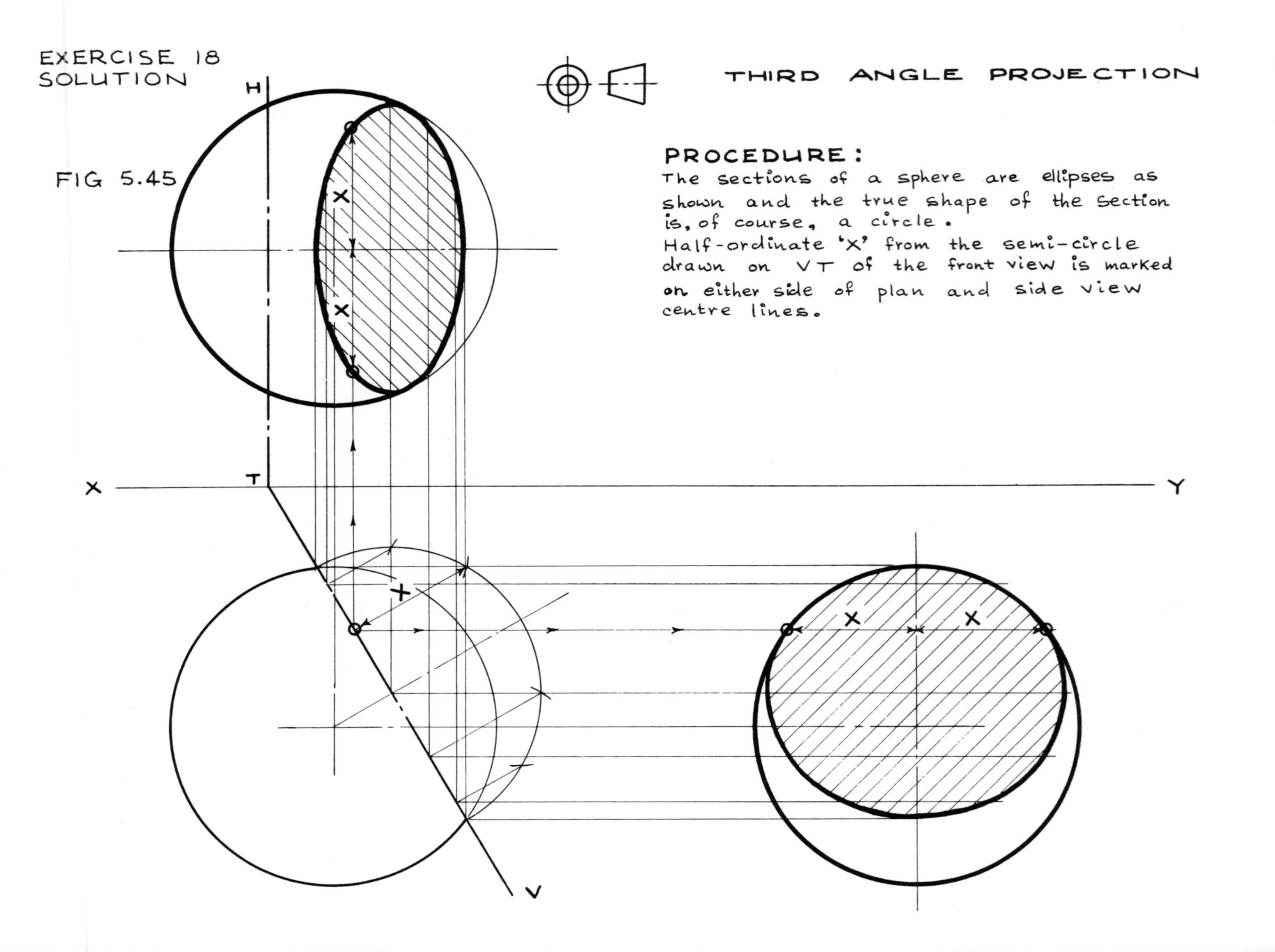
EXERCISE 18
SOLUTION
FIG 5.45
THIRD ANGLE PROJECTION
PROCEDURE:
The sections of a sphere are ellipses as shown and the true shape of the section is, of course, a circle.
Half-ordinate 'X' from the semi-circle drawn on VT of the front view is marked on either side of plan and side view centre lines.
H
T
X
Y
V
X

EXERCISE 19 SOLUTION

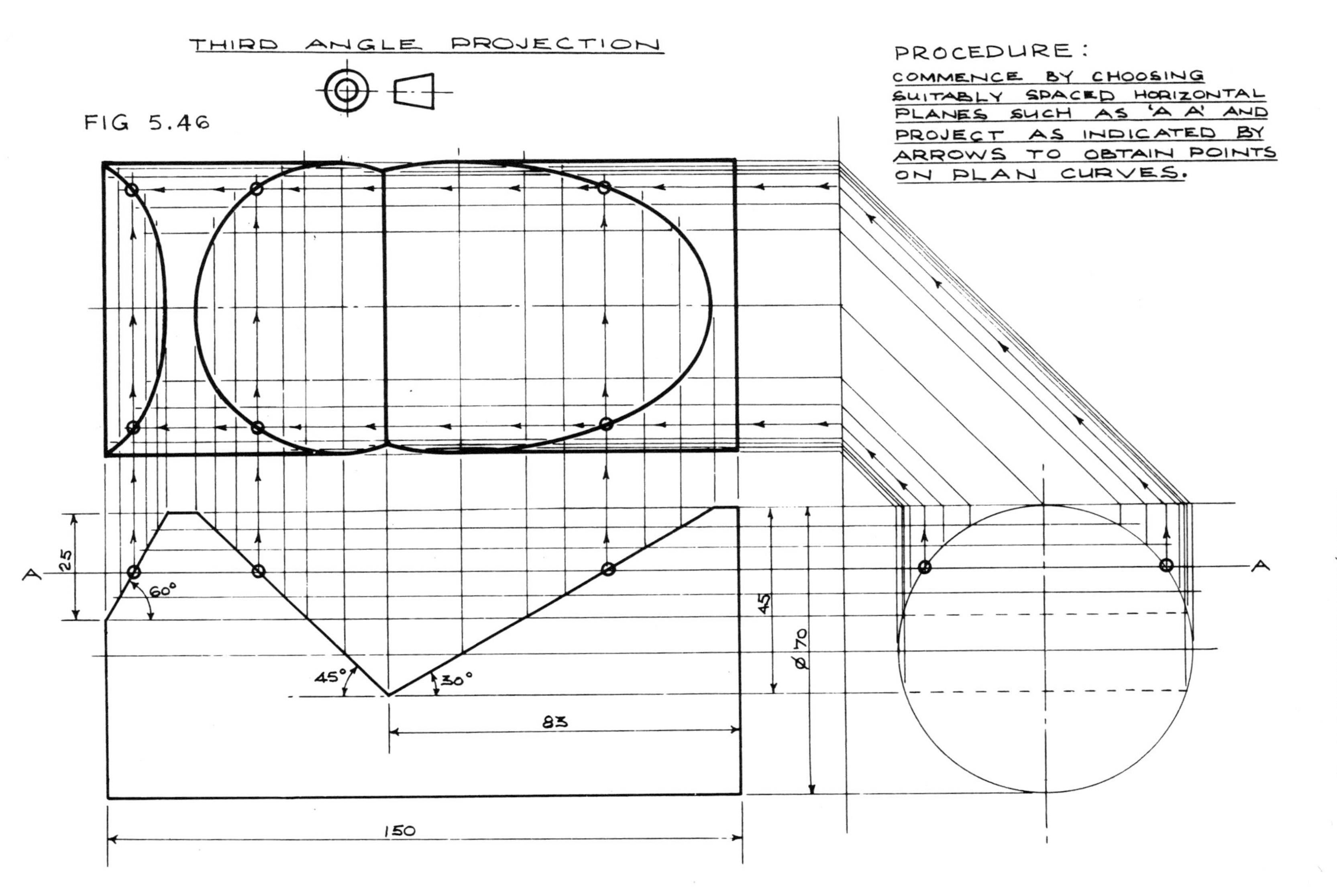
THIRD ANGLE PROJECTION
FIG 5.46
PROCEDURE:
COMMENCE BY CHOOSING SUITABLY SPACED HORIZONTAL PLANES SUCH AS 'A A' AND PROJECT AS INDICATED BY ARROWS TO OBTAIN POINTS ON PLAN CURVES.
A
A
25
60°
45°
30°
45
⌀ 70
83
150

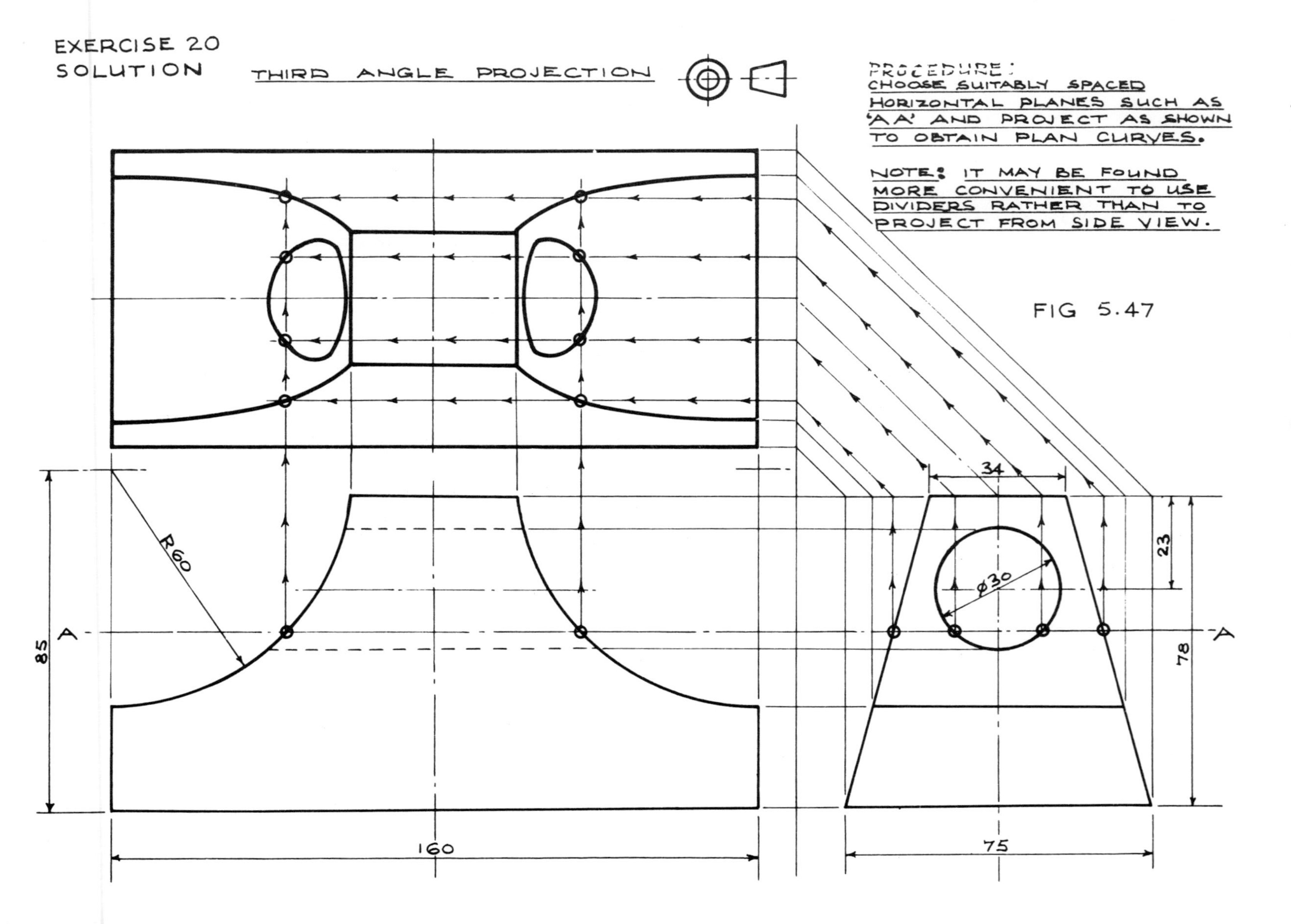
EXERCISE 20
SOLUTION
THIRD ANGLE PROJECTION
PROCEDURE:
CHOOSE SUITABLY SPACED HORIZONTAL PLANES SUCH AS 'A A' AND PROJECT AS SHOWN TO OBTAIN PLAN CURVES.
NOTE: IT MAY BE FOUND MORE CONVENIENT TO USE DIVIDERS RATHER THAN TO PROJECT FROM SIDE VIEW.
34
23
Ø30
R60
A
A
85
78
160
75

FIG 5.47

EXERCISE 21 SOLUTION

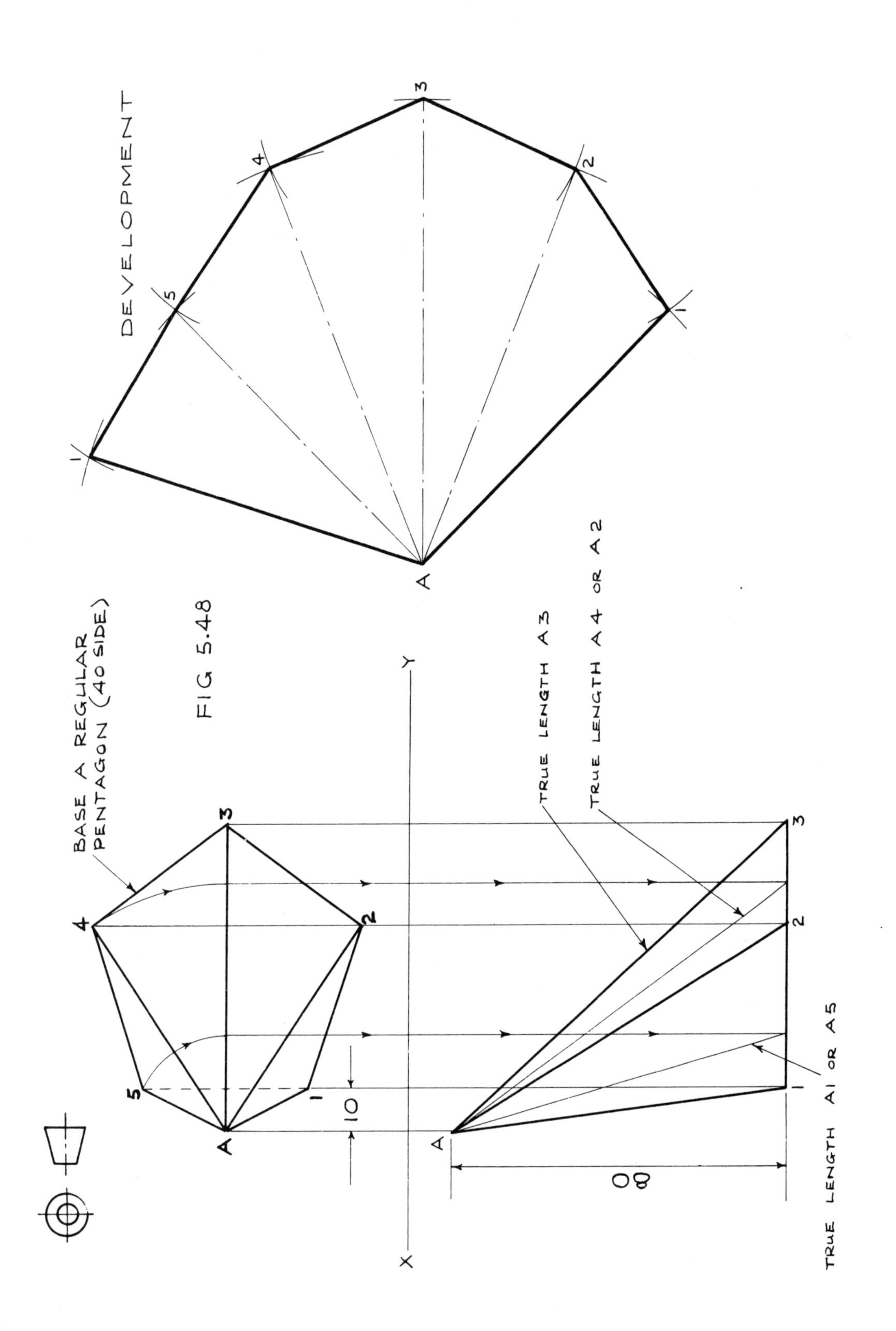
BASE A REGULAR
PENTAGON (40 SIDE)
DEVELOPMENT
TRUE LENGTH A3
TRUE LENGTH A4 OR A2
TRUE LENGTH A1 OR A5
10
80
X
Y
A
1
2
3
4
5

FIG 5.48

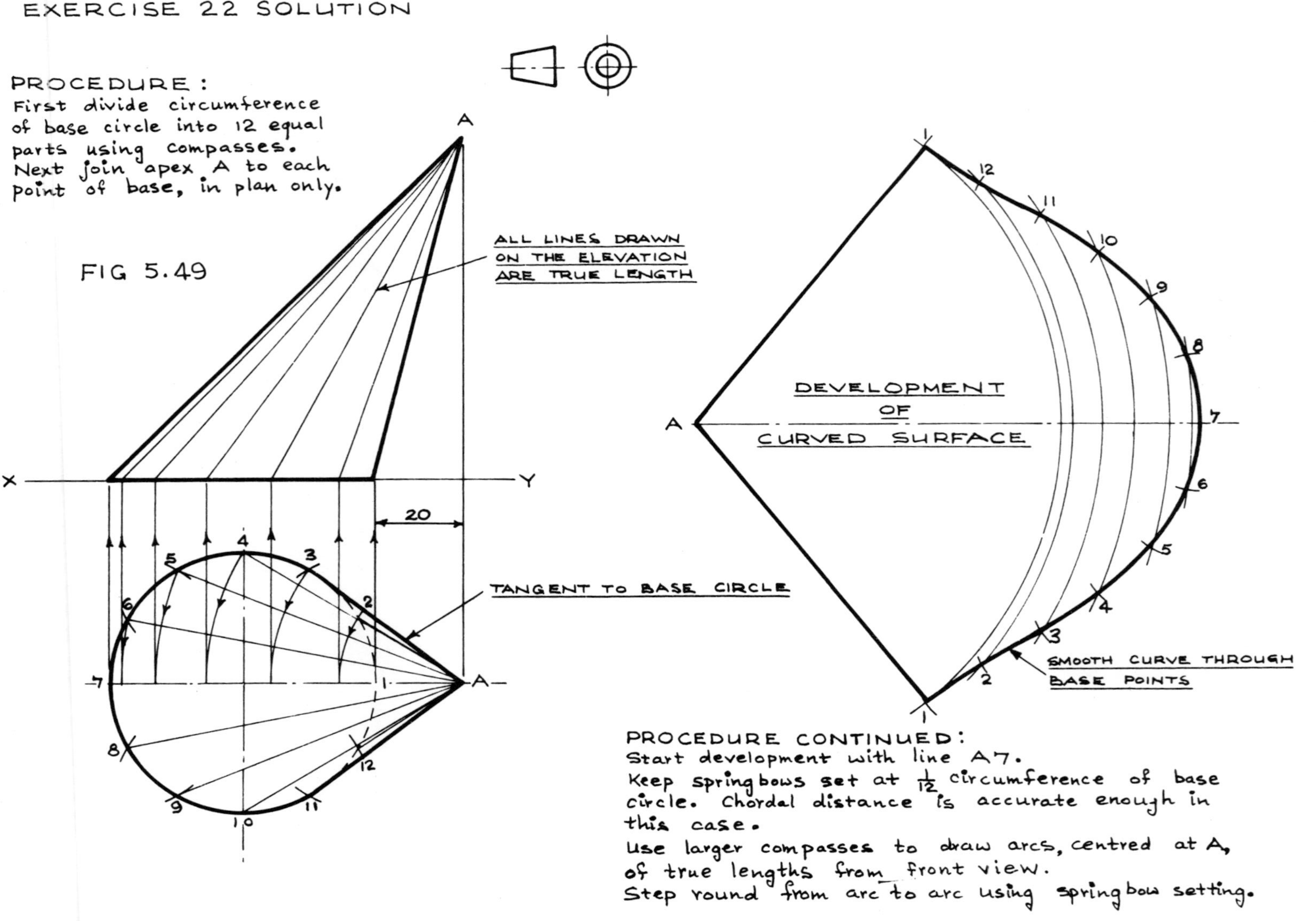

EXERCISE 22 SOLUTION
PROCEDURE:
First divide circumference of base circle into 12 equal parts using compasses.
Next join apex A to each point of base, in plan only.
ALL LINES DRAWN ON THE ELEVATION ARE TRUE LENGTH
TANGENT TO BASE CIRCLE
DEVELOPMENT OF CURVED SURFACE
SMOOTH CURVE THROUGH BASE POINTS
PROCEDURE CONTINUED:
Start development with line A7.
Keep springbows set at 1/12 circumference of base circle. Chordal distance is accurate enough in this case.
Use larger compasses to draw arcs, centred at A, of true lengths from front view.
Step round from arc to arc using springbow setting.
A
X
Y
20
1
2
3
4
5
6
7
8
9
10
11
12

FIG 5.49

EXERCISE 23 SOLUTION

FIG 5.50

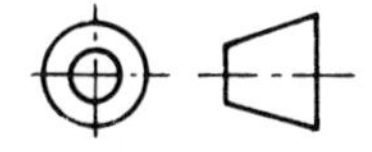

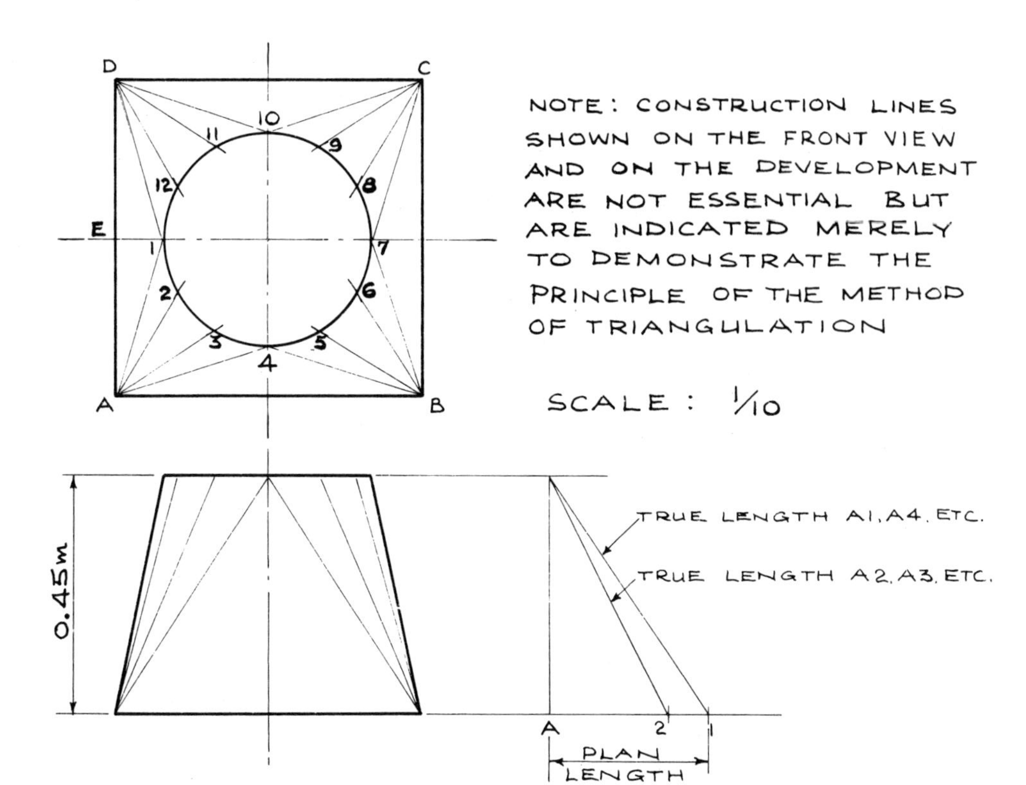

When constructing the various triangles which make up the development, using true lengths and starting with BC 7, set springbows permanently at $\frac{1}{12}$ circumference from plan.

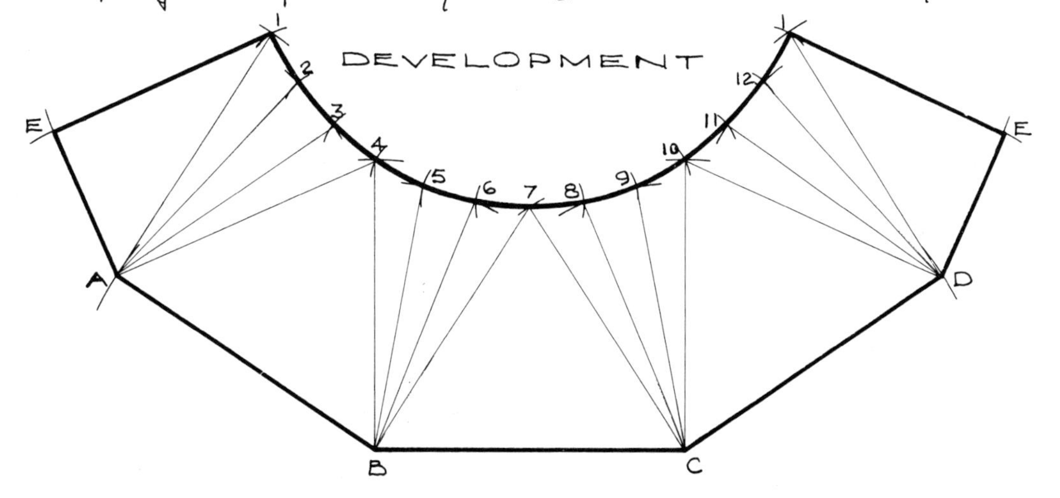

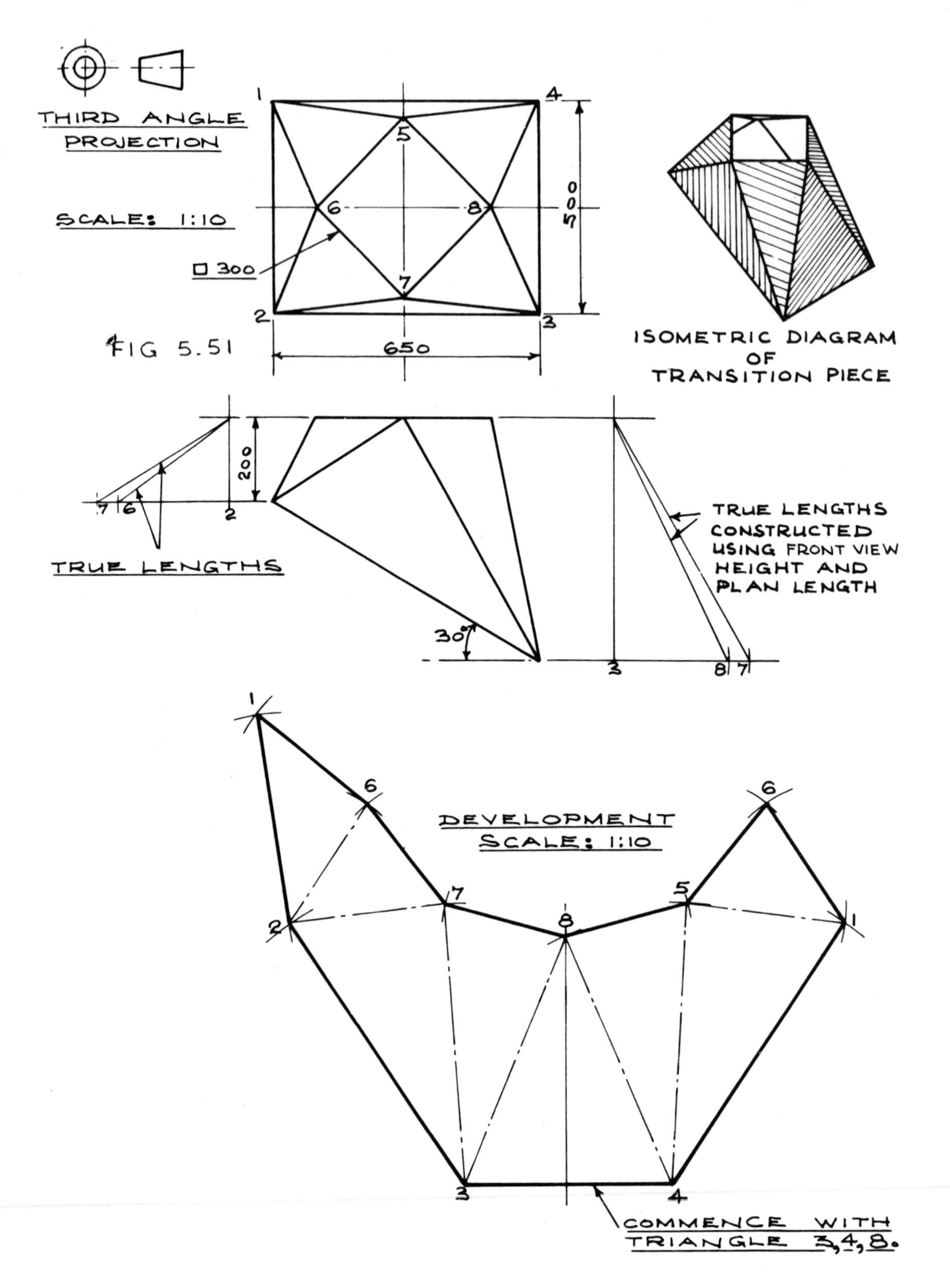
EXERCISE 24 SOLUTION
THIRD ANGLE PROJECTION
SCALE: 1:10
□ 300
FIG 5.51
650
500
1
2
3
4
5
6
7
8
ISOMETRIC DIAGRAM OF TRANSITION PIECE
200
TRUE LENGTHS
30°
TRUE LENGTHS CONSTRUCTED USING FRONT VIEW HEIGHT AND PLAN LENGTH
DEVELOPMENT
SCALE: 1:10
COMMENCE WITH TRIANGLE 3,4,8.

EXERCISE 25 SOLUTION

FIG 5.52

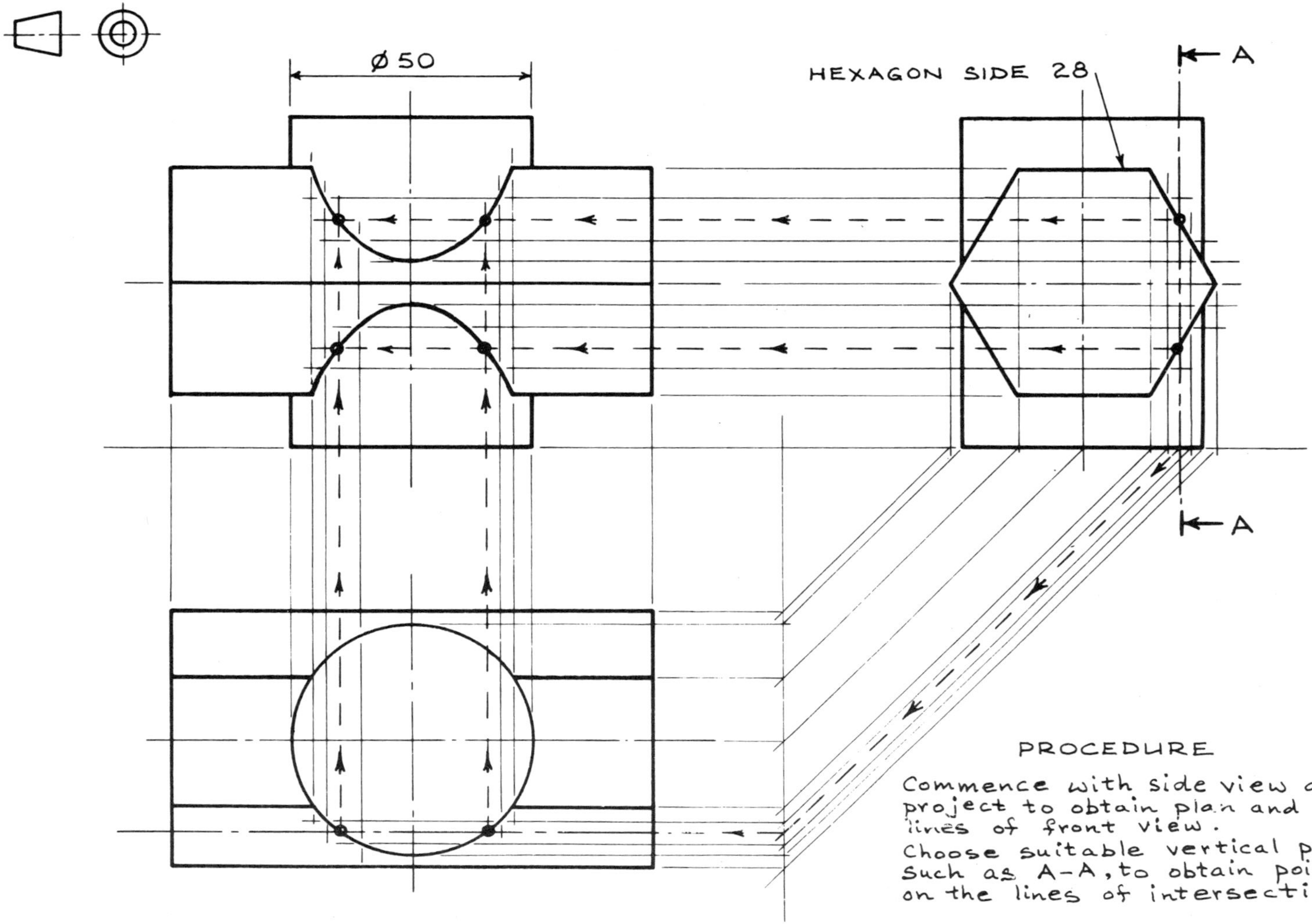

Ø50

HEXAGON SIDE 28

A

A

PROCEDURE

Commence with side view and project to obtain plan and main lines of front view.
Choose suitable vertical planes, such as A–A, to obtain points on the lines of intersection.

EXERCISE 26 SOLUTION

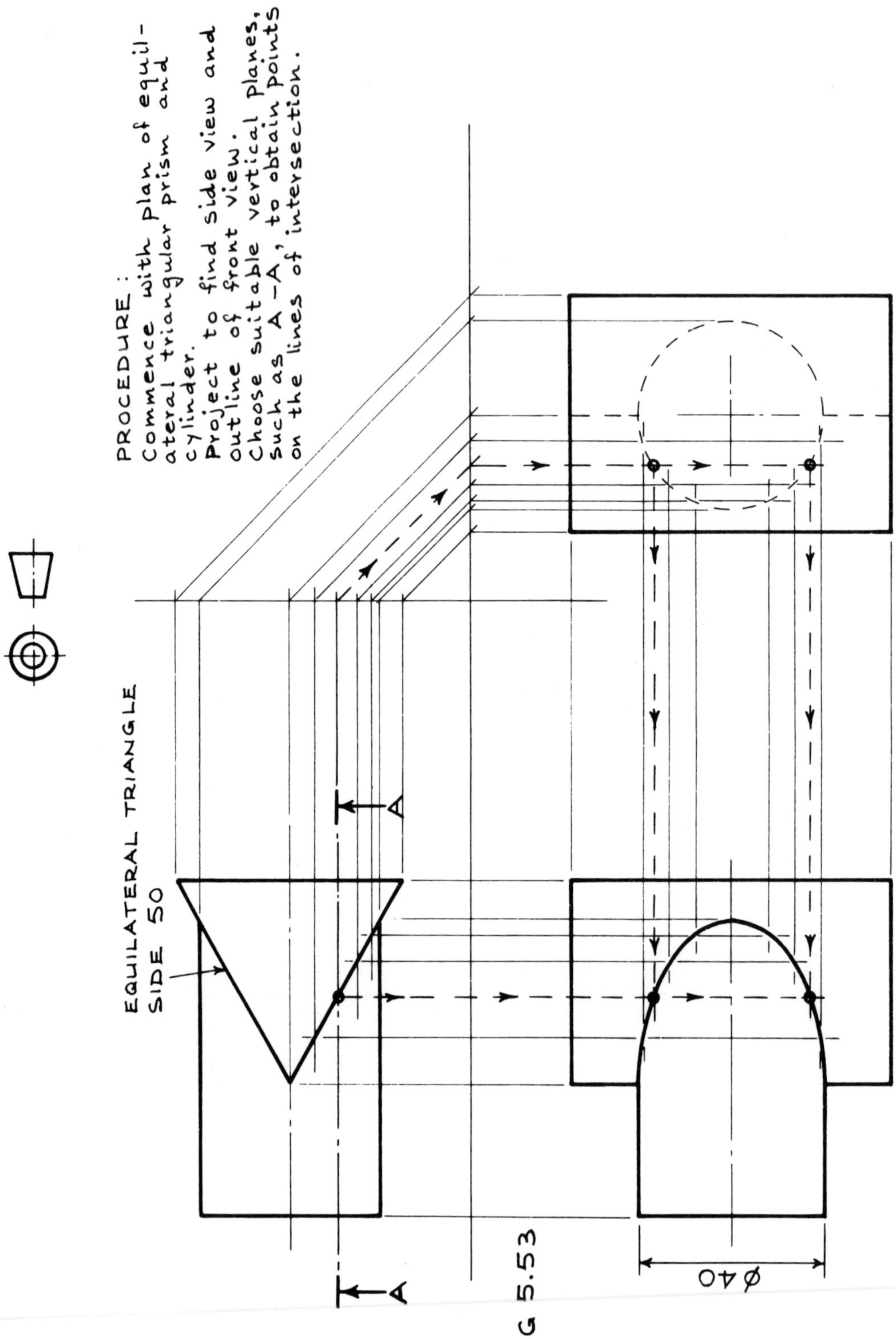

EQUILATERAL TRIANGLE
SIDE 50

A

A

PROCEDURE :
Commence with plan of equilateral triangular prism and cylinder.
Project to find side view and outline of front view.
Choose suitable vertical planes, such as A – A, to obtain points on the lines of intersection.

FIG 5.53

Ø40

EXERCISE 27 SOLUTION

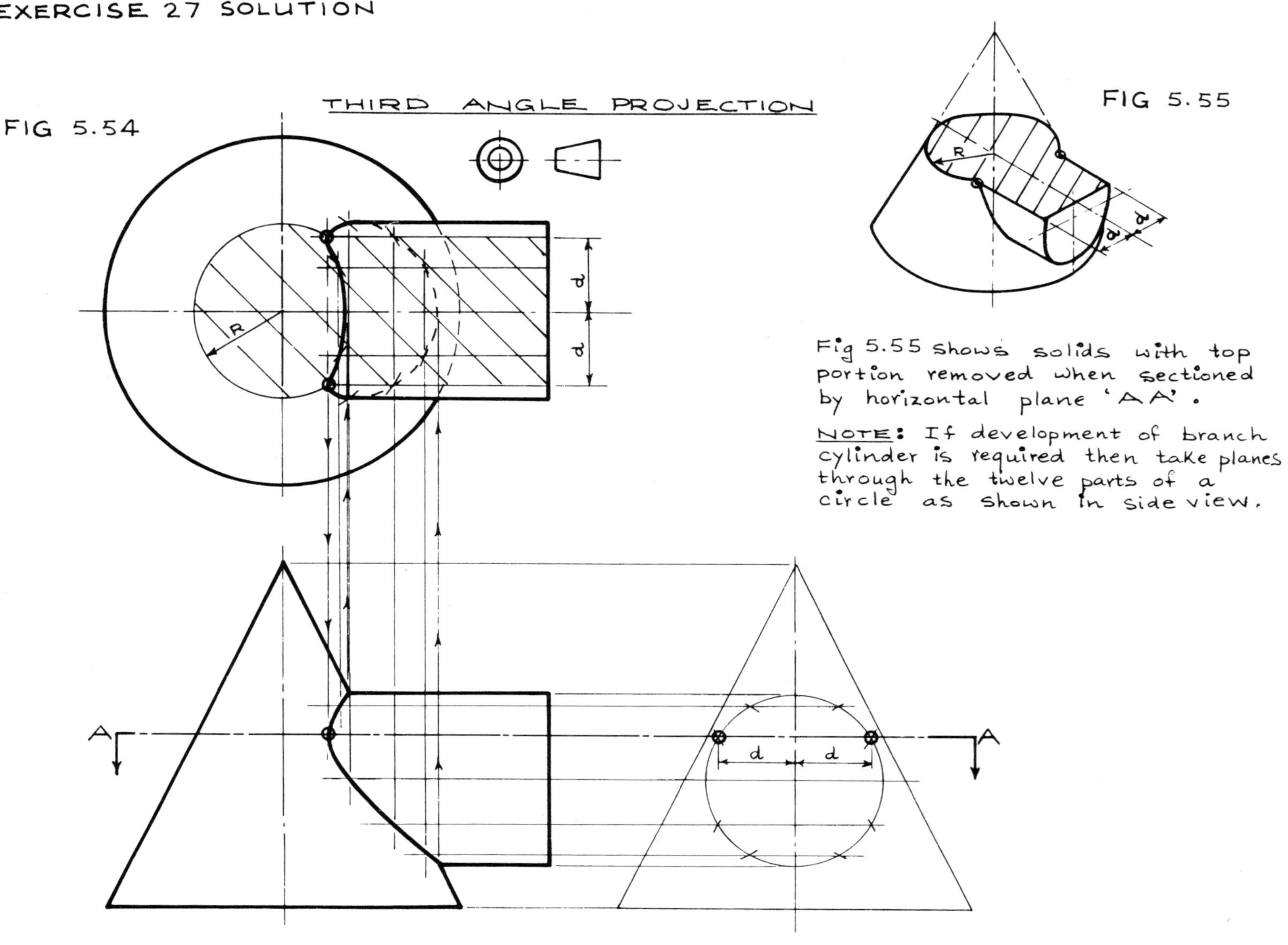
THIRD ANGLE PROJECTION
FIG 5.54
R
d
d
A
A
d
d
FIG 5.55
R
d
d
Fig 5.55 shows solids with top portion removed when sectioned by horizontal plane 'AA'.
NOTE: If development of branch cylinder is required then take planes through the twelve parts of a circle as shown in side view.

EXERCISE 28 SOLUTIONS

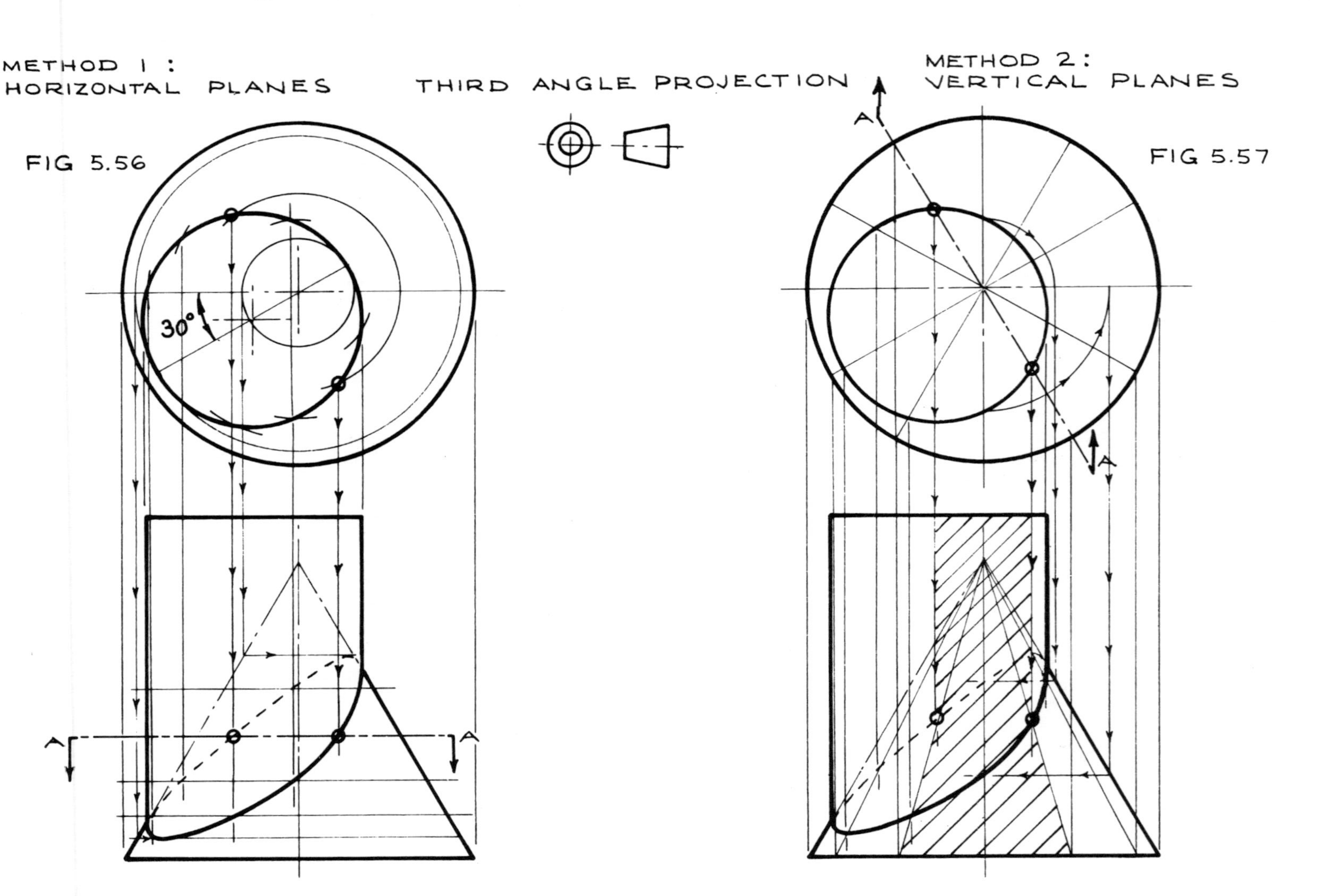
METHOD 1 :
HORIZONTAL PLANES
THIRD ANGLE PROJECTION
METHOD 2 :
VERTICAL PLANES
FIG 5.56
30°
A
A
FIG 5.57
A
A

EXERCISE 29 SOLUTION

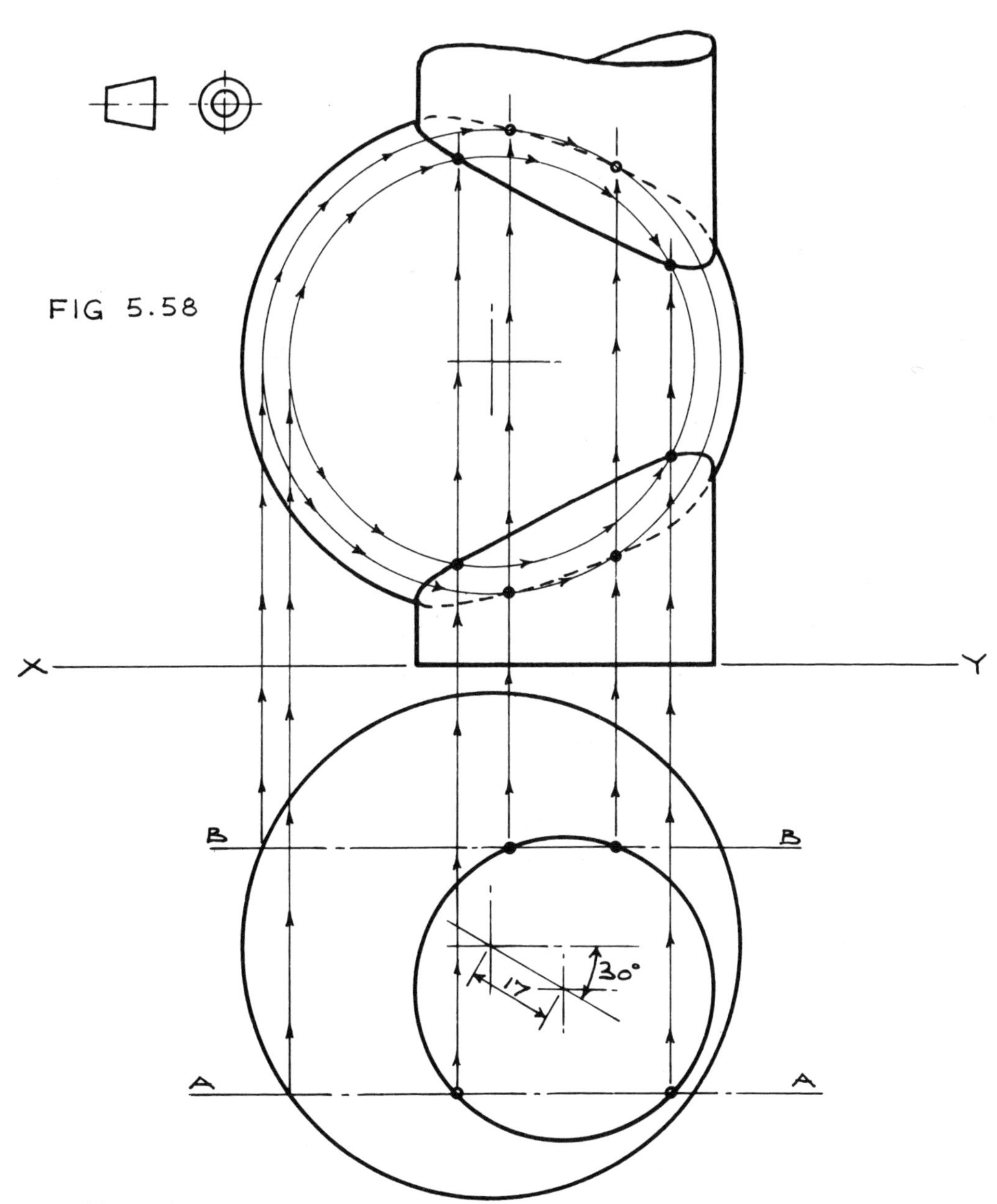

PROCEDURE:

use a series of vertical cutting planes, resulting in concentric circles on the sphere and parallel lines on the cylinder in front view.

A typical plane is 'A-A' which gives four intersection line points on the front of the front view and also 'B-B' which results in four points at the rear.

Special planes giving particular points are through centres of sphere and cylinder and to touch the cylinder.

EXERCISE 30 SOLUTION

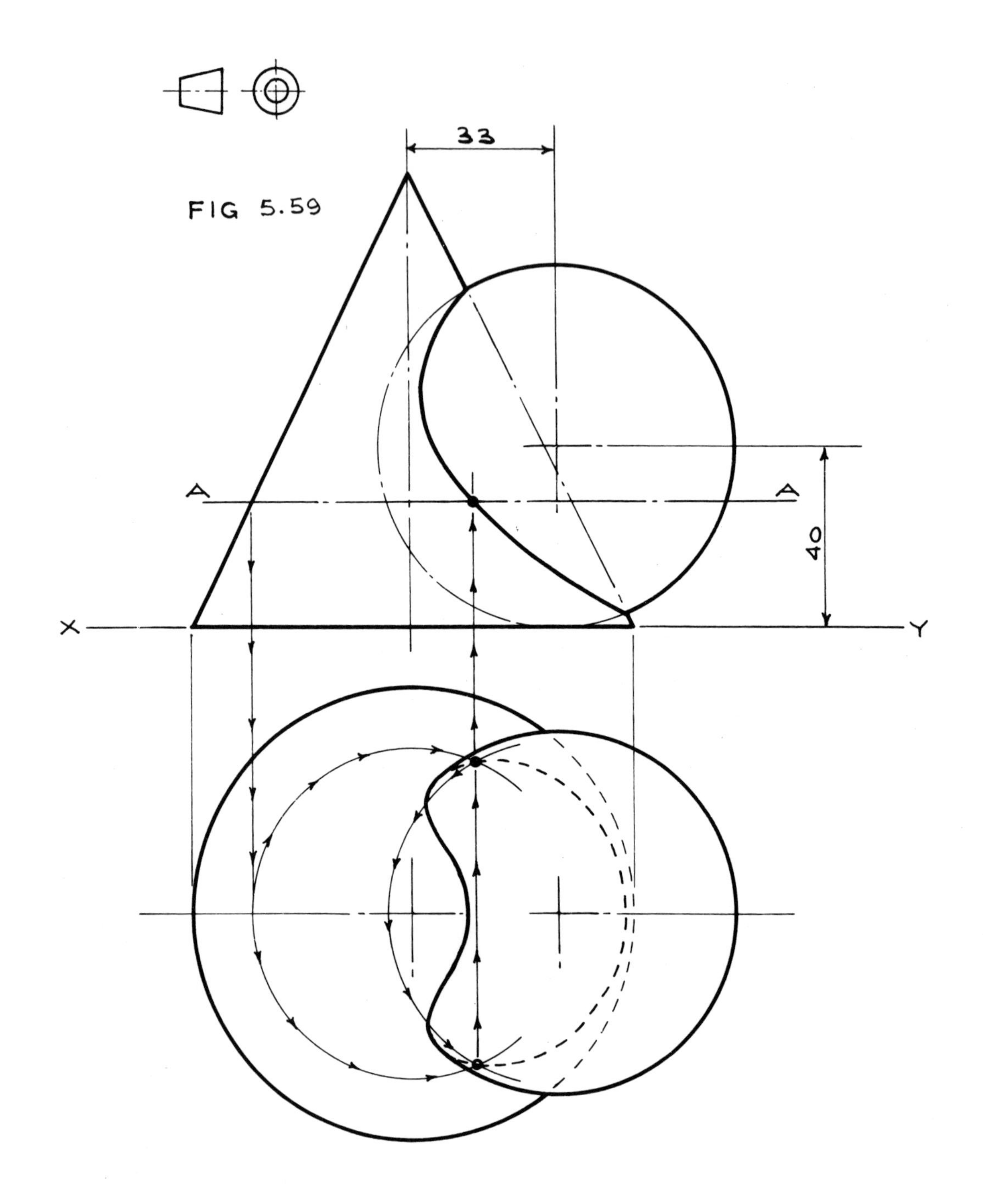

PROCEDURE:
Use horizontal section cutting planes to produce a circle in both cone and sphere.
Where the two circles, or arcs only, intersect in plan view two points are obtained which by projection to 'AA' give their front view.

6
LOCI

A locus is the path of a point which moves and obeys certain laws regarding its position at any time.

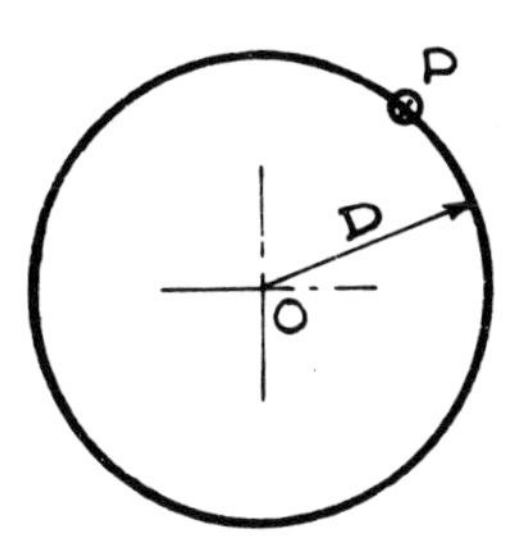

FIG 6.1

The locus of a point P, which is always a certain distance D from a fixed point O, is a circle, centre O and radius equal to D.

FIG 6.2

The locus of a point P, which is always a fixed distance D from a given straight line AB, is either of the parallel lines CD or EF.

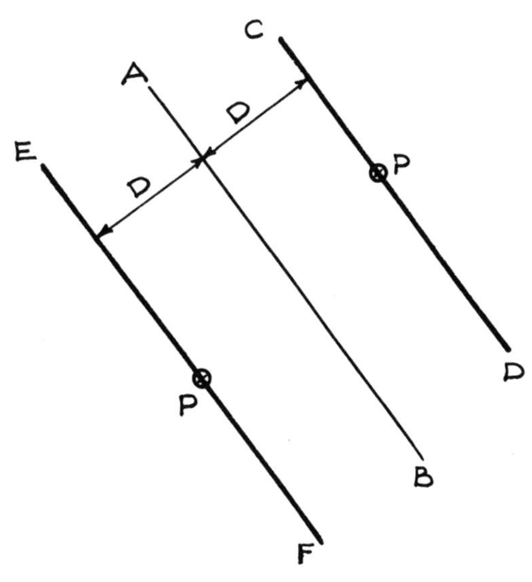

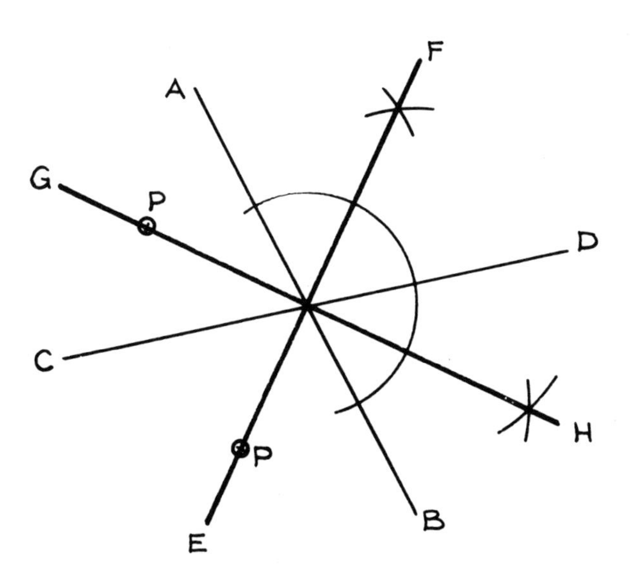

FIG 6.3

The locus of a point P which is equi-distant from two intersecting straight lines AB and CD is either EF or GH, the bisectors of the angles.

Exercise 1. Plot the locus of a point P which is equi-distant from a given circle, centre O, and a given straight line AB. Make the circle diameter 40 mm and the distance of its centre O from line AB, 50 mm.

Exercise 2. Two circles centred at A and B have diameters of 50 mm and 40 mm respectively and AB is 70 mm. Obtain the locus of a point P which moves so that it is equi-distant from the two circles.

Exercise 3. Point P is on the line AB which is 70 mm long and AP is 10 mm initially. At the same time as AB rotates in a horizontal plane about A, point P moves uniformly along the line towards B. If P moves at the rate of 5 mm for each 30° of turn of AB plot its locus as the line makes one complete turn counter-clockwise.

Exercise 4. Suppose that the condition of movement of point P in the previous exercise is that the ratio of two adjacent radius vectors is always the same. This means that AP/A1 = A1/A2 = A2/A3, etc. Plot the locus of P for one complete turn of AB counter-clockwise being given that the constant ratio is 11/14.

Exercise 5. Thread is unwound from a cotton reel 40 mm diameter, the thread being held taut and in a plane perpendicular to the axis of the reel. Draw the locus of the end of the thread for one turn unwound.

Exercise 6. Two straight lines AB and CD intersect at right-angles at E. F is a point on CD such that EF is 40 mm. Plot the locus of point P given that the distance of P from F is always equal to the perpendicular distance of P from line AB.

Exercise 7. Suppose that point P in the previous exercise moves so that the ratio of the distance P to F and the perpendicular distance of P from line AB is always constant and greater than unity. Construct the locus of P for a ratio of 5:4 and use the distance EF 30 mm.

Exercise 8. If the ratio of the distances referred to in the previous exercise is always constant and less than unity, suppose 6:7, plot the complete locus of P.

Exercise 9. Construct an ellipse of major axis 130 mm and minor axis 80 mm using the method of Intersecting Arcs.

Exercise 10. Draw an ellipse having major axis 120 mm and minor axis 70 mm using a paper trammel.

EXERCISE 1
SOLUTION

THE LOCUS OF A POINT P WHICH IS EQUI-DISTANT FROM A GIVEN LINE AB AND A GIVEN CIRCLE

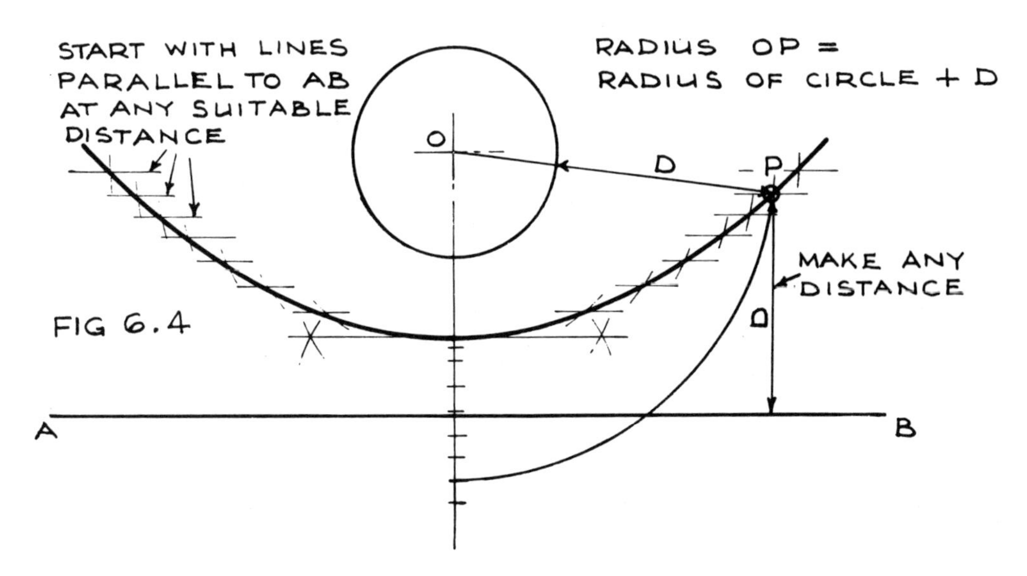

FIG 6.4

EXERCISE 2
SOLUTION

THE LOCUS OF A POINT P WHICH IS EQUI-DISTANT FROM TWO GIVEN CIRCLES

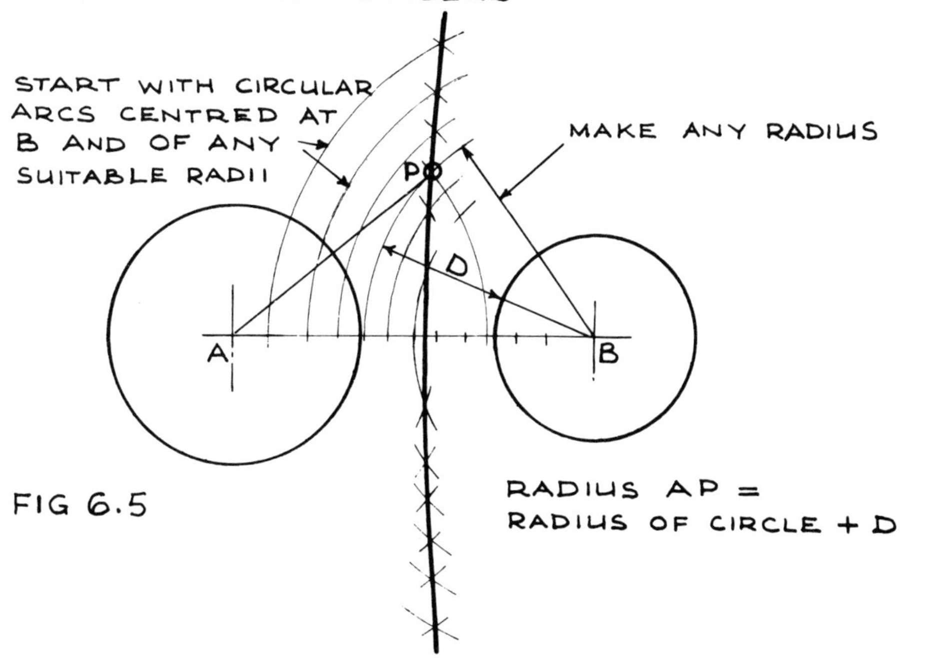

FIG 6.5

EXERCISE 3 SOLUTION

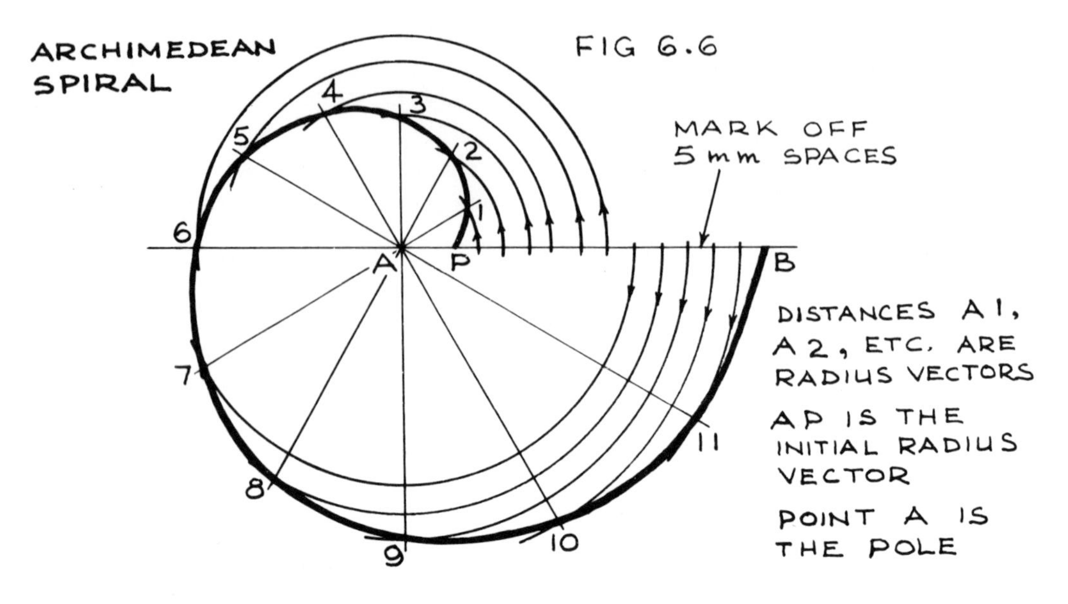

EXERCISE 4 SOLUTION

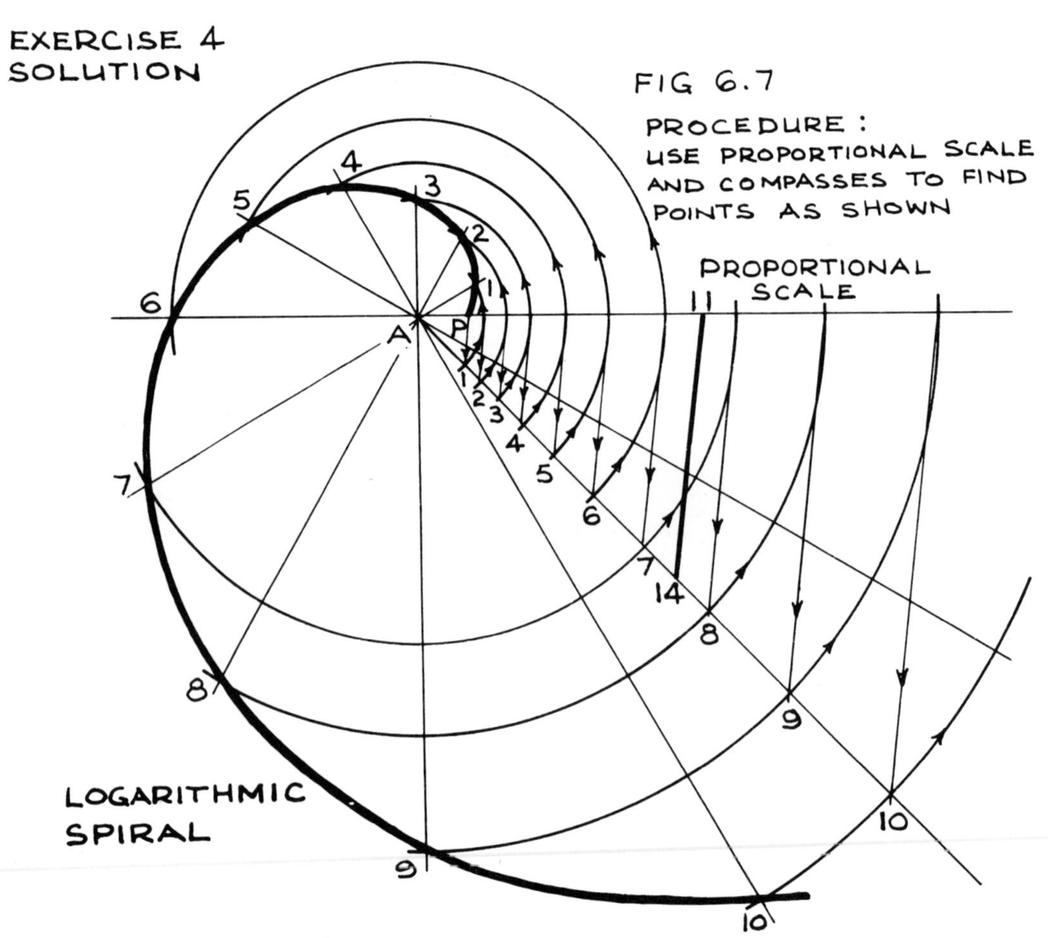

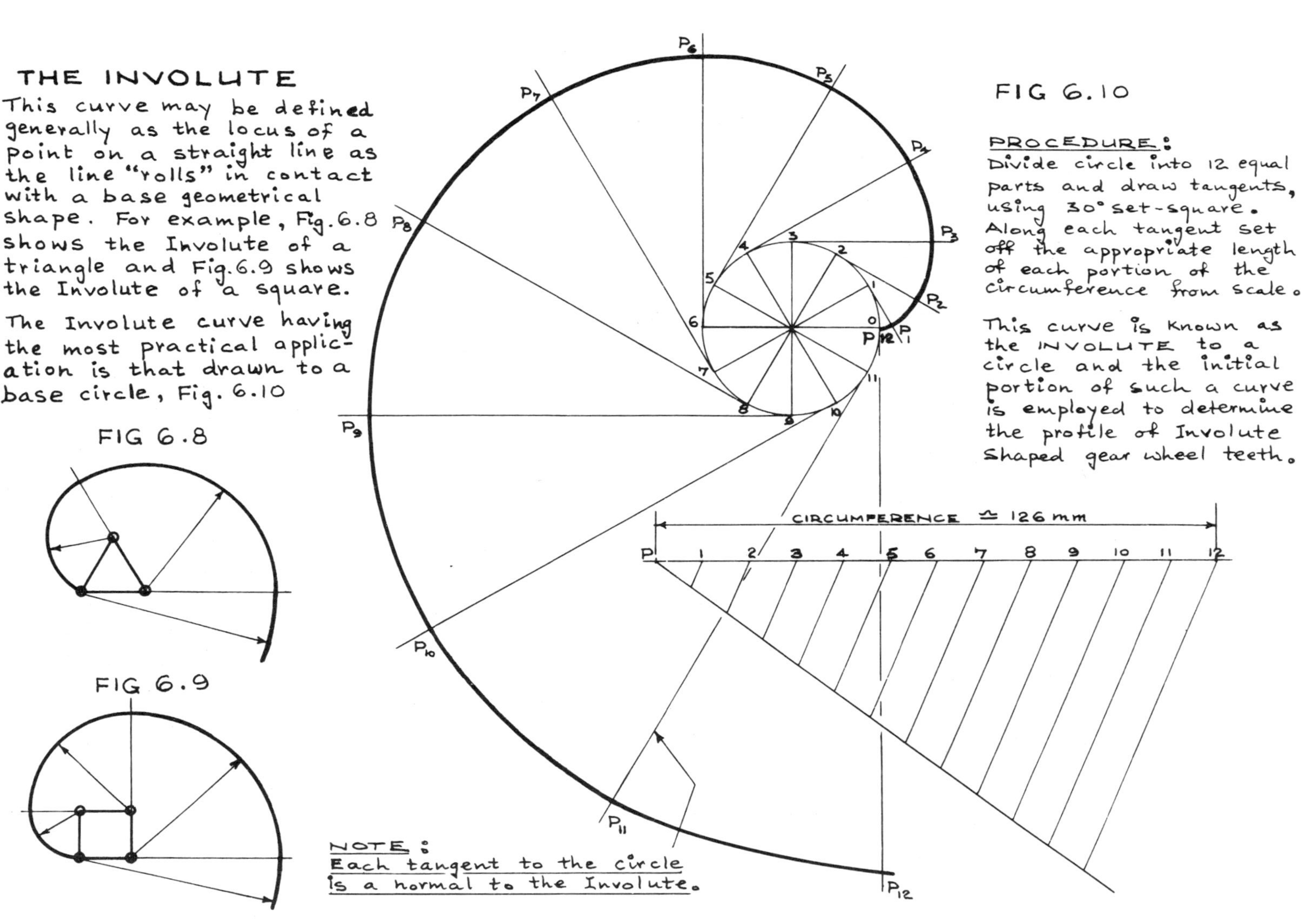
EXERCISE 5 SOLUTION
THE INVOLUTE
This curve may be defined generally as the locus of a point on a straight line as the line "rolls" in contact with a base geometrical shape. For example, Fig. 6.8 shows the Involute of a triangle and Fig. 6.9 shows the Involute of a square.
The Involute curve having the most practical application is that drawn to a base circle, Fig. 6.10
FIG 6.8
FIG 6.9
FIG 6.10
PROCEDURE:
Divide circle into 12 equal parts and draw tangents, using 30° set-square. Along each tangent set off the appropriate length of each portion of the circumference from scale.
This curve is known as the INVOLUTE to a circle and the initial portion of such a curve is employed to determine the profile of Involute shaped gear wheel teeth.
CIRCUMFERENCE ≃ 126 mm
P 1 2 3 4 5 6 7 8 9 10 11 12
P1 P2 P3 P4 P5 P6 P7 P8 P9 P10 P11 P12
NOTE:
Each tangent to the circle is a normal to the Involute.

EXERCISE 6 SOLUTION

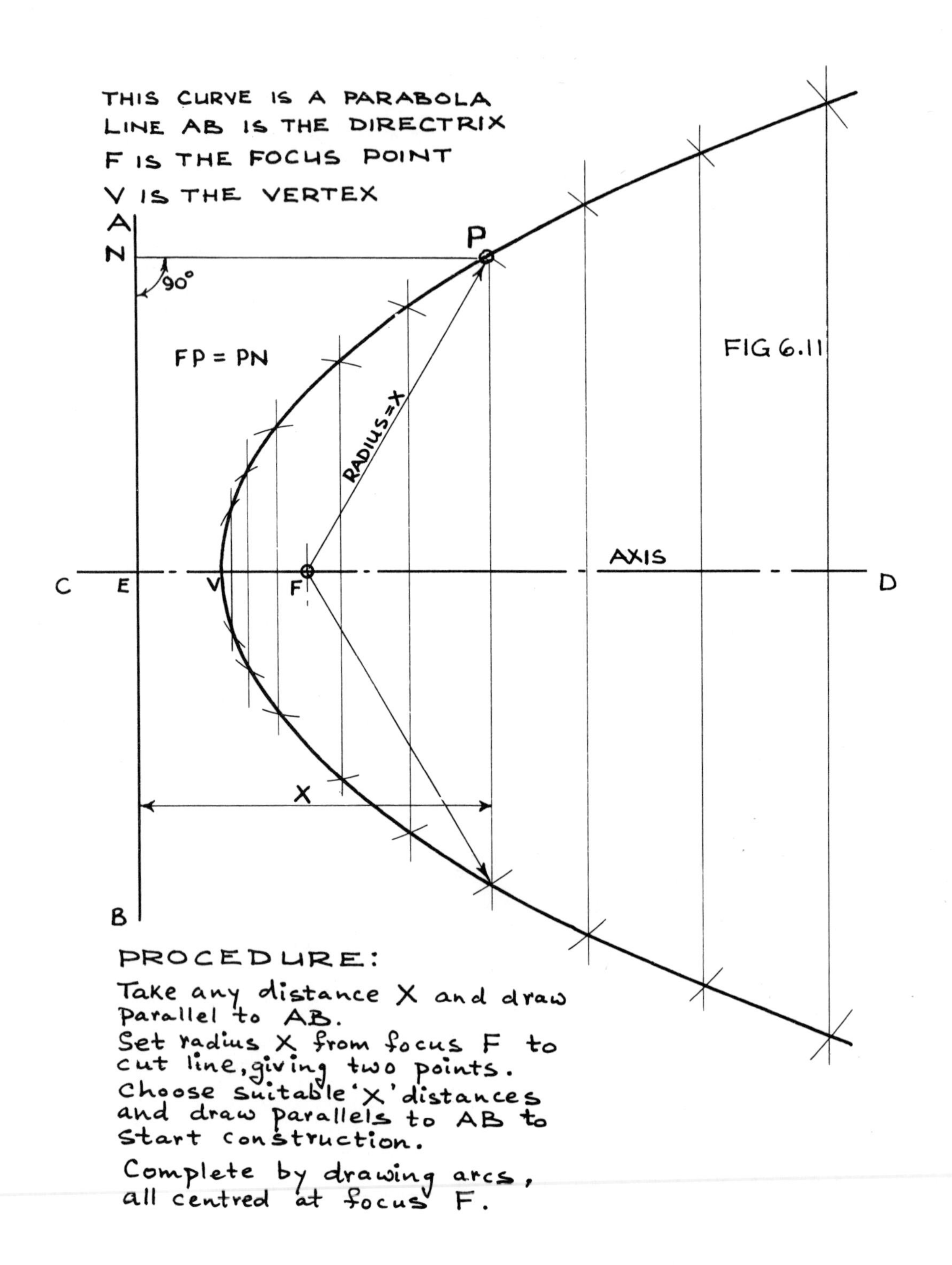

EXERCISE 7 SOLUTION

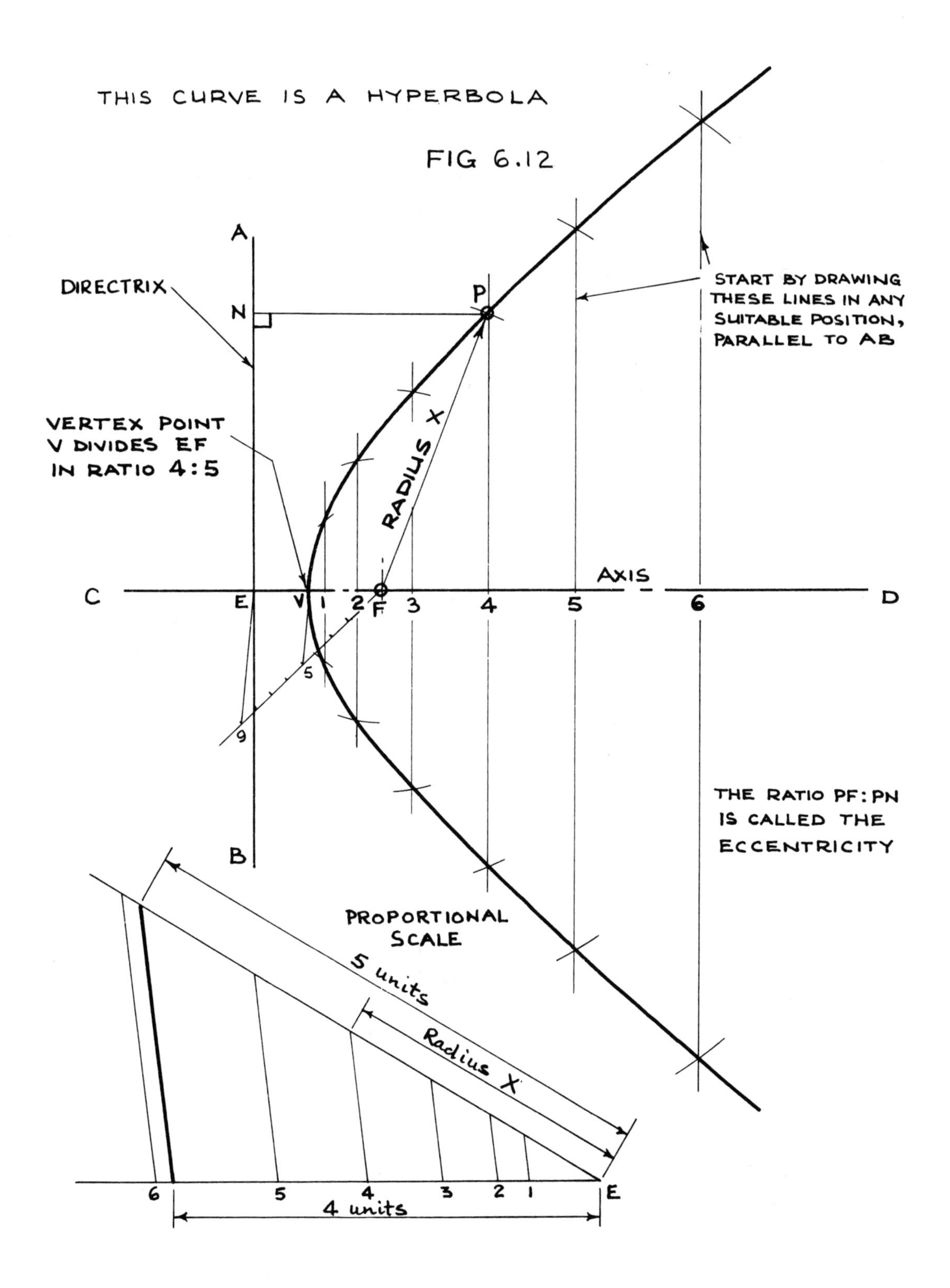

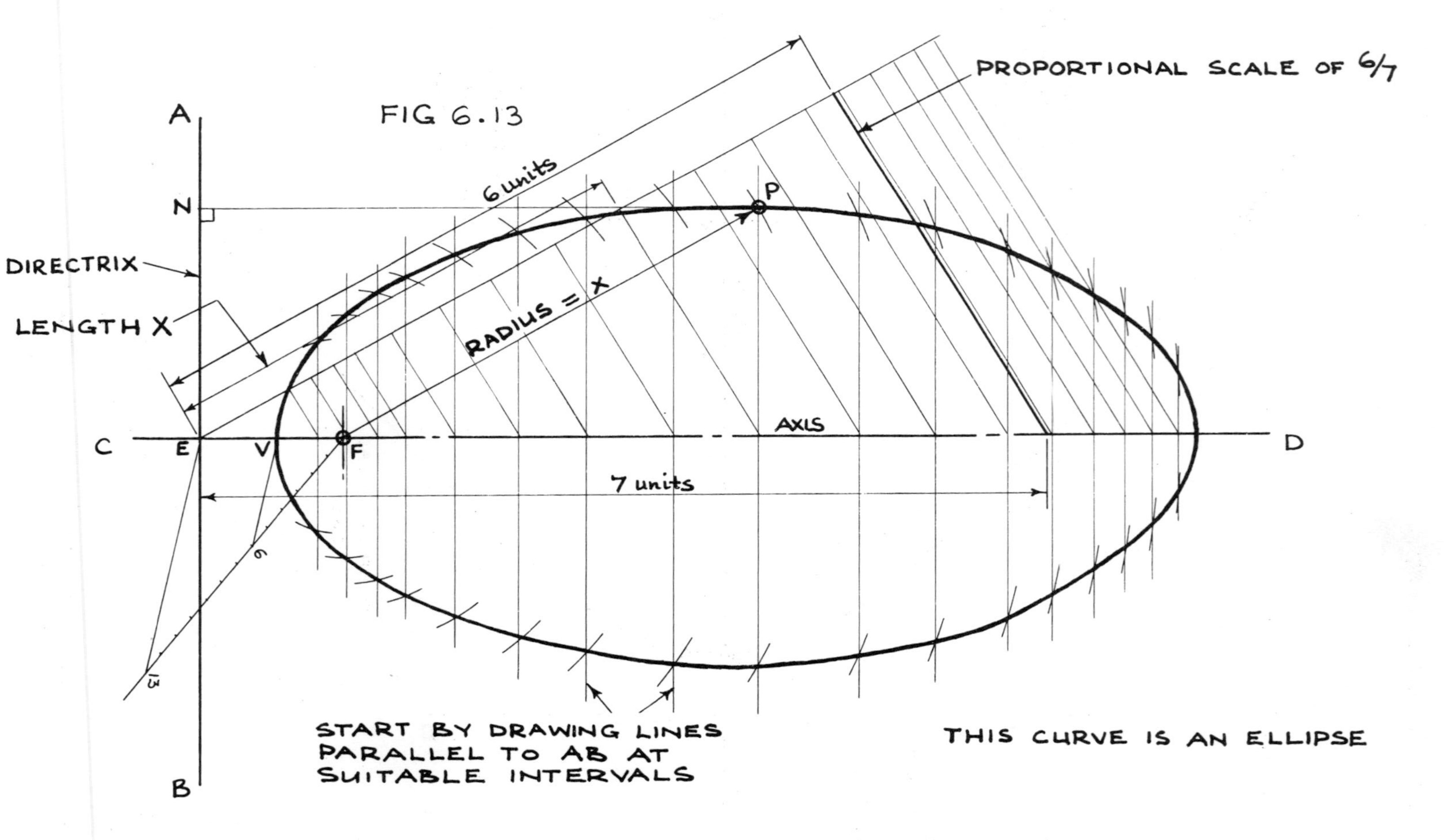

NOTE: Various methods for drawing the ellipse will be given.
These will be found much simpler than the above method.

SOME BASIC FACTS ABOUT THE ELLIPSE

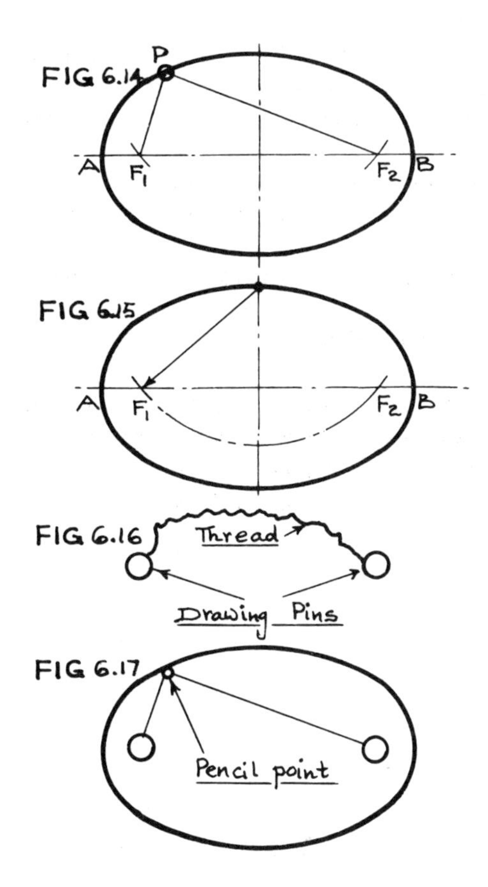

For any point on an ellipse, the sum of the distances to the two focus points is equal to the length of the major axis.

$$PF_1 + PF_2 = AB$$

Thus, if the major and minor axes are known, this gives a method of obtaining the focus points.

A quick method, though not an accurate one due to stretch in the thread, of drawing an ellipse is shown by Figs. 6.16 and 6.17. Thread is tied to two drawing pins and kept taut as a pencil point is used to trace out the curve.

An ellipse may be defined as the locus of a point which moves so that the sum of its distances from two fixed points is constant.

EXERCISE 9 SOLUTION

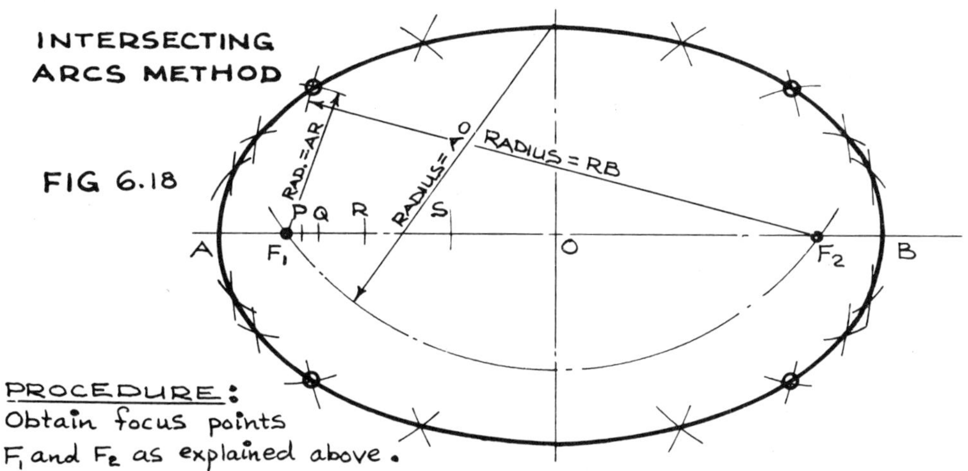

PROCEDURE:
Obtain focus points F_1 and F_2 as explained above.
P, Q, R and S are any points to the right of F_1. Four arcs are drawn for one compass setting, always centred on focus points.

Exercise 11. Use the Auxiliary Circles method to construct an ellipse similar to the one given in the previous exercise.

Exercise 12. Draw two parallelograms having adjacent sides 120 mm and 90 mm including a 60° angle. Show two separate methods of inscribing an ellipse in a parallelogram.

Note. The methods shown in figures 6.27 and 6.28 apply equally to a rectangle which is just a special type of parallelogram.

Exercise 13. Construct an ellipse, by the Trammel method, having major axis 140 mm and minor axis 80 mm. At a point P on the curve, 37 mm from the minor axis, construct a tangent and normal.

Exercise 14. By Trammel method and using 120 mm and 70 mm for major and minor axes respectively, construct an ellipse. From a point P, 37 mm from the major axis and 72 mm from the minor axis, construct the two possible tangents to the curve.

Exercise 15. Draw two similar rectangles 160 mm x 100 mm and inscribe a parabola in each using different methods.

Exercise 16. Draw a parabola by an accurate method and find the position of the focus point and the directrix.

Exercise 17. Construct a parabola having the focus 20 mm from the directrix by the method of *Exercise 6*. Find a point P on the curve, 82 mm from the directrix, and draw a tangent to the parabola at this point.

Exercise 18. The asymptotes to a hyperbola intersect one another at 70°. A point P on a branch of the curve is 20 mm from one asymptote and 32 mm from the other. Construct one branch of this curve.

SOME FURTHER BASIC METHODS FOR ELLIPSE CONSTRUCTION

TRAMMEL METHOD

FIG 6.19

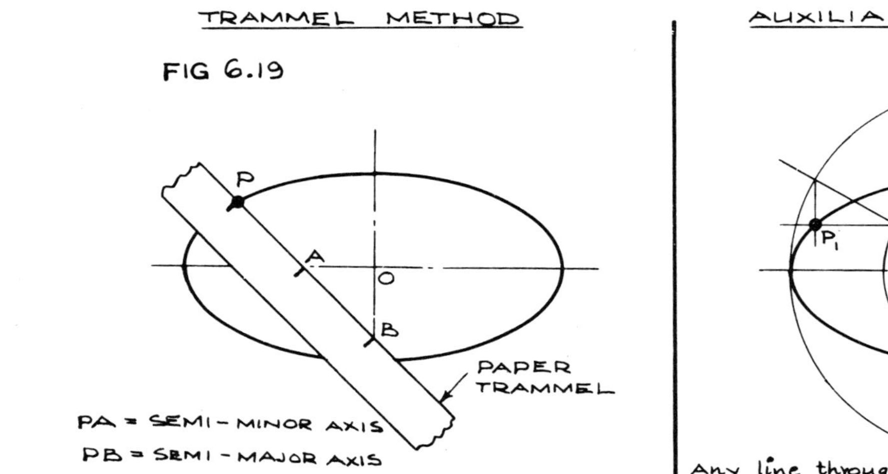

AUXILIARY CIRCLES METHOD

FIG 6.20

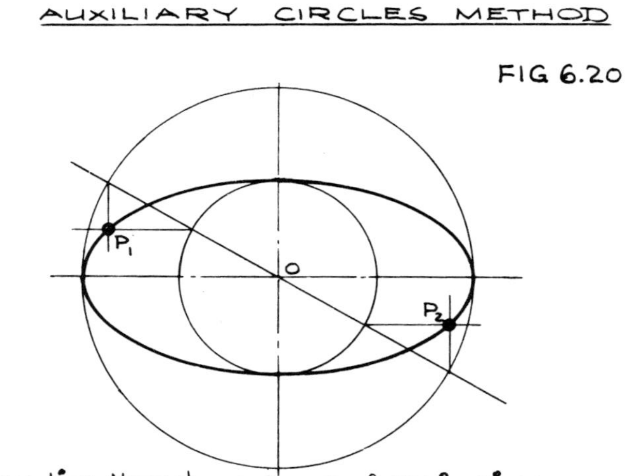

Any line through centre O of conic gives rise to two points, P_1 and P_2, on curve.

The TRAMMEL method, Fig. 6.19, is an exceedingly quick and accurate method for ellipse construction and is based on a sound mathematical property of the curve. No construction lines are necessary but care must be exercised in marking the position of P, using a sharp-point pencil, and A and B must be exactly positioned on the major and minor axes respectively. Exercise 10 solution, Fig. 6.25, illustrates the procedure.

The AUXILIARY CIRCLES method is also soundly based mathematically but requires rather a lot of construction. However, the method is very simple and the construction may be quickly executed. The method is shown completely by Exercise 11 solution, Fig. 6.26

Two methods are shown, in Figs. 6.21 and 6.22 for inscribing an ellipse inside a parallelogram, which includes the rectangle, and approximately equal amounts of construction are involved in each ease. The second method, Fig. 6.22 is perhaps the simpler of the two, both of which are demonstrated in detail by Exercises 12 and 13.

Tangents to the ellipse, Figs 6.23 and 6.24 are explained in detail by Exercises 13 and 14 solutions in Figs. 6.29 and 6.30 respectively.

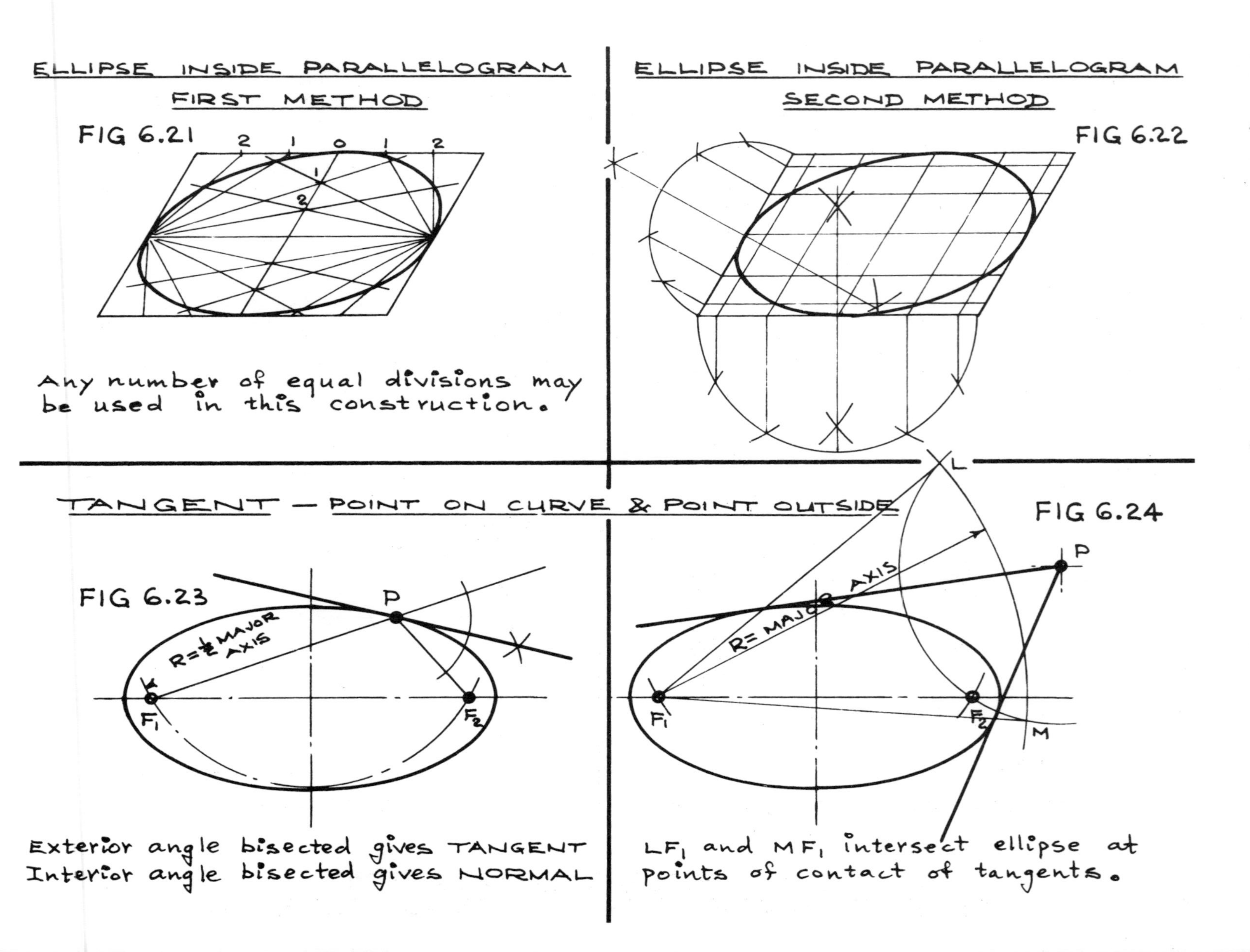
ELLIPSE INSIDE PARALLELOGRAM
FIRST METHOD
FIG 6.21
2 1 0 1 2
1
2
Any number of equal divisions may be used in this construction.
ELLIPSE INSIDE PARALLELOGRAM
SECOND METHOD
FIG 6.22
TANGENT – POINT ON CURVE & POINT OUTSIDE
FIG 6.23
P
R=½ MAJOR AXIS
F1
F2
Exterior angle bisected gives TANGENT
Interior angle bisected gives NORMAL
FIG 6.24
L
P
R= MAJOR AXIS
F1
F2
M
LF1 and MF1 intersect ellipse at points of contact of tangents.

EXERCISE 10 SOLUTION

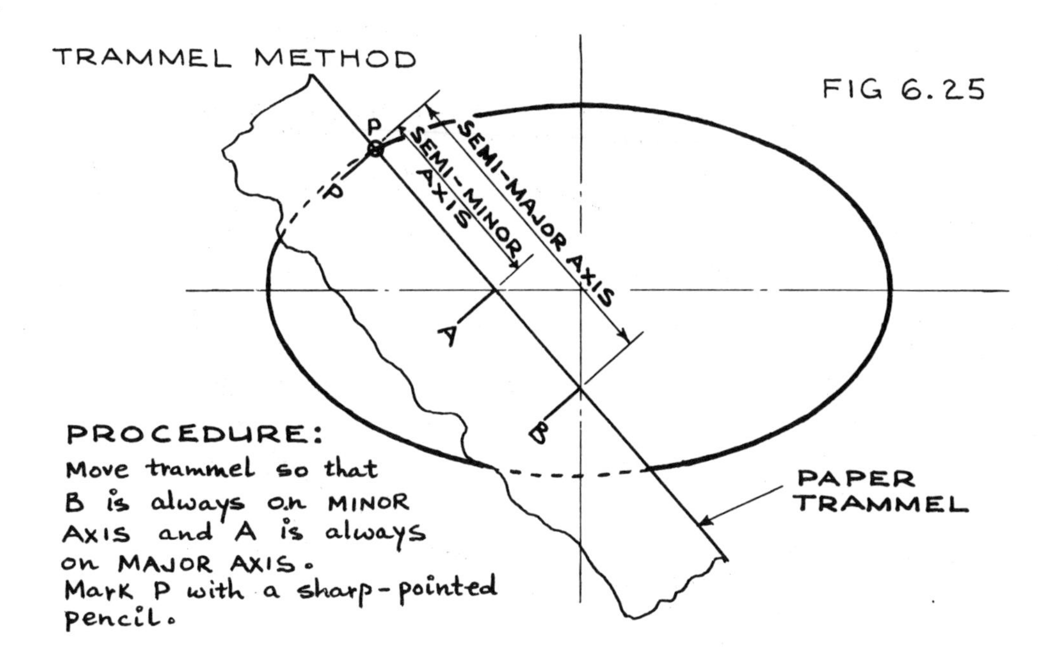

EXERCISE 11 SOLUTION

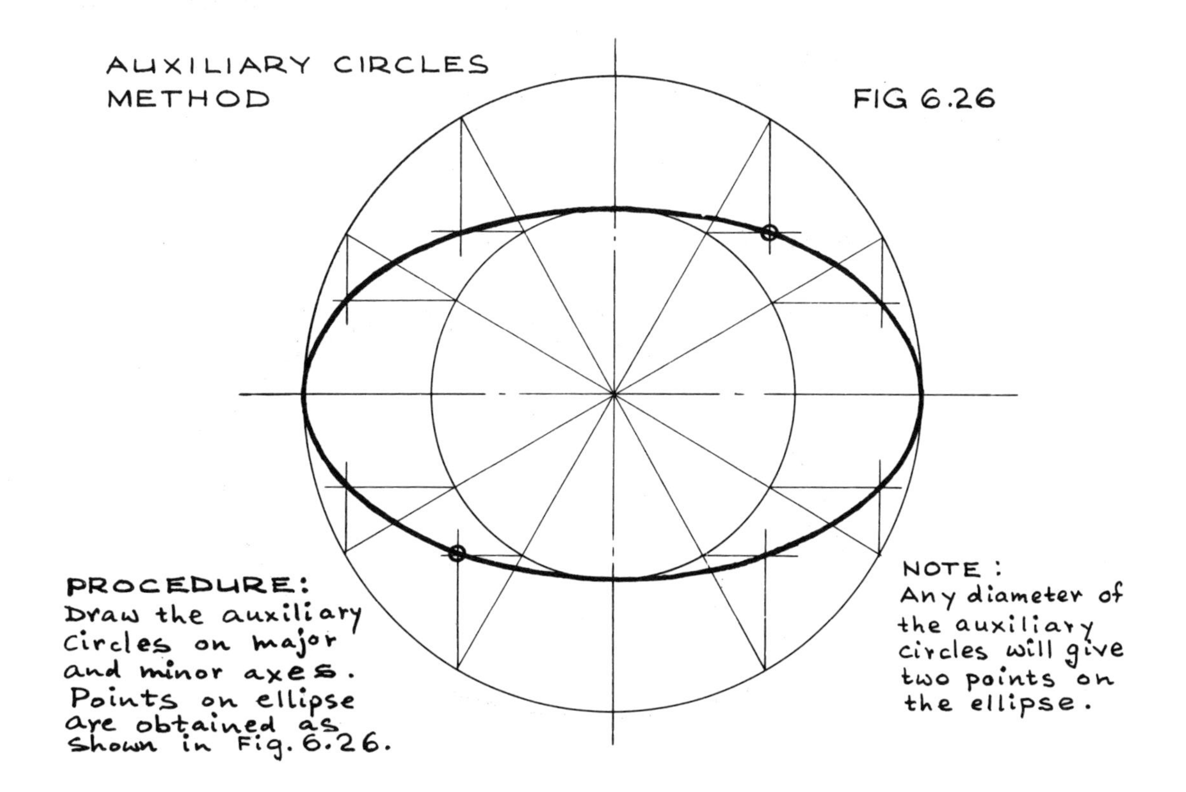

EXERCISE 12 SOLUTION

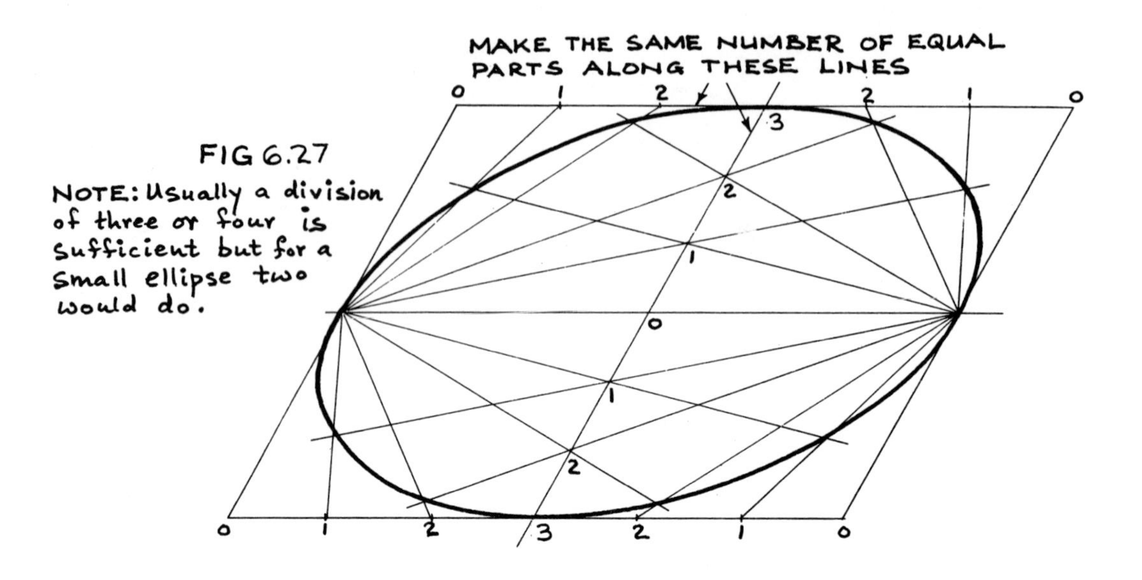

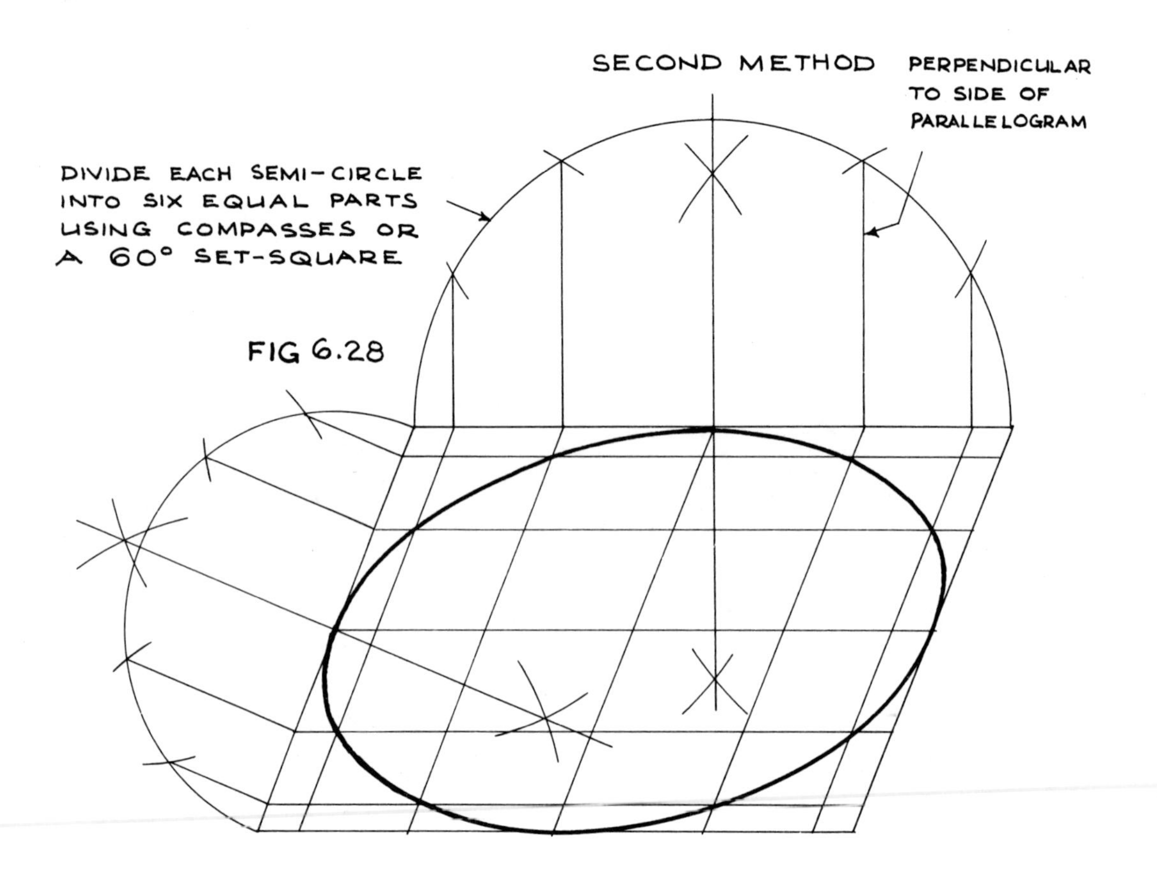

EXERCISE 13 SOLUTION

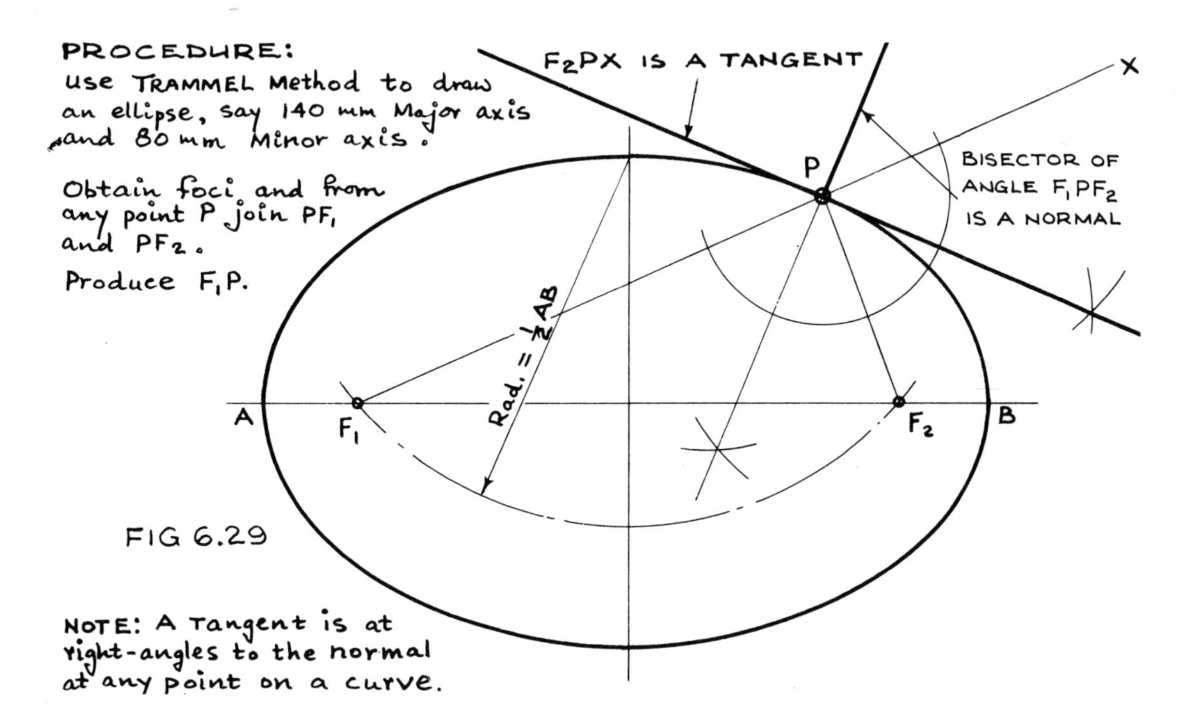

FIG 6.29

EXERCISE 14 SOLUTION

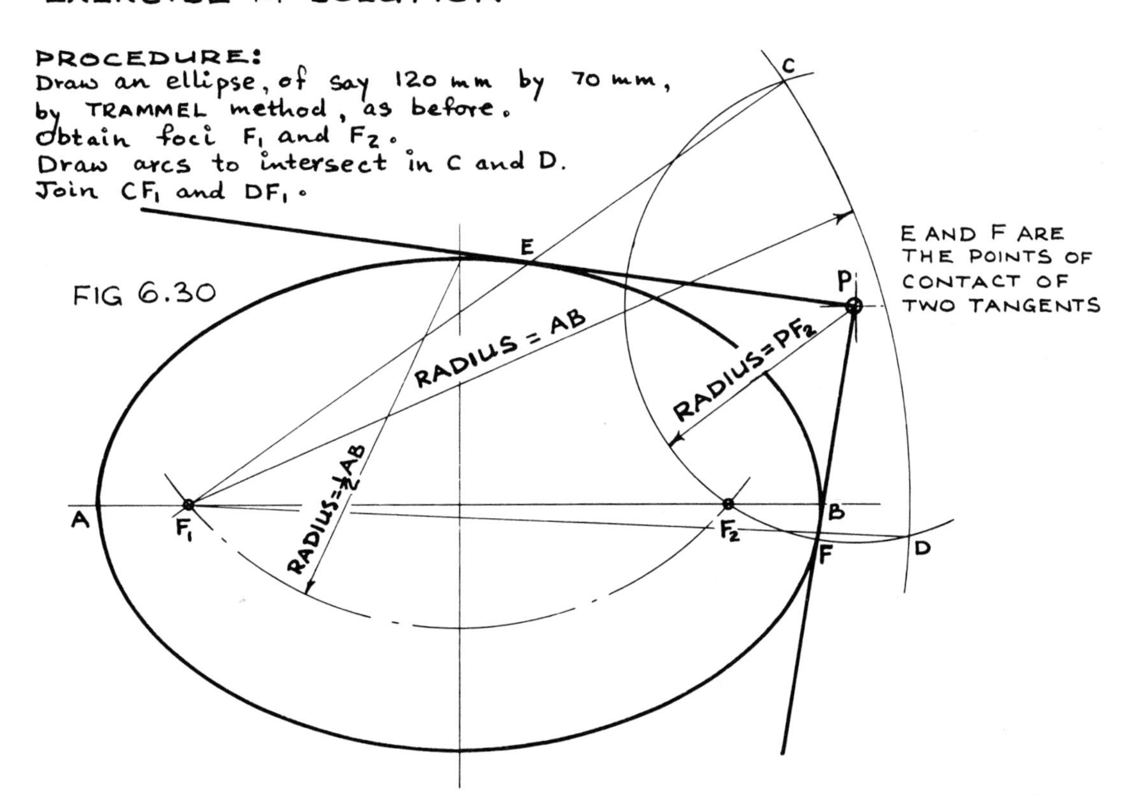

FIG 6.30

EXERCISE 15 SOLUTION

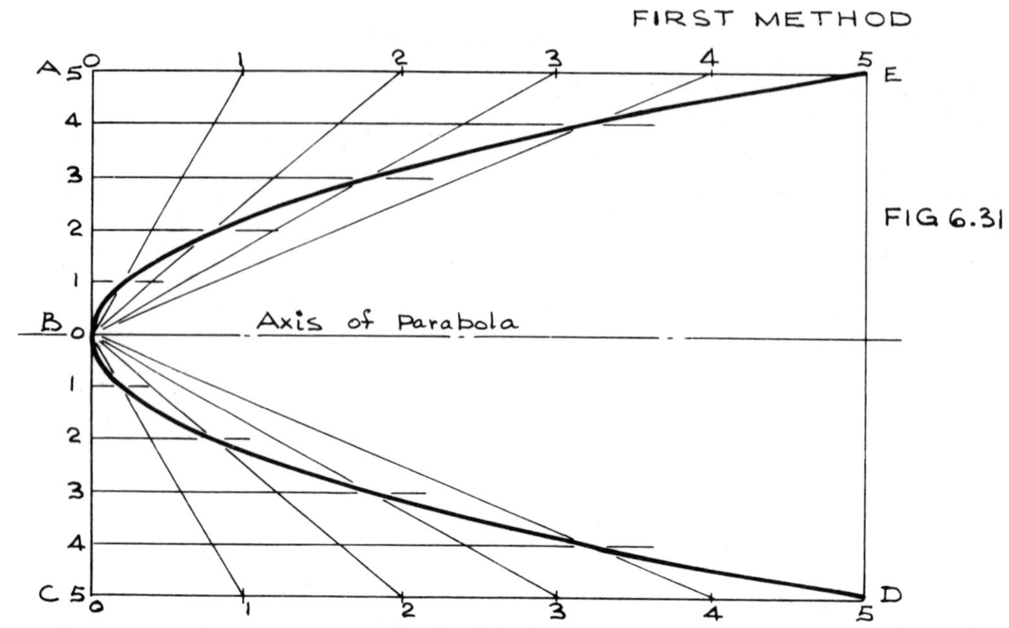

Mark off a convenient number of equal parts, having the same number along AB as along AE. Use ruler directly. (or a geometrical method)

NOTE: Bending Moment diagrams are sometimes Parabolic in shape. A Student of Engineering or Architecture may find **this construction useful.**

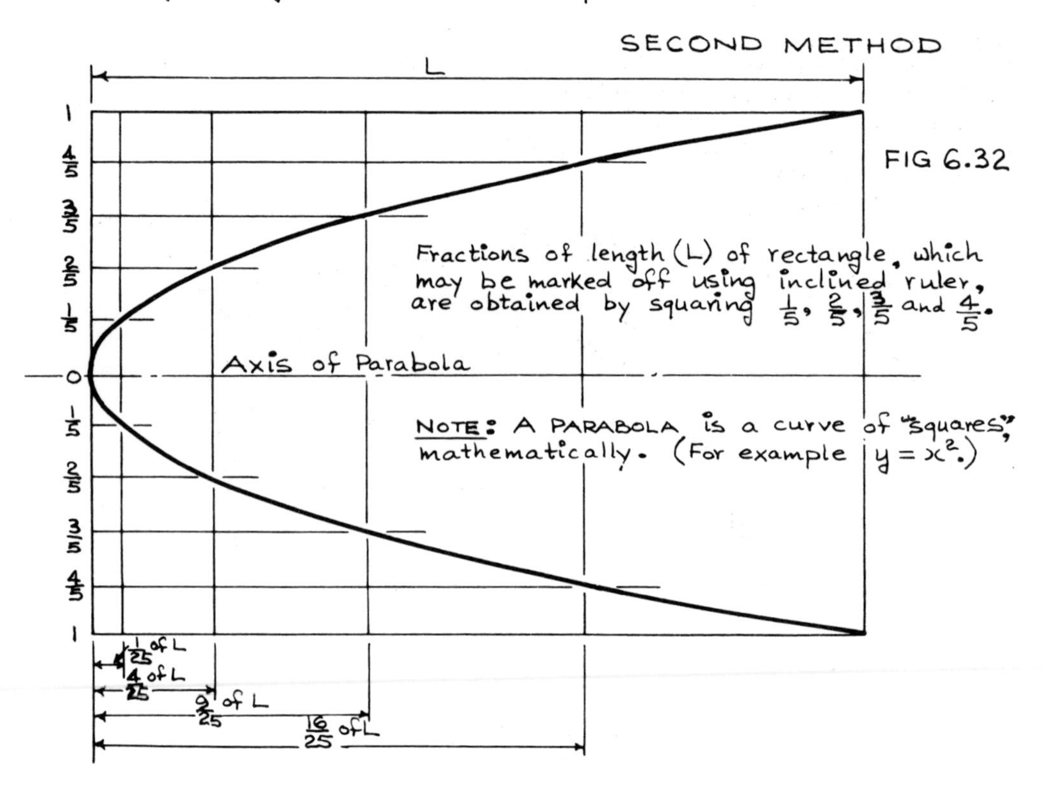

AN IMPORTANT PROPERTY OF THE PARABOLA

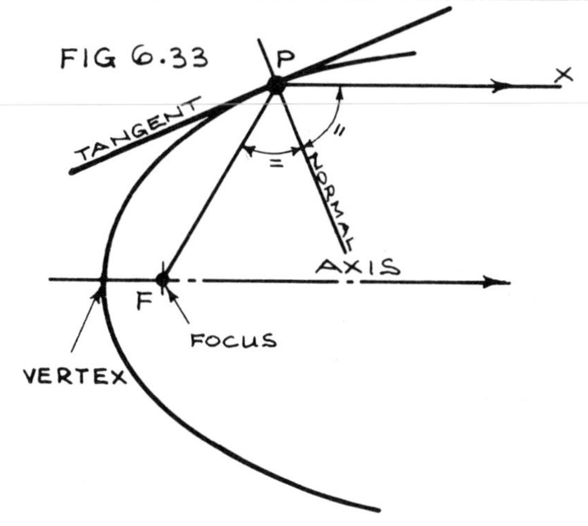

If any point P on the curve is joined to the focus F and if PX is drawn parallel to the AXIS, then FP and PX make equal angles with the NORMAL to the curve at P.

This fact is important, for example, in the design of search-light or head lamp reflectors. Since the incident light and reflected light make equal angles with the normal to the surface, thus a light source placed at the Focus will throw out a parallel beam if the reflector is paraboloid in shape.

EXERCISE 16 SOLUTION

PROCEDURE:

First construct a parabola inside a rectangle, of say 120 x 90 mm, by the method of Fig 6.31.

Through any point P draw a line perpendicular to the axis.

Make NX = 2 NV and join XV. Through Q draw parallel to NX to obtain focus F.

For a parabola, the vertex V is mid-way between focus and directrix

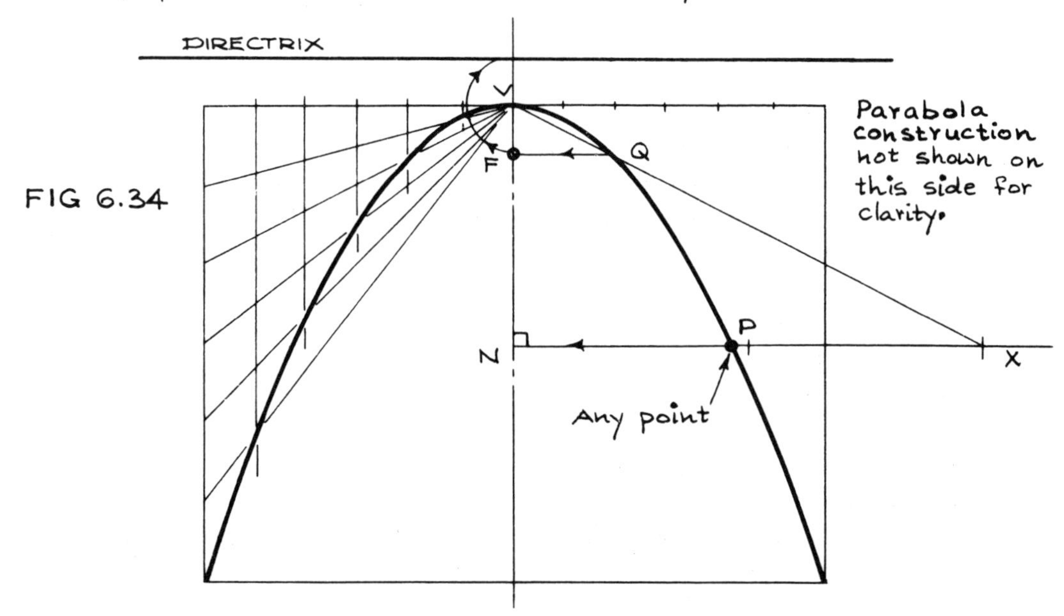

EXERCISE 17 SOLUTION

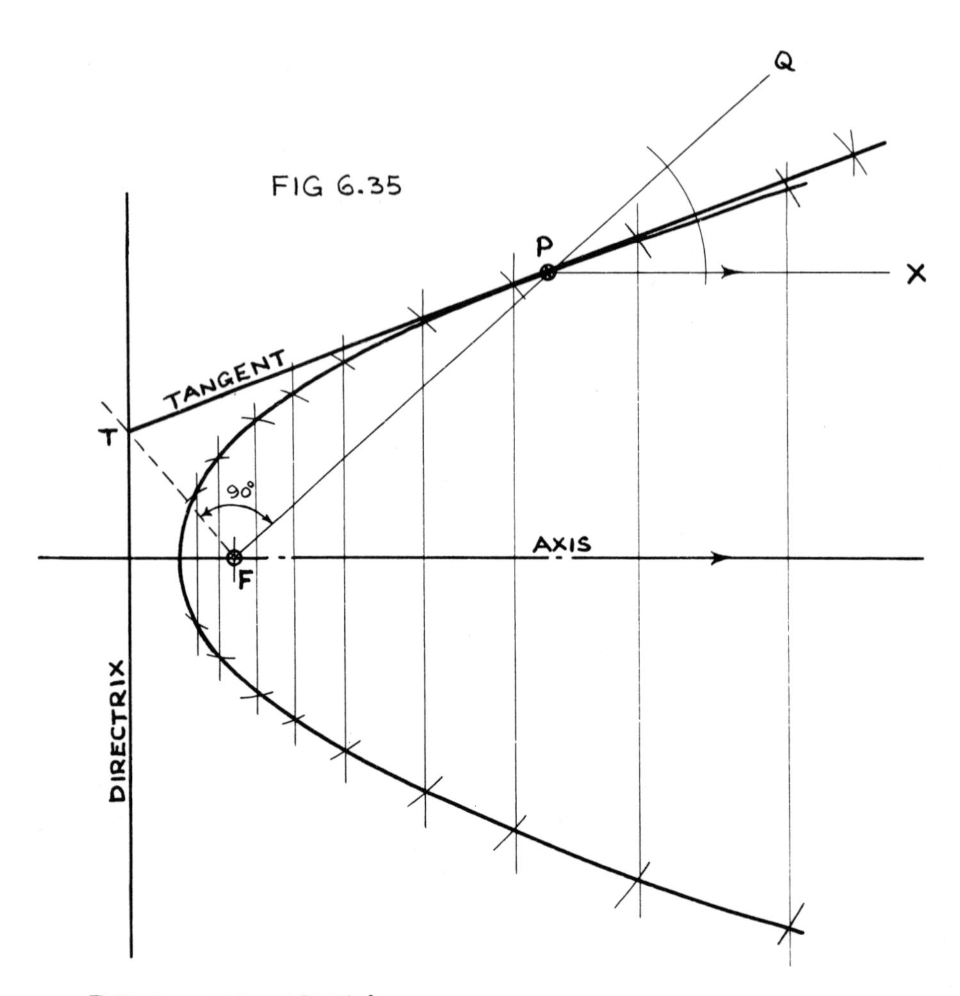

PROCEDURE:

First draw any parabola; this time suppose the focus is 20 mm from the directrix. See Exercise 6 solution for the method of constructing a parabola, being given the distance of focus from directrix.

Join focus F to any point P and produce to Q. Through P draw PX parallel to the axis of the parabola. Bisect the exterior angle QPX to obtain the tangent.

NOTE: For a normal at P bisect angle FPX. Also notice that PT subtends a right-angle at the focus and this is true also for the hyperbola. T is the point where the tangent produced meets the directrix.

EXERCISE 18 SOLUTION

PROCEDURE:
Through P draw CD and EF parallel to OA and OB respectively.
Through O draw any radial lines to intersect CD and EF. For example OG intersects CD in H and EF in K.
Parallels to asymptotes drawn through H and K give another point Q on the curve.
Other points are obtained in a similar manner.

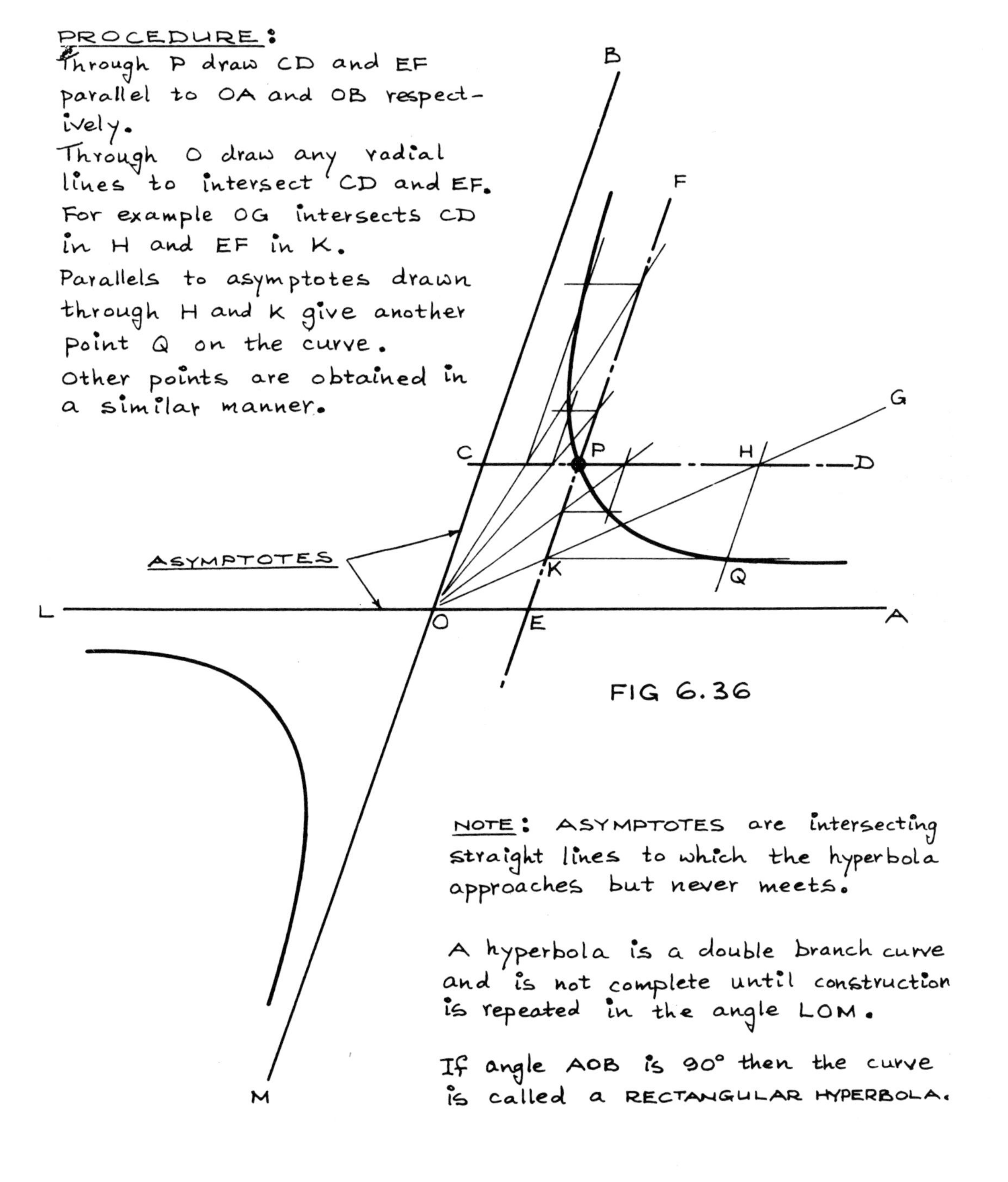

FIG 6.36

NOTE: ASYMPTOTES are intersecting straight lines to which the hyperbola approaches but never meets.

A hyperbola is a double branch curve and is not complete until construction is repeated in the angle LOM.

If angle AOB is 90° then the curve is called a RECTANGULAR HYPERBOLA.

Exercise 19. In a simple slider crank chain mechanism, figure 6.38, the crank OA is 30 mm and the connecting rod AB is 100 mm. Piston B is constrained to slide along OX. Plot the locus of a point P on the connecting rod, where AP is 30 mm, for one turn of the crank. Use two different methods for drawing the locus.
Exercise 20. A crank O_2B of length 20 mm turns clockwise about O_2. Another crank O_1A of length 35 mm is caused to oscillate about O_1 when linked to crank O_2B by connecting rod AB. The mechanism is shown in figure 6.39. Plot the locus of a point P on AB where BP is 35 mm.
Exercise 21. A crank OA of length 25 mm rotates clockwise about O. PQ is a rod which is freely pin-jointed to point A, where AP is 20 mm, and is constrained to slide through a fixed point B. The mechanism is shown in figure 6.40. Plot the locus of each end of the rod PQ as crank OA makes one complete revolution.
Exercise 22. Figure 6.41 shows a mechanism consisting of a crank OA which rotates about O, a connecting rod AB hinged to the crank at A and to a piston at B which slides in the line OB. CD is a link hinged to the connecting rod at C and constrained to move through the fixed point E. Obtain the locus of end D of the rod for one turn of the crank OA.
Exercise 23. A wheel rolls one half turn clockwise from the initial position shown, figure 6.43, along a horizontal straight rail. Point O_1 is a fixed pivot for rod O_1C and rods CD and PF may slide through the fixed pivots E and G respectively. Points A, P, B and C are pin-jointed. Plot loci for P, F and D of the mechanism and find the maximum angle turned through by rod O_1C.
Exercise 24. A thin circular disc 50 mm diameter rests in a vertical plane with point P of the circumference in contact with a thin, straight, horizontal rail AB. Draw the locus of P as the disc rolls, axis horizontal, for one complete turn along AB.
Exercise 25. Plot the locus of P for a similar disc to that described for the previous exercise but in this case make P a point in the disc 5 mm from the rolling edge.
Exercise 26. Construct an epicycloid curve, as shown in figure 6.46, being given that the base circle has a radius of 80 mm and the generating circle radius is 20 mm.
Exercise 27. Plot the locus of a point on the circumference of a 40 mm diameter circle which is assumed to roll on the inside of a base circle of radius 80 mm. The solution is shown in figure 6.47.

LOCI : MECHANISMS

In order to trace the path of a point on a moving mechanism there are two methods commonly employed in engineering graphics. A point P on the connecting rod AB is chosen for the locus shown in Figs. 6.37 and 6.38.

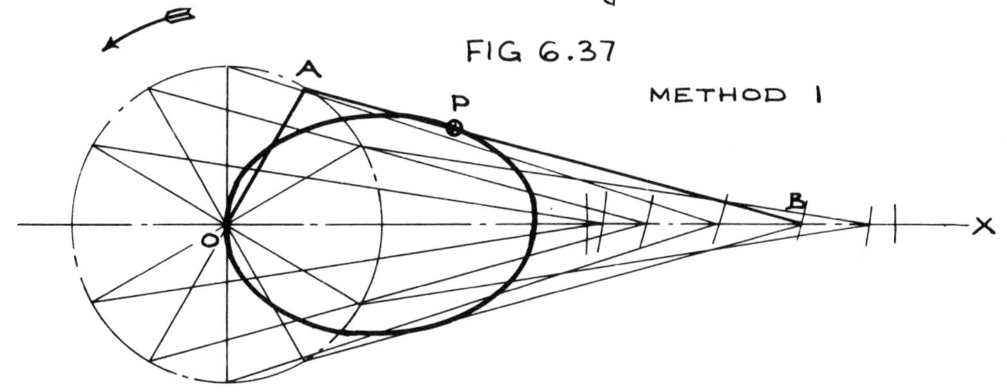

FIG 6.37

In Figs. 6.37 and 6.38, OA is a crank rotating about O, AB is a connecting rod pin-jointed at A and with end B free to move in the line OX. This simple mechanism is known as the SLIDER CRANK CHAIN MECHANISM and has practical application in the piston type engine.

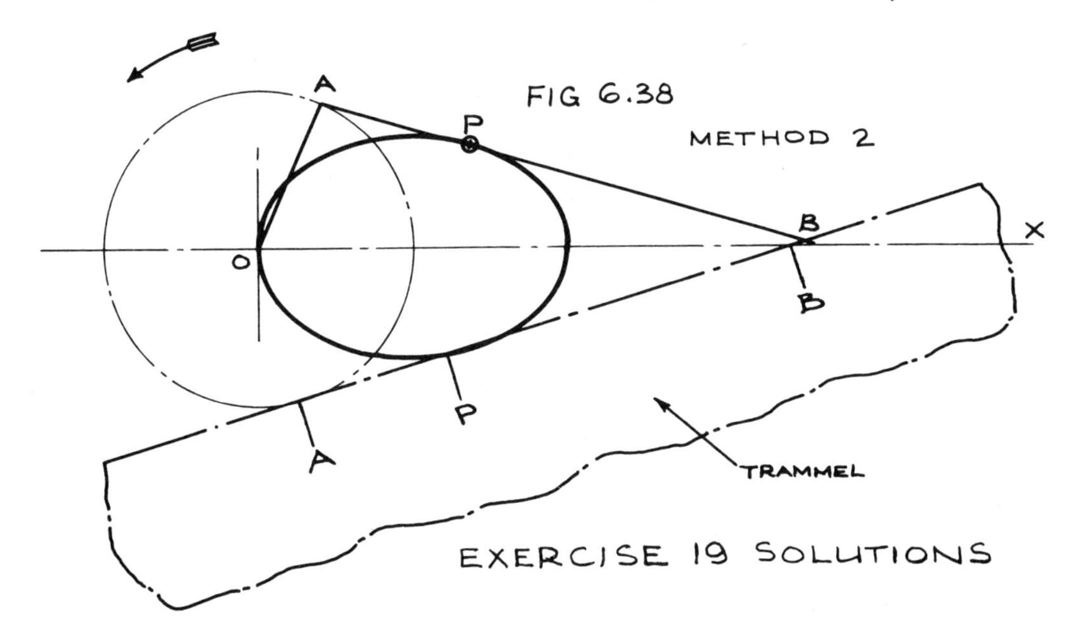

FIG 6.38

EXERCISE 19 SOLUTIONS

METHOD 1 consists of drawing the mechanism in a number of different positions and marking point P in each position.

METHOD 2 requires the use of a paper straight edge, called a trammel, on which has been marked the connecting rod length and the position of P. This method is quick and also has the advantages that the student may more readily visualise the motion and choose points in greater numbers on arcs of higher curvature.

EXERCISE 20 SOLUTION

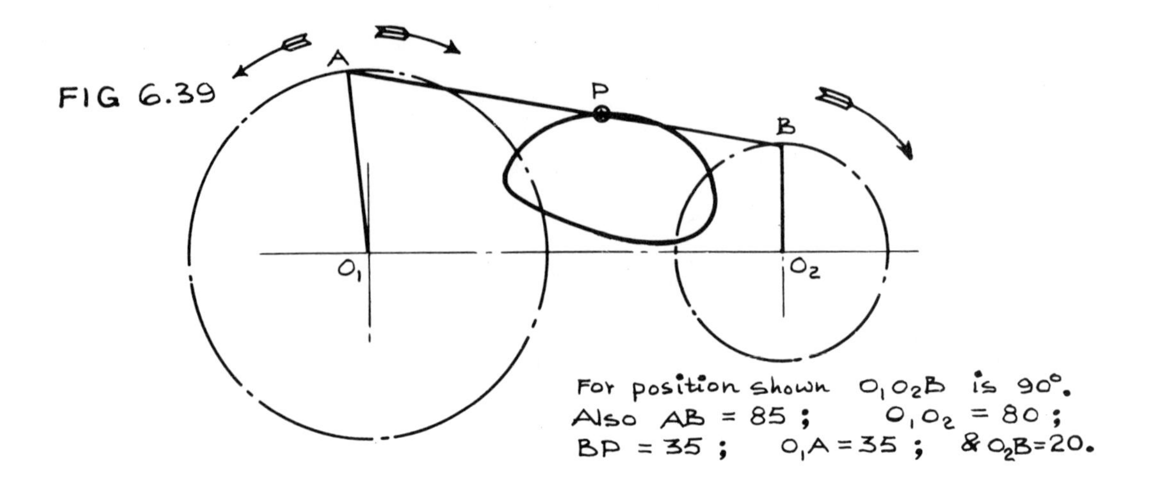

EXERCISE 21 SOLUTION

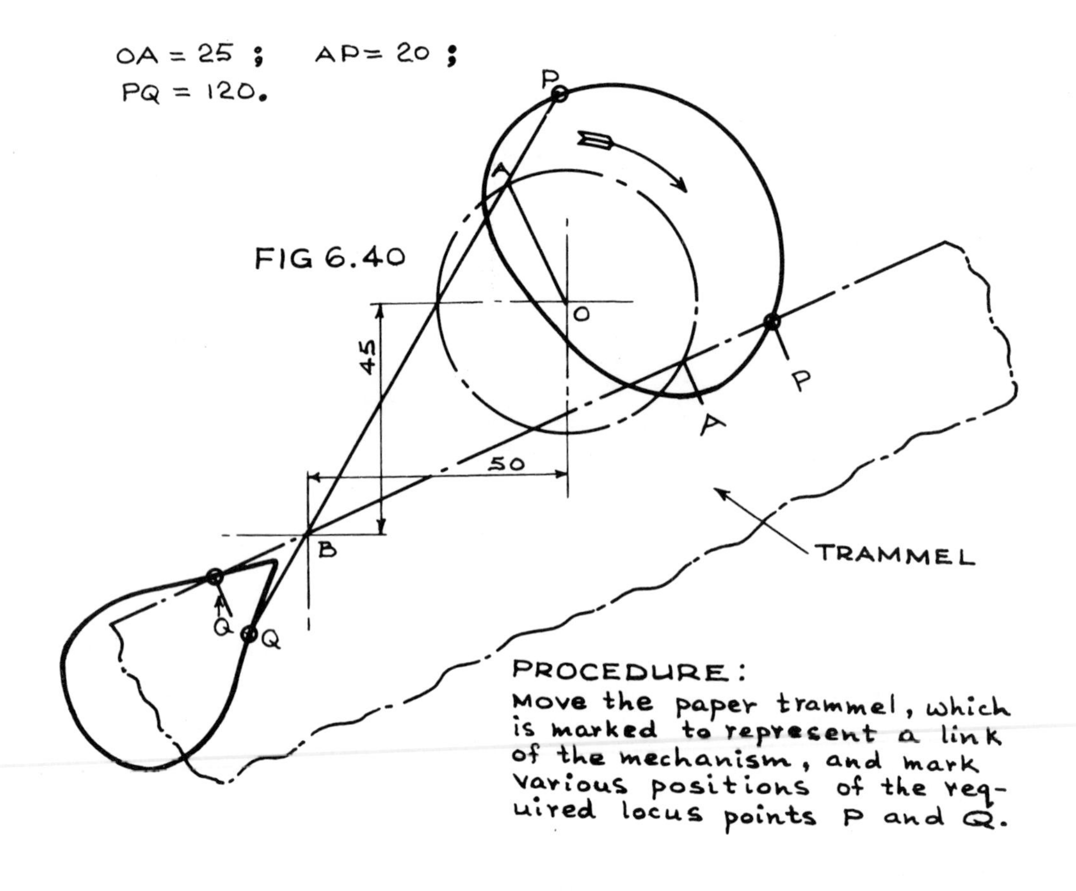

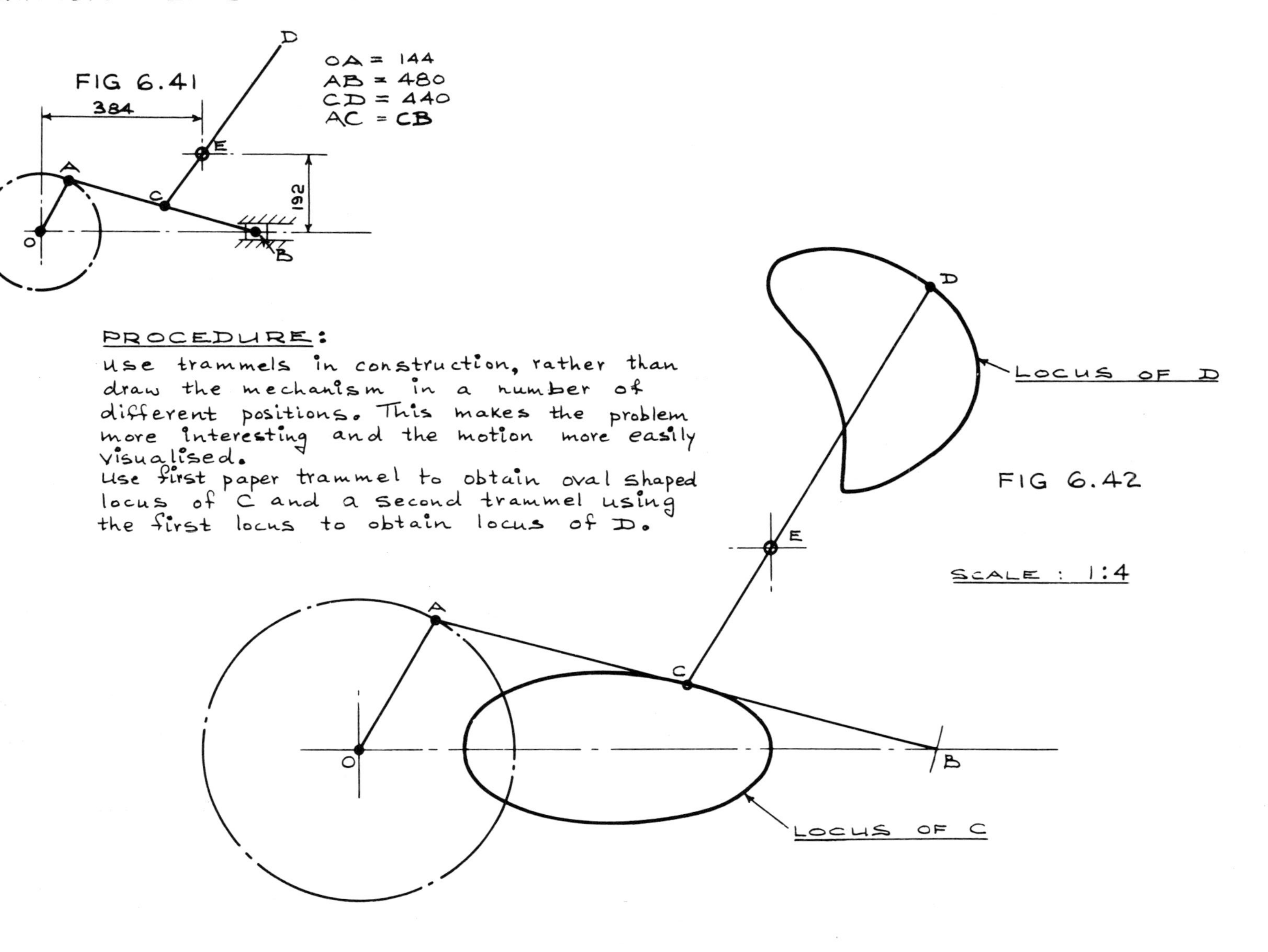
EXERCISE 22 SOLUTION
FIG 6.41
384
192
OA = 144
AB = 480
CD = 440
AC = CB
O
A
C
E
D
B
PROCEDURE:
Use trammels in construction, rather than draw the mechanism in a number of different positions. This makes the problem more interesting and the motion more easily visualised.
Use first paper trammel to obtain oval shaped locus of C and a second trammel using the first locus to obtain locus of D.
LOCUS OF D
FIG 6.42
SCALE : 1:4
LOCUS OF C

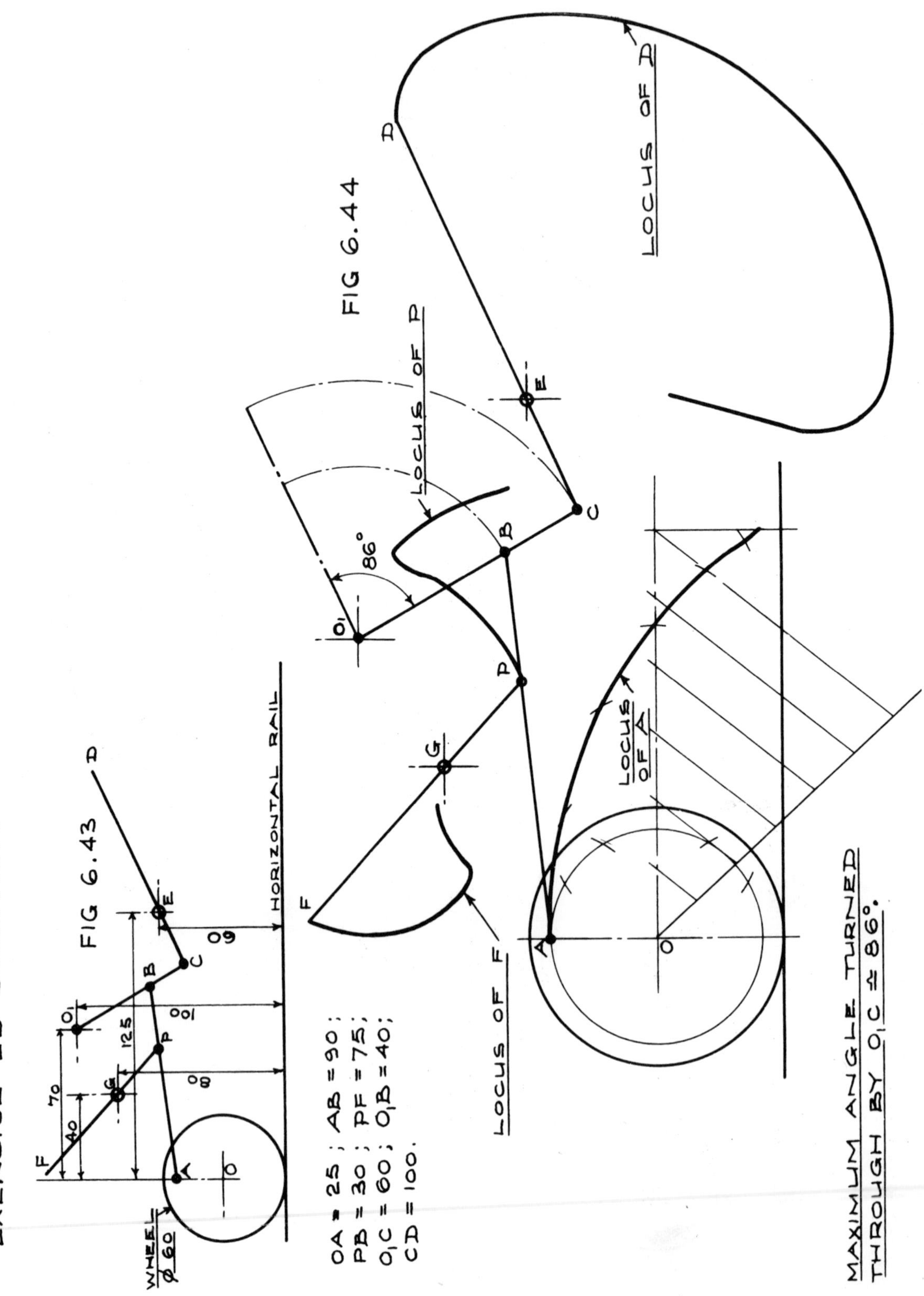

EXERCISE 23 SOLUTION
FIG 6.43
FIG 6.44
WHEEL
Ø 60
HORIZONTAL RAIL
60
100
80
125
70
40
OA = 25 ; AB = 90 ;
PB = 30 ; PF = 75 ;
O1C = 60 ; O1B = 40 ;
CD = 100.
86°
LOCUS OF D
LOCUS OF P
LOCUS OF F
LOCUS OF A
MAXIMUM ANGLE TURNED
THROUGH BY O1C ≏ 86°.

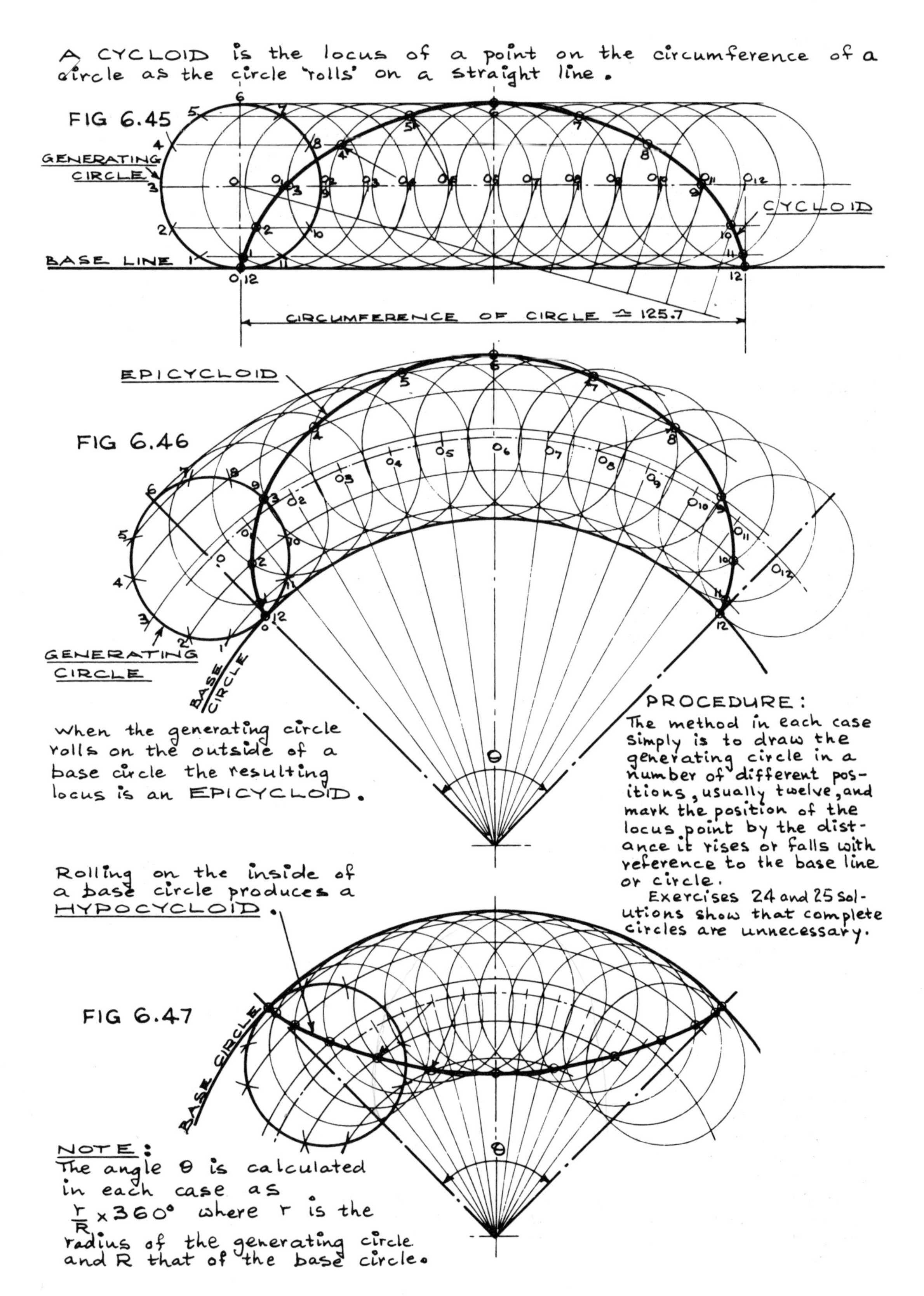
A CYCLOID is the locus of a point on the circumference of a circle as the circle 'rolls' on a straight line.
FIG 6.45
GENERATING CIRCLE
BASE LINE
CYCLOID
CIRCUMFERENCE OF CIRCLE ≏ 125.7
EPICYCLOID
FIG 6.46
GENERATING CIRCLE
BASE CIRCLE
When the generating circle rolls on the outside of a base circle the resulting locus is an EPICYCLOID.
PROCEDURE:
The method in each case simply is to draw the generating circle in a number of different positions, usually twelve, and mark the position of the locus point by the distance it rises or falls with reference to the base line or circle.
Exercises 24 and 25 solutions show that complete circles are unnecessary.
Rolling on the inside of a base circle produces a HYPOCYCLOID.
FIG 6.47
BASE CIRCLE
NOTE:
The angle θ is calculated in each case as
r/R x 360° where r is the radius of the generating circle and R that of the base circle.

EXERCISE 24 SOLUTION

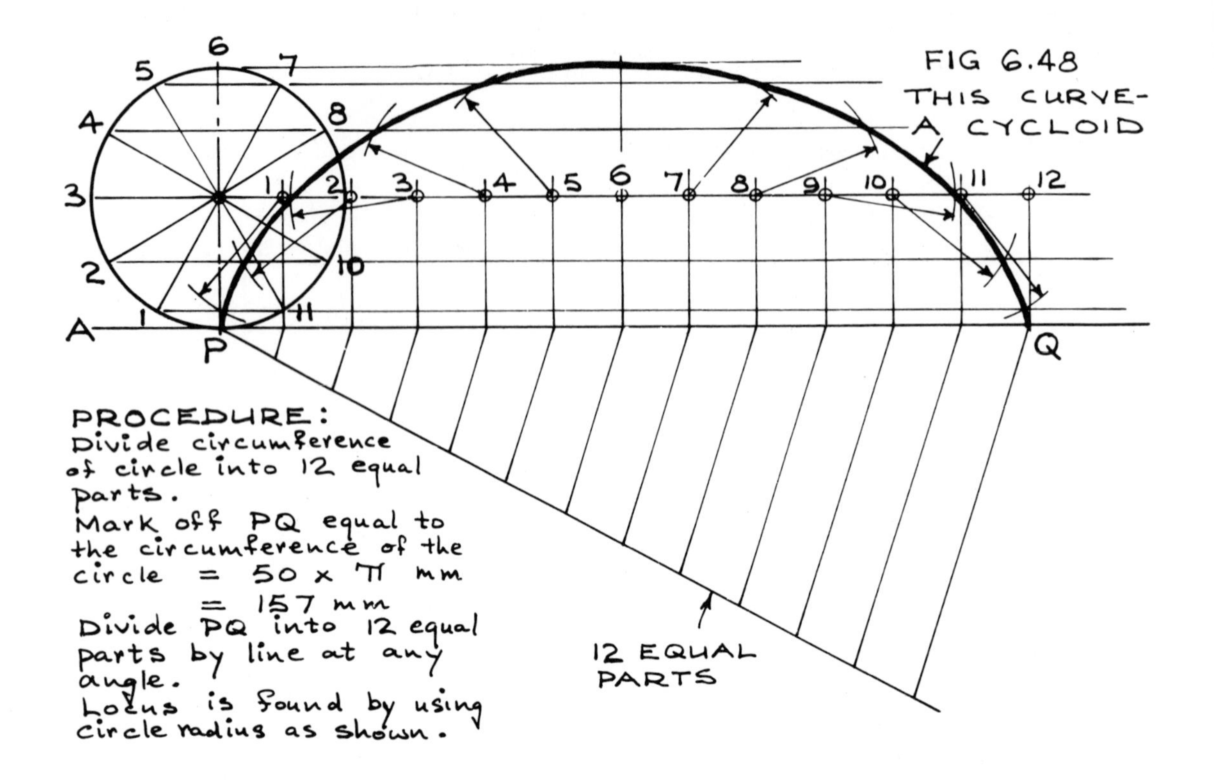

EXERCISE 25 SOLUTION

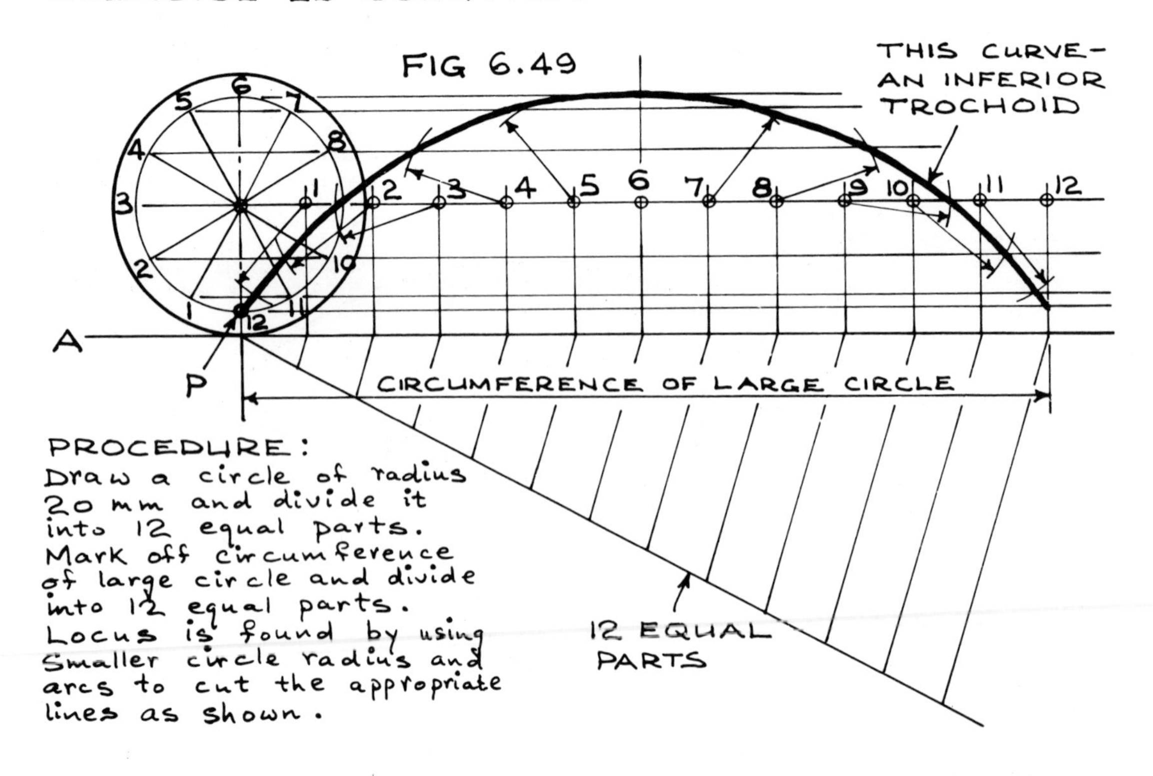

THE HELIX

By far the most common example of the three-dimensional curve called the *helix* is to be seen in the ordinary screw thread.

The procedure for obtaining a basic helix curve in front view, as shown in figure 6.50, is as follows:

(i) Draw a semi-circle representing half the plan view of the given helix and divide the arc into six equal parts. This is quicker than drawing a complete circle divided into twelve parts and also saves space. Number the points as shown in figure 6.50.

(ii) Set off a few pitch lengths as required according to the given information for the helix.

(iii) Divide each pitch length into twelve equal parts, i.e. the same number of parts as for the complete circular plan of the helix.

(iv) Plot the points in front view as shown in figure 6.50, remembering that the law of the curve locus is: advancement in a direction parallel to the axis is proportional to the angular movement around circumference of helix plan. The numbered points projected from the half circle plan should agree with the numbered horizontal lines in elevation.

By use of spring bows, as shown in figure 6.51, this basic helix can easily be transformed into the elevation of a right-handed coil helical spring made from rod of circular cross-section, figure 6.52.

Exercise 28. Construct a helix 70 mm diameter, pitch 50 mm and convert it to a right-handed coil circular cross-section spring of mean diameter 70 mm and wire cross-section diameter 10 mm.

Exercise 29. Draw a few turns of a right-handed single start square thread 80 mm outside diameter and 30 mm pitch.

Procedure note. Figure 6.53 shows basic helices for the outside diameter and thread root diameter for the given pitch. Use spring bows, set to half the pitch, to mark parallel helices as shown in figure 6.54. Simply move the steel point of the compasses along the initial curves, figure 6.53, keeping the compass width parallel to the axis and make pencil dots at suitable intervals. Finish off as in figure 6.55 by emphasising lines sloping upwards towards the right and complete to show solid core of screw.

Exercise 30. Draw a front view showing three complete turns of a left-handed helical groove 15 mm wide and 15 mm deep cut in a cylinder 80 mm diameter. The pitch of the helix is to be 40 mm.

Procedure note. Draw basic helices as shown in figure 6.56 and parallel helices as in figure 6.57. Complete as in figure 6.58 by emphasising lines sloping upward to the left. Finish off to show solid core as for previous exercise.

Exercise 31. Show in front view, three complete turns of a square sectioned helical spring 80 mm outside diameter. The spring is made from bar 10 mm square in cross-section and the pitch of the helix is 40 mm.

Procedure note. Similar to *Exercise 29* except that, unlike the screw, the spring has no solid core.

Exercise 32. The form of a simple Buttress thread is shown in figure 6.62. Project a front view showing a few turns of the thread cut on a bar 80 mm diameter using a pitch of 16 mm.

Exercise 33. A double-start square thread has a lead of 40 mm and an outside diameter of 80 mm. Obtain a front view showing 120 mm length of screwed bar.

EXERCISE 28 SOLUTION

THE HELIX

This is a three dimensional curve which is common in everyday life in the form of the screw thread. The elevation of a helix is easily obtained if it is remembered that a moving point on the curve advances parallel to the axis by equal amounts for equal angles turned through about the axis. The axial advance for one complete turn is called the PITCH.

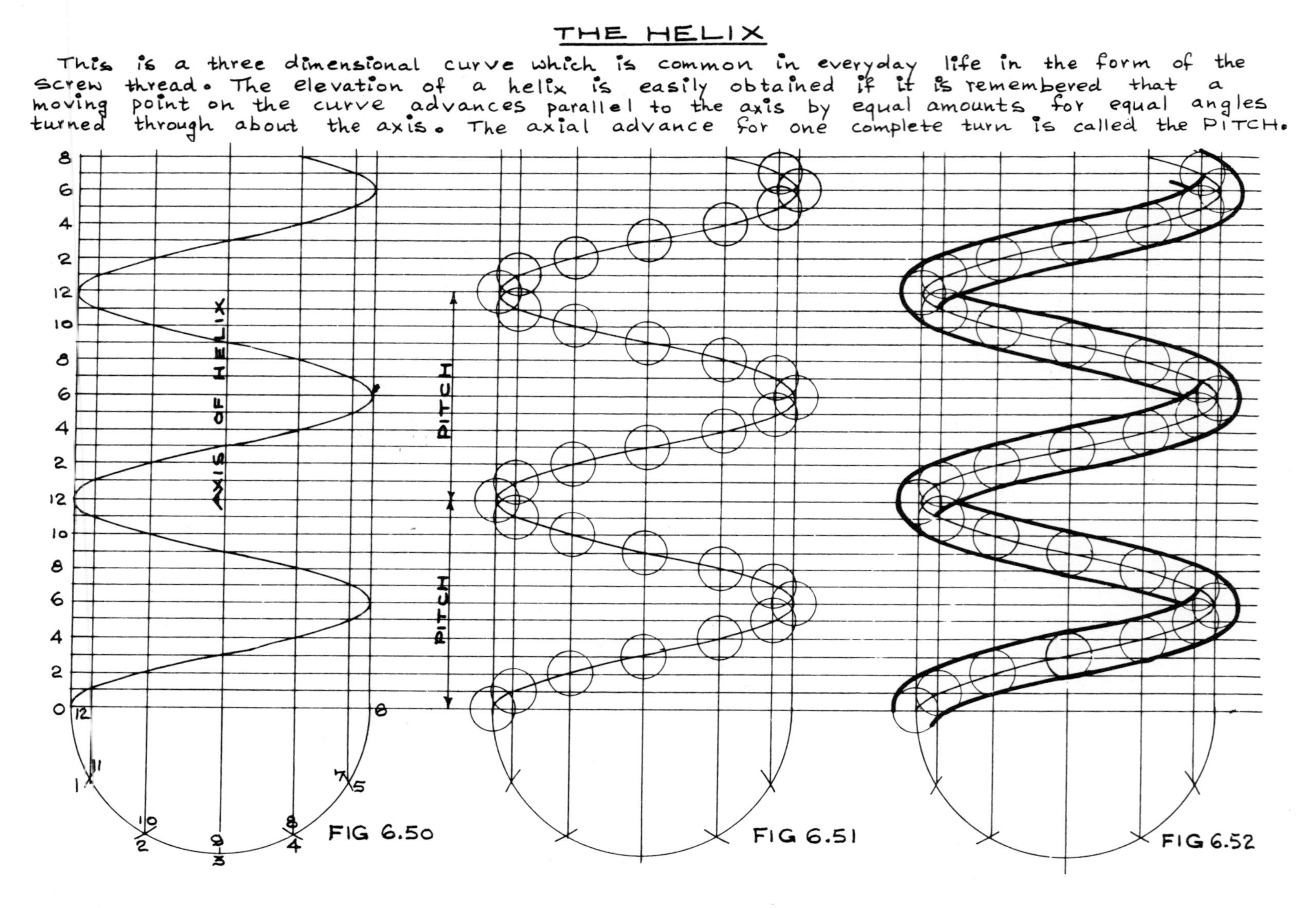

EXERCISE 29 SOLUTION

SCREW THREADS NOTE

In the case of a single-start screw thread, the common type, in one turn of* the nut it advances along the bolt by a distance equal to the pitch of the thread. In a multi-start thread the nut advances a distance known as the LEAD, in one complete turn. Thus in a double-start thread the lead is twice the pitch and in a triple-start thread, three times, etc. Multi-start threads are used to speed up the relative axial movement between nut and screw. In the foregoing the PITCH is defined as the distance, measured parallel to the axis, between a point on one thread and the corresponding point on the next thread. This sometimes causes a little confusion in the mind of the student on meeting multi-start threads for the first time. It may be helpful to remember that the lead of a multi-start thread is in fact the pitch of ONE, of the separate threads which make up the multi-start thread.

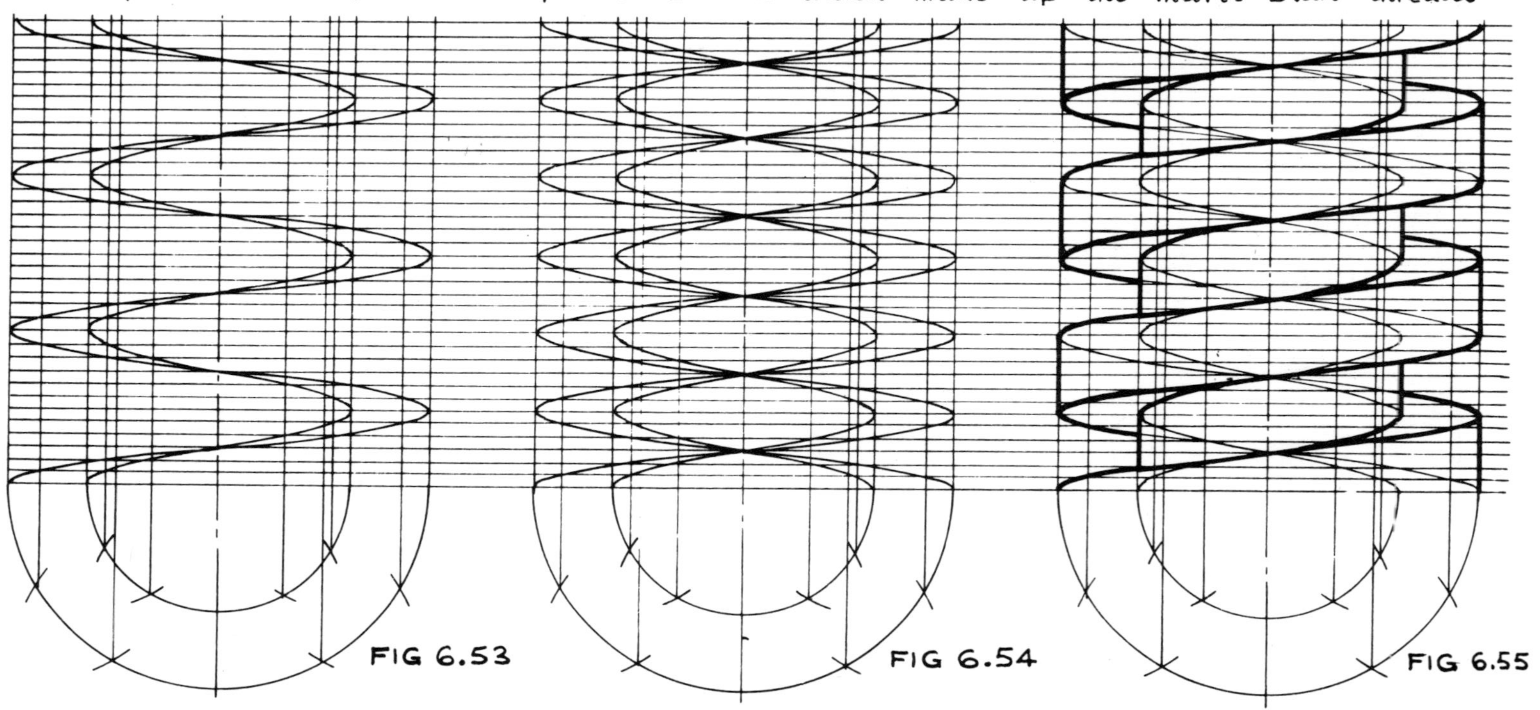

FIG 6.53 FIG 6.54 FIG 6.55

EXERCISE 30 SOLUTION

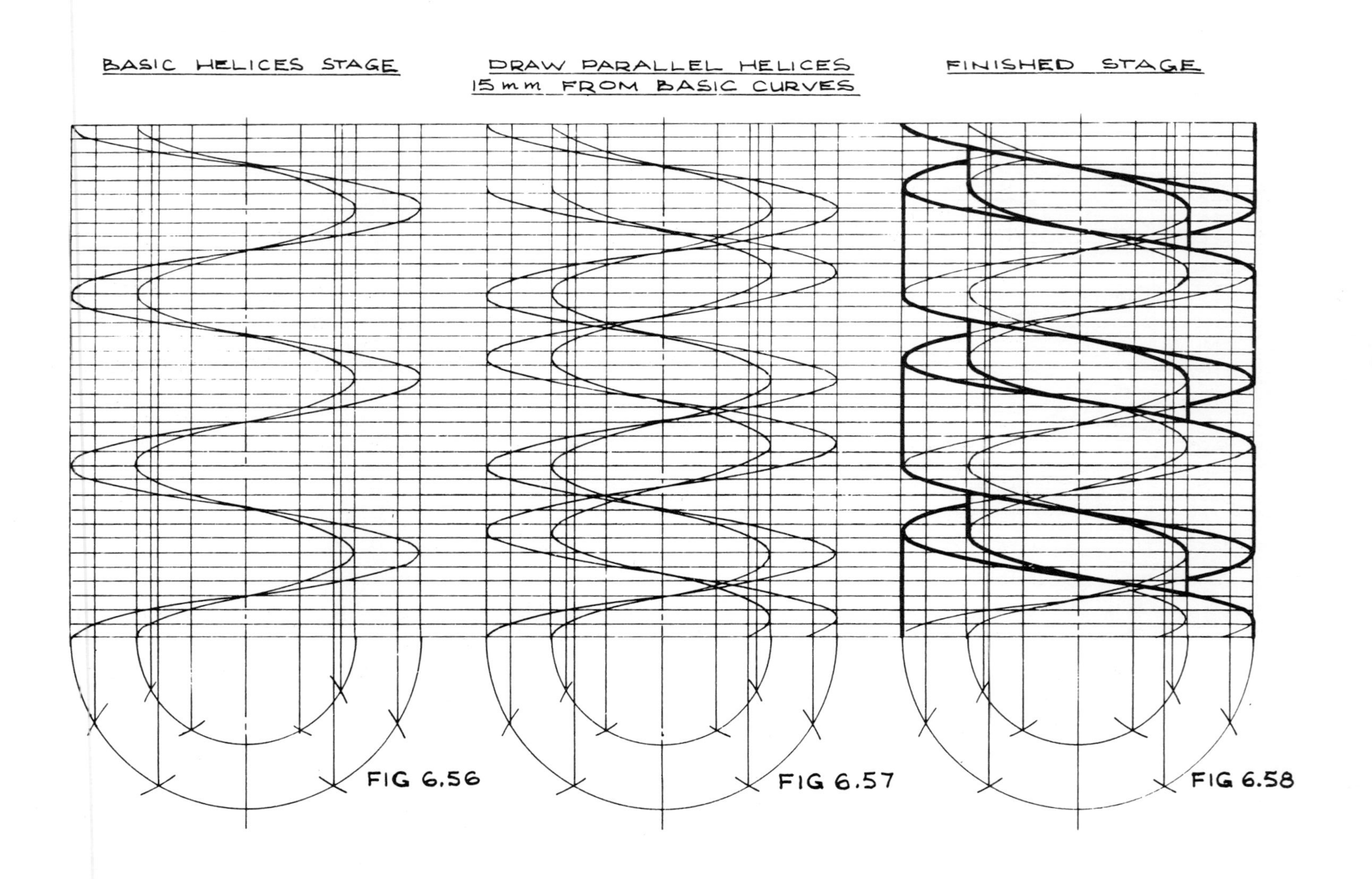
BASIC HELICES STAGE
DRAW PARALLEL HELICES
15mm FROM BASIC CURVES
FINISHED STAGE
FIG 6.56
FIG 6.57
FIG 6.58

EXERCISE 31 SOLUTION

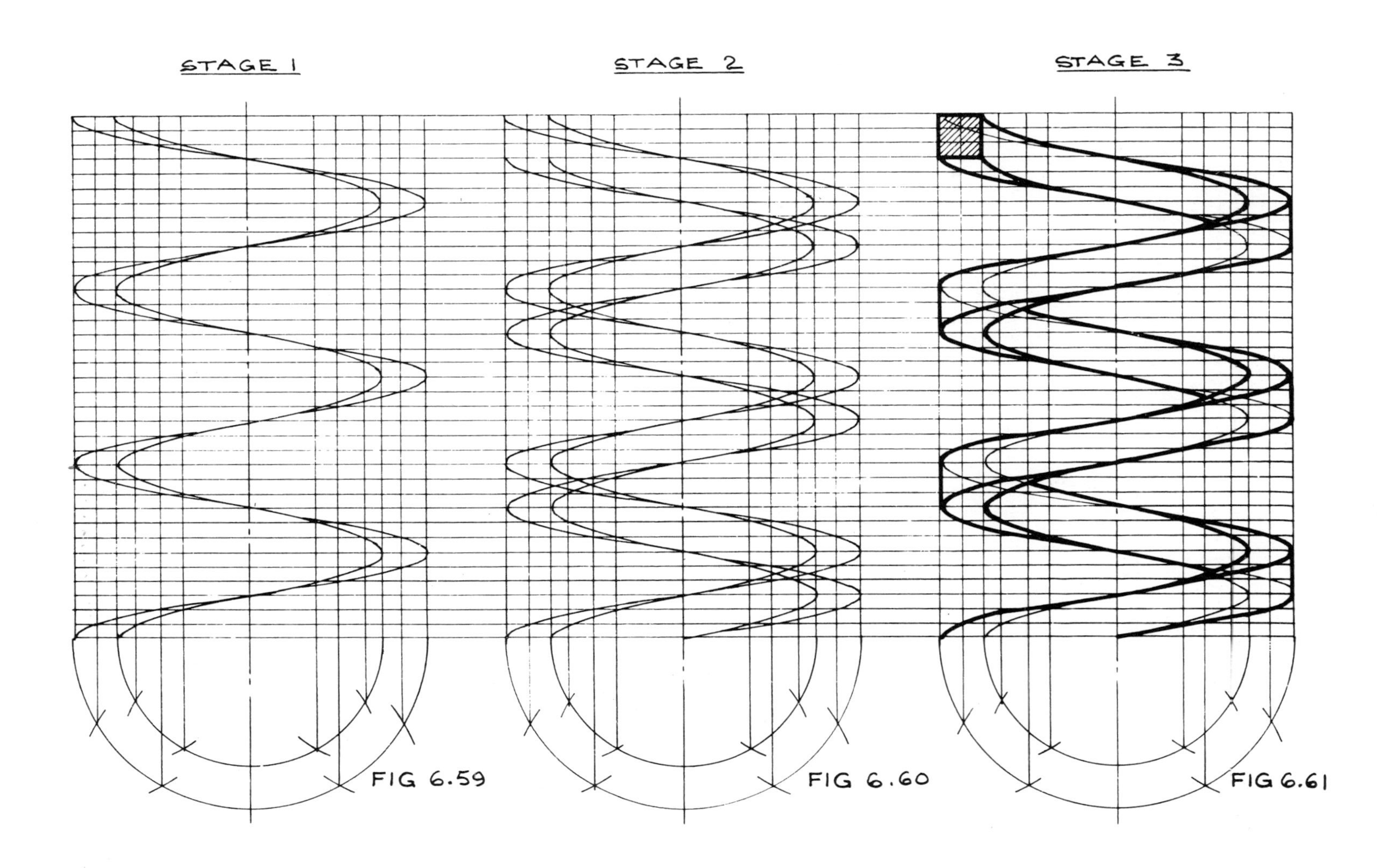

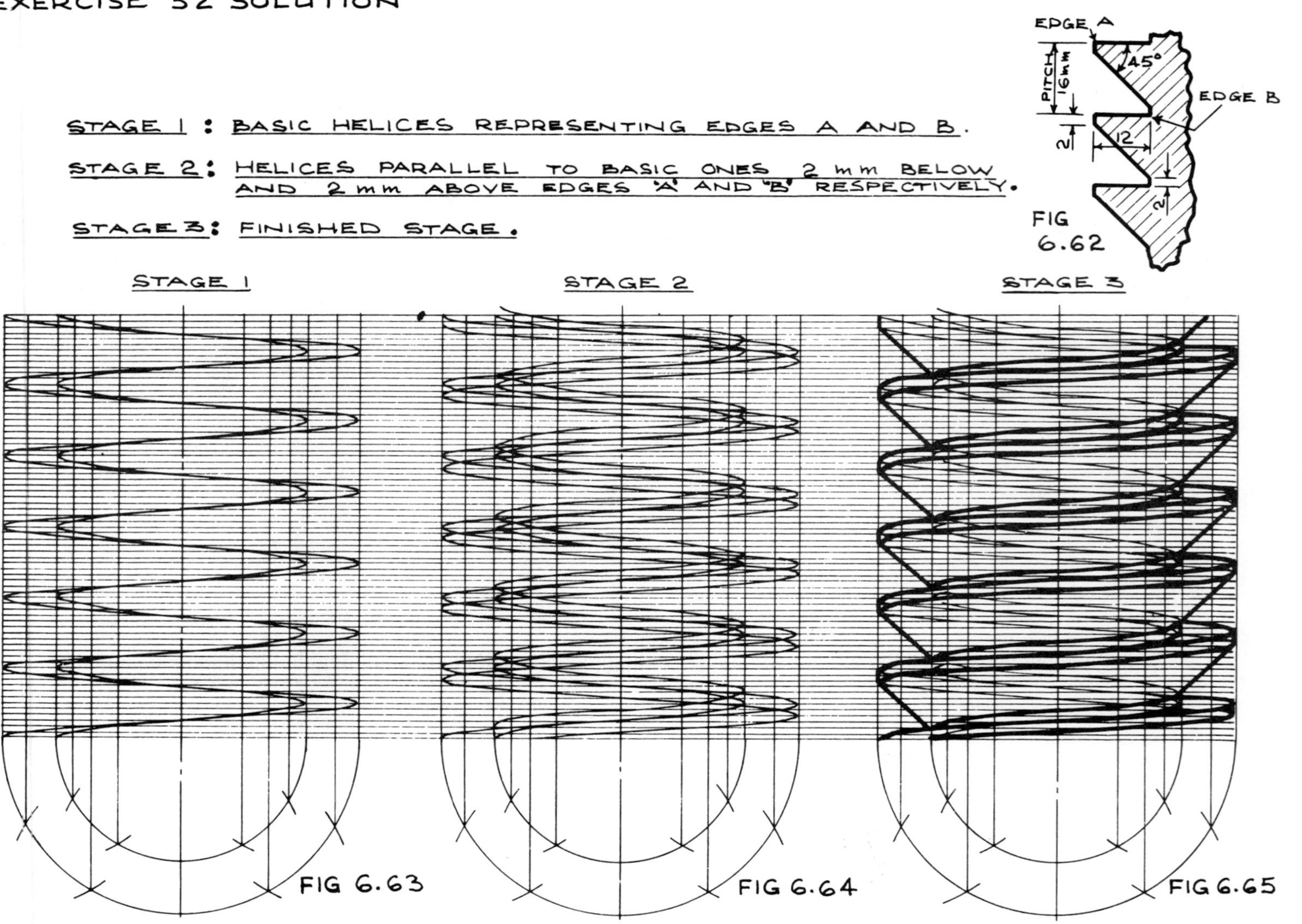

FIG 6.62

FIG 6.63

FIG 6.64

FIG 6.65

EXERCISE 33 SOLUTION

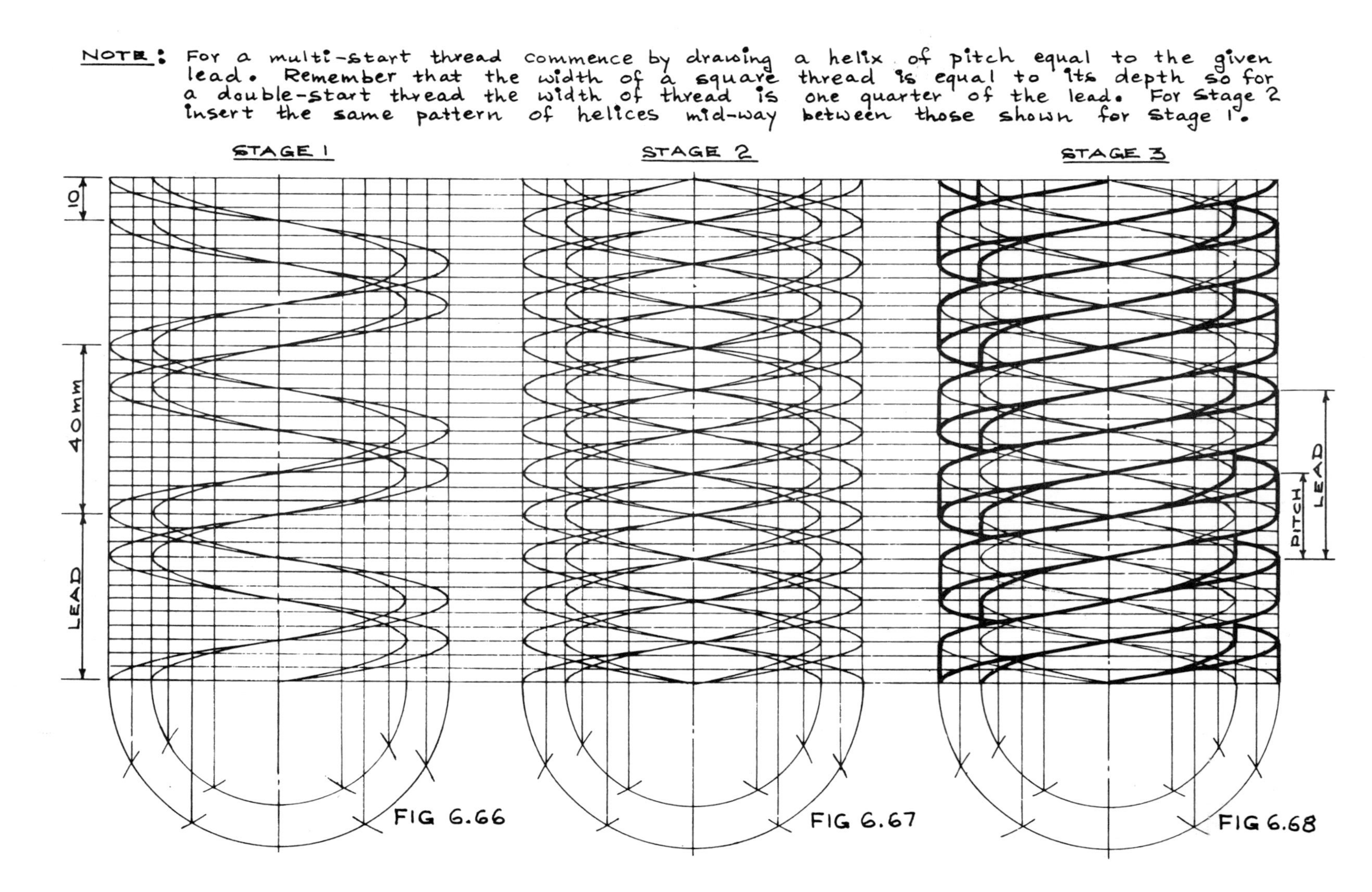

FIG 6.66

FIG 6.67

FIG 6.68

7
ISOMETRIC, AXONOMETRIC AND PERSPECTIVE PROJECTION

ISOMETRIC PROJECTION

Isometric projection is used to portray a solid by means of a single diagram. In chapter 3, isometric 'drawing' of simple objects is treated without attempting to introduce the *isometric scale*. When consideration is given to the foreshortening effect on the dimensions of an object which is portrayed isometrically this is termed isometric 'projection'. In this chapter some more advanced exercises will be demonstrated, all of which make use of the isometric scale.

In the case of simple rectangular solids, for example the cube shown in figure 7.1, the edges OA, OB and OC are assumed to be equally inclined to the direction of viewing, that is, these edges are equally inclined to the principal vertical plane.

The isometric view is drawn with the aid of a 30° set-square and this is the left-hand view in figure 7.1. An end view, projected from this isometric view, is shown to the right in figure 7.1.

The student should notice the following important points about isometric projection:

(i) Edges of cube parallel to the isometric axes OA, OB and OC are foreshortened by the same amount (reduction ratio 1:0.8165).

(ii) The amount of foreshortening is shown by the end view in figure 7.1, where edge OC is found in true length.

(iii) In conventional isometric drawing no allowance is made for this foreshortening effect and, for convenience, lengths parallel to the isometric axes are made full size. A conventional isometric drawing thus appears larger than the actual size of the object.

(iv) Notice that the true angle of 'tilt' of the cube base, angle α in the side view, is approximately 35°. The exact value of this angle α is 35° 16′ (35.27°).

Note. In actual isometric projection the student will not require to use this angle but this is given for information only.

(v) When it is desired to draw a true isometric projection, that is, taking into account the foreshortening effect, an isometric scale may be easily constructed as follows.

Consider the top face OBDA of a cube shown in figure 7.1. Its true shape is the square OBDA but its isometric shape is a rhombus, using the same horizontal diagonal AB. The isometric length may be obtained as indicated in figure 7.2. Since all measurements parallel to the isometric axes are foreshortened in the same proportion, an isometric scale may be simply constructed as shown in figure 7.3. The true isometric projection of an object is therefore similar to that obtained by using isometric scale on all measurements parallel to the three isometric axes.

Conventional isometric drawing, although quicker to execute, suffers from the following disadvantages. First of all, the object appears to be larger than it really is. For example, an isometric drawing of a cylinder has an ellipse major axis appreciably larger than the given cylinder diameter. Secondly, there is the problem of drawing a sphere or a portion thereof. A sphere, of course, ought to appear the same size that is, full size, no matter in which direction it is viewed. Thus, when using full size measurements as in conventional isometric drawing a sphere would require to have an *increased* radius in order to make it look correct in comparison to other geometrical solids which may comprise the object. For example a hemispherical cap on the cylinder, figure 7.4, would require to have a radius of X/2, centred at the ellipse centre. The increased radius for use in the portrayal of a sphere in conventional isometric drawing may be obtained by using an isometric scale *in reverse*, as shown in figure 7.5.

It is for the foregoing reasons that some examining authorities are careful to exclude the sphere from elementary syllabuses. However, it is the opinion of the author that it would be much better to tell the whole truth right from the start and use nothing but isometric scale. Thus, diameter X of cylinder in figure 7.4 would measure its actual diameter and there would be no need to ignore the sphere in elementary work. Indeed, rulers containing an isometric scale could be quite cheaply manufactured or the keen student could even make up such a scale ruler. The author has found that a paper isometric scale attached to the hypotenuse edge of a 30° set-square, and covered with transparent adhesive tape, is a very useful instrument for isometric projections.

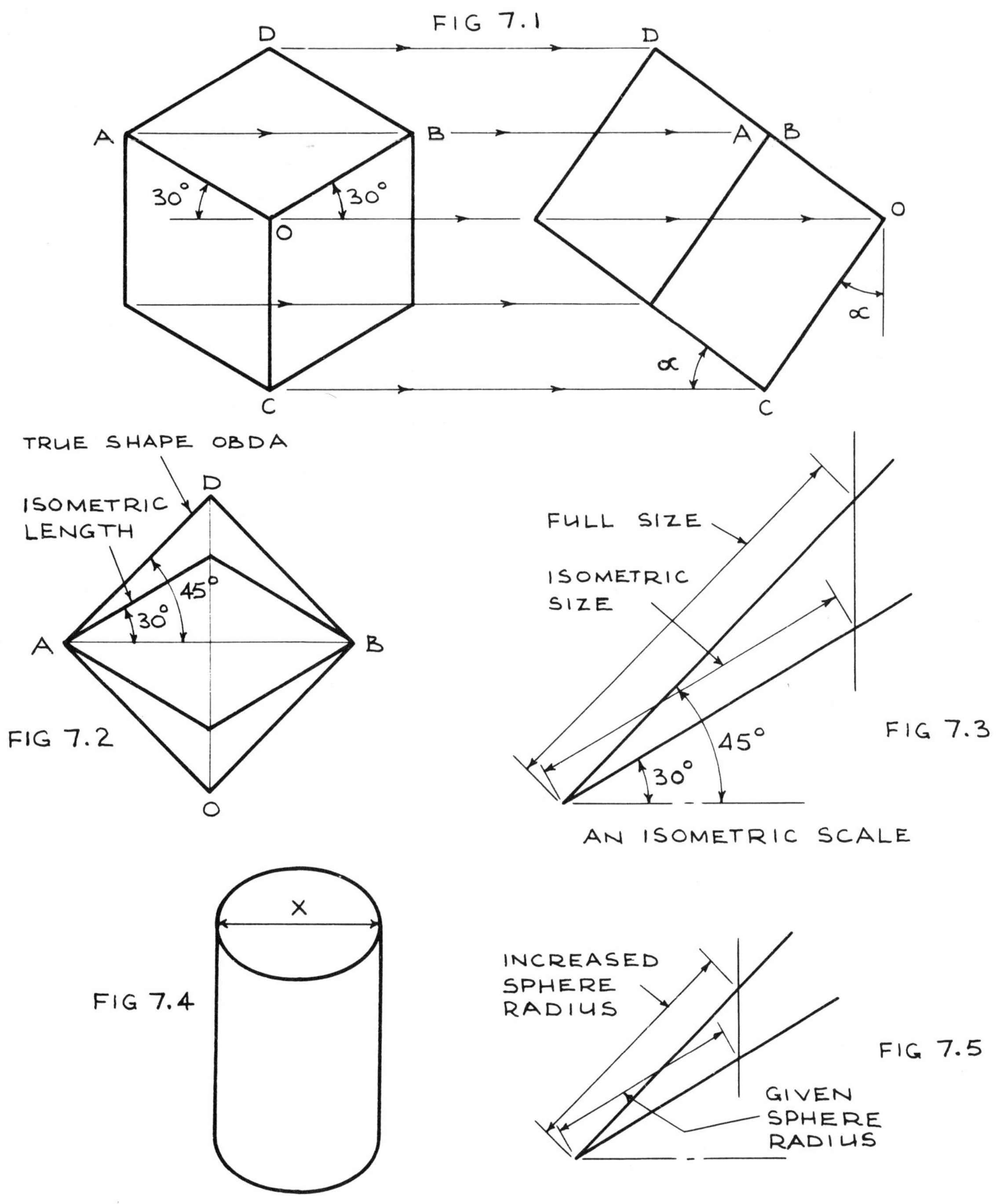

EXERCISES

Exercise 1. Two views of a Pipe Support Block are shown in third angle orthographic projection in figure 7.6. Using isometric scale make an isometric projection of the block with PQ vertical and A as the centre of isometric axes.

Exercise 2. Obtain an isometric projection of the special bracket, two views of which are shown in first angle orthographic projection in figure 7.9 (N.I.G.C.E.).

Exercise 3. Particulars of a thrust rod are shown in figure 7.12. Make an isometric projection of the rod so that the axis is in the given position (N.I.G.C.E.).

Exercise 4. A cranked bracket is shown by two views in first angle orthographic projection in figure 7.14. Make an isometric projection of the bracket as indicated by the arrows in the orthographic views.

Exercise 5. The front view and plan of a small tool-holder are shown in first angle orthographic projection in figure 7.17. Using an isometric scale draw an isometric view when the quarter bounded by vertical planes through AB and BC is removed (N.I.G.C.E.).

Exercise 6. Using a suitable scale make an axonometric projection of either (a) a book case, cocktail cabinet, cupboard unit, or (b) a music centre on a stand to include speakers and record storage compartment.

Exercise 7. Using a suitable scale make an axonometric projection of either (a) part of the interior arrangement of your kitchen at home, or (b) part of the interior arrangement of your lounge.

Exercise 8. Two views of a building outline are given in figure 7.31, the dimensions being in metres. Draw these views to a suitable scale and then obtain a two-point perspective view.

Exercise 9. Draw two views of a house, or bungalow, to scale, as in the example shown in figure 7.33. Using these views, and a plan view, make a perspective drawing of the dwelling.

EXERCISE 1 SOLUTION

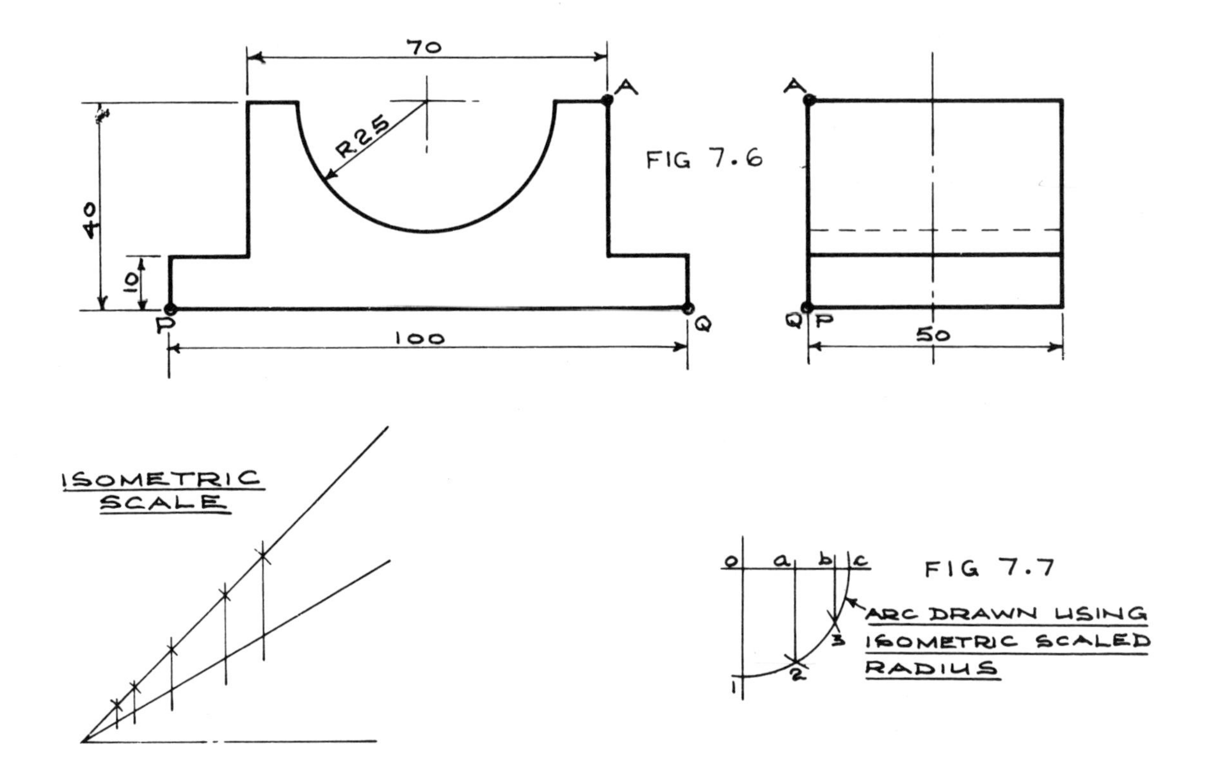

NOTE: Where circles or parts of circles are involved it is best to reduce the given radius by isometric scale first of all. The half-ordinates then taken from the construction arc, figure 7.7, may be used directly for the finished drawing. Thus much less work is involved by this method than by drawing a construction arc full-size and then proceeding to reduce each of the half-ordinates for use on the isometric scale drawing.

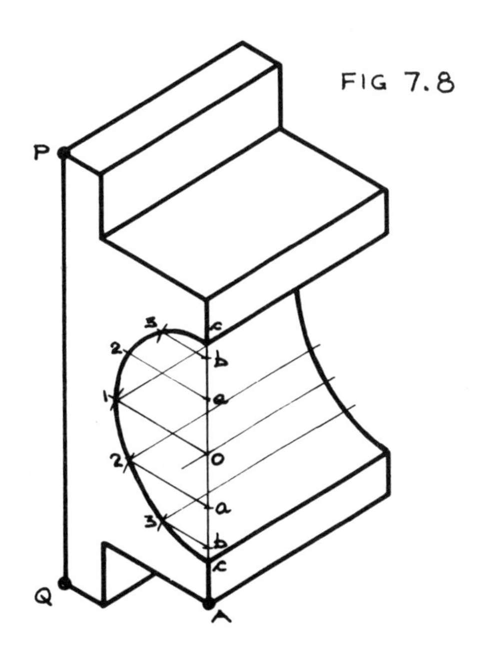

EXERCISE 2 SOLUTION

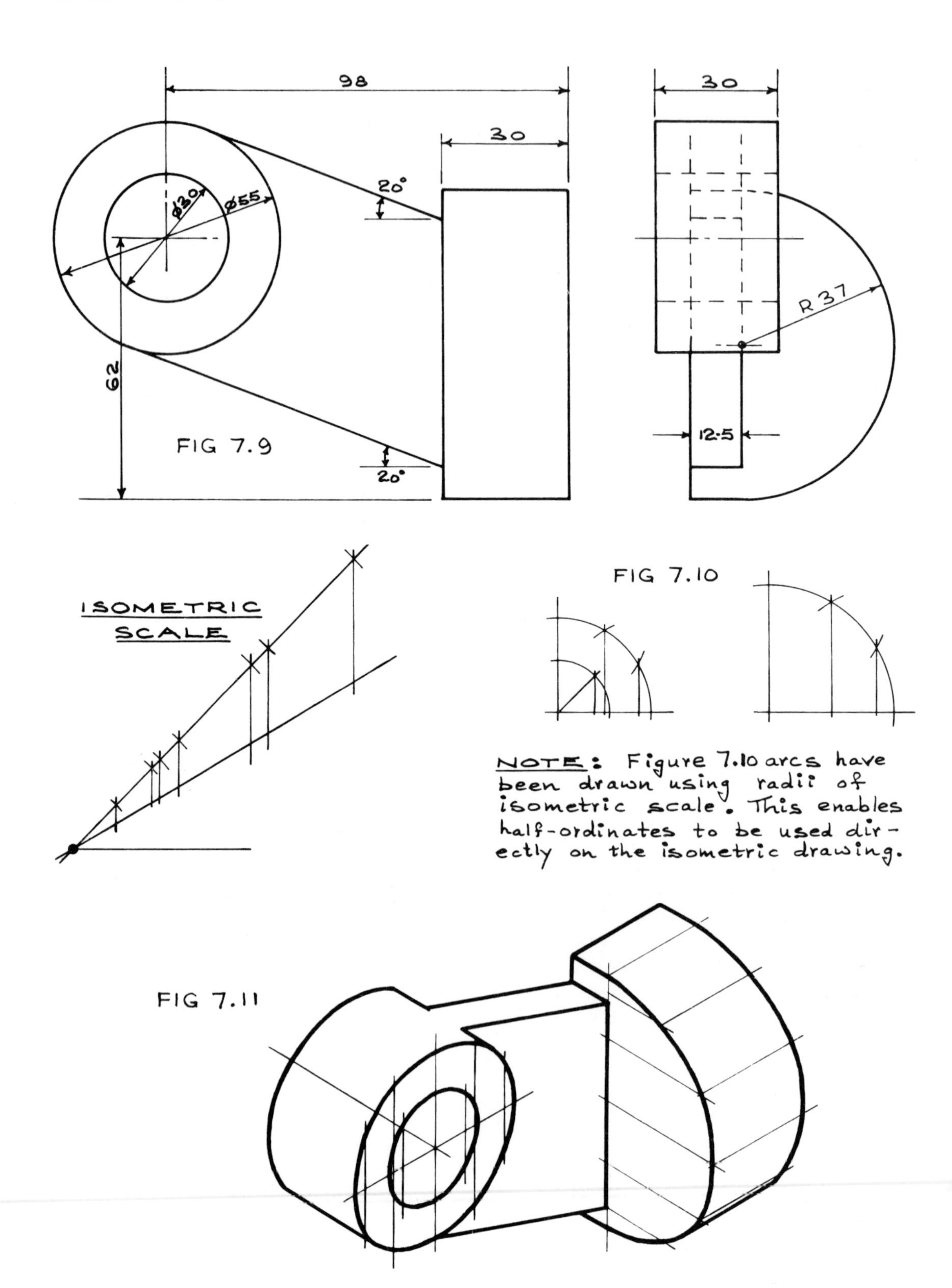

EXERCISE 3 SOLUTION

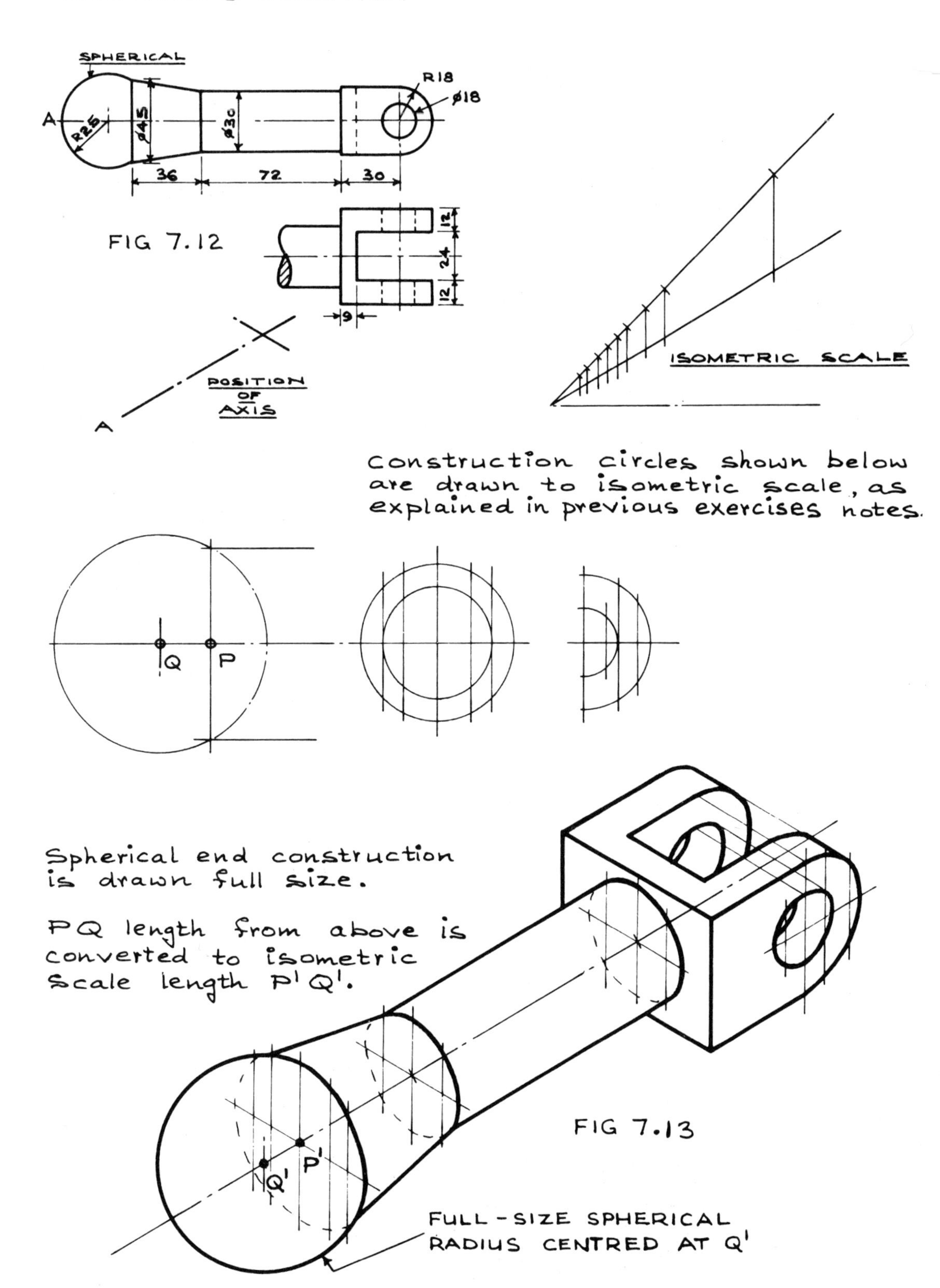

FIG 7.12

FIG 7.13

EXERCISE 4 SOLUTION

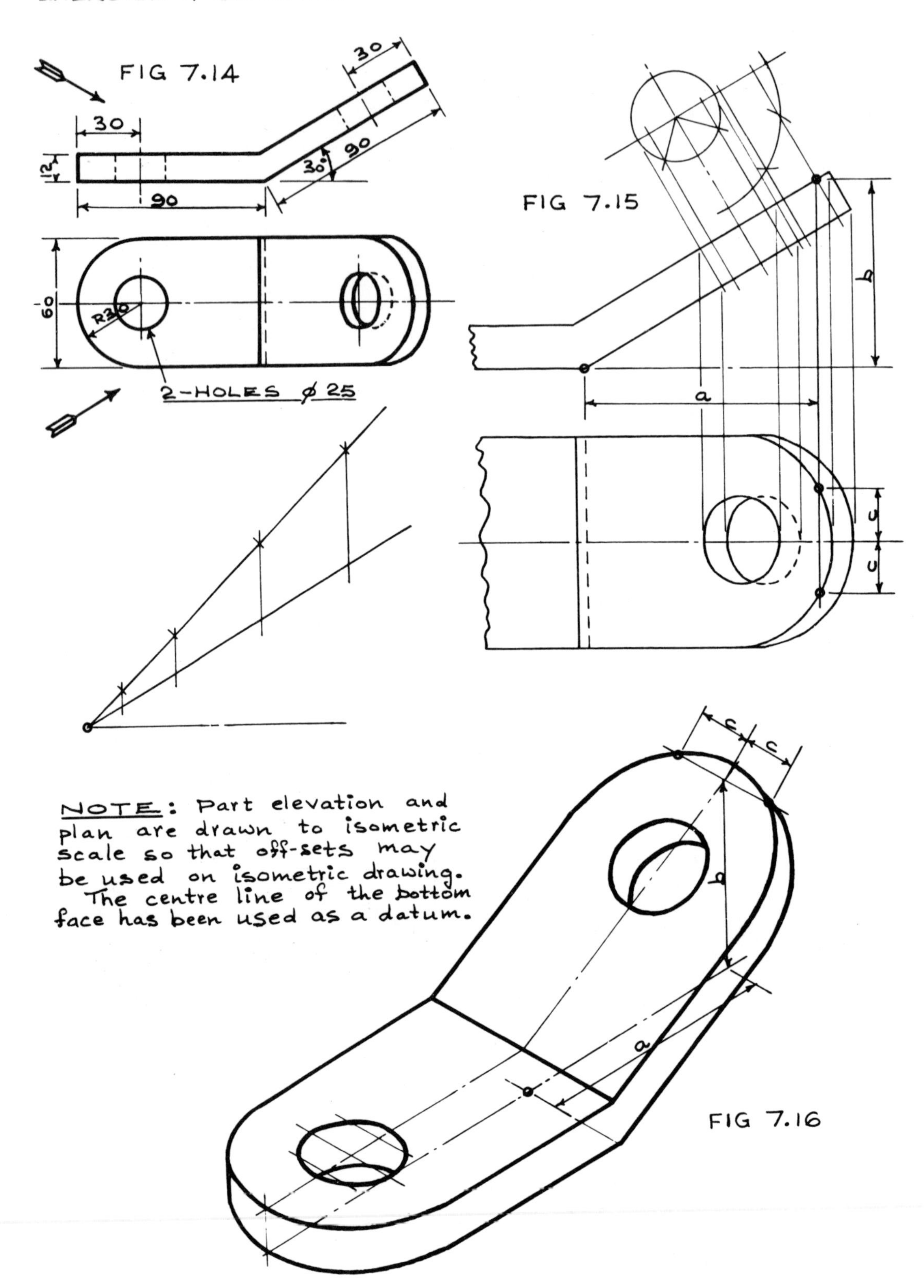

EXERCISE 5 SOLUTION

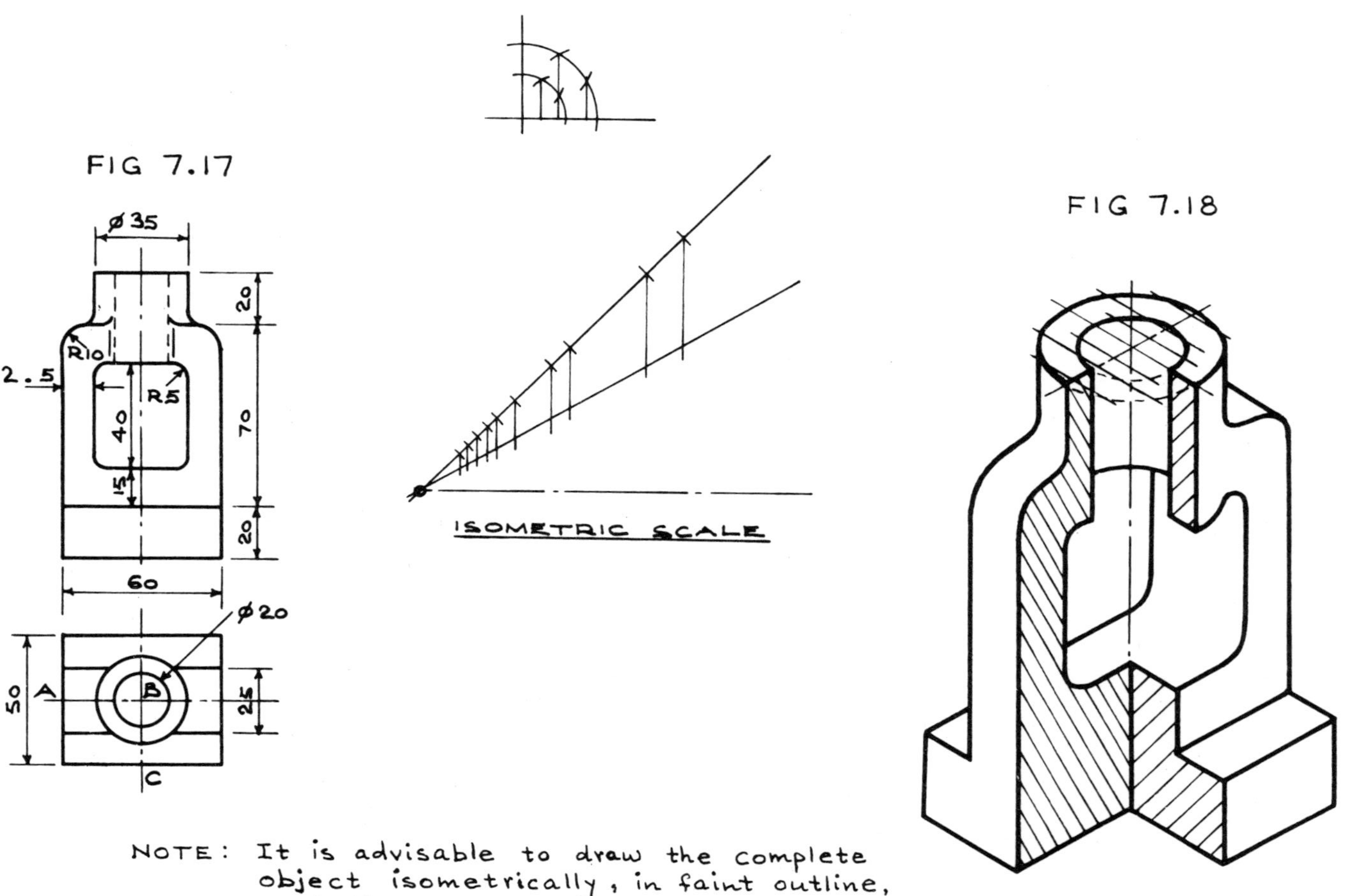

FIG 7.17

FIG 7.18

NOTE: It is advisable to draw the complete object isometrically, in faint outline, and then remove the quarter.

AXONOMETRIC PROJECTION

Axonometric projection is a commonly used metric projection system which is particularly suitable for showing layouts of buildings, plant and equipment, furniture, etc. Architects, factory planners, or interior designers may use this system.

The impression given by this method is usually that of a view looking down on the objects from a high view-point. All horizontal surfaces are drawn as in a true plan view and for this reason the system is sometimes referred to a planometric drawing.

The angles used are complementary, for example, 45° and 45°, or 30° and 60° are commonly employed, but any combination adding up to 90° is suitable.

Vertical lines are made verticals in axonometric projection and all horizontals and verticals are made full-size. Inclined lines, however, must be treated as in isometric projection and inclined surfaces must be drawn using a grid. An inclined circular surface would thus appear elliptical, as in isometric projection.

A simple example of a box, figure 7.19, with lid having a circular hole and opened 180°, though not a good example for axonometric use, serves to illustrate the method of drawing. In this case the very common and convenient equal complementary angles of 45° are used. The circular hole in the lid is drawn entirely with compasses.

A second simple example is of a stool, figure 7.20, and in this case the complementary angles used are 30° and 60°, which are also very convenient.

Part of a kitchen interior layout is shown in figure 7.21 and in this case angles of 45° have been used. As an exercise the student may find it of interest to make a planometric layout of part of a classroom, laboratory, or workshop.

Note. In the examples shown in figures 7.19 to 7.21 the rectangular top surfaces of objects do not appear to be rectangular. In fact they are truly rectangular and this is merely an optical illusion.

FIG 7.19

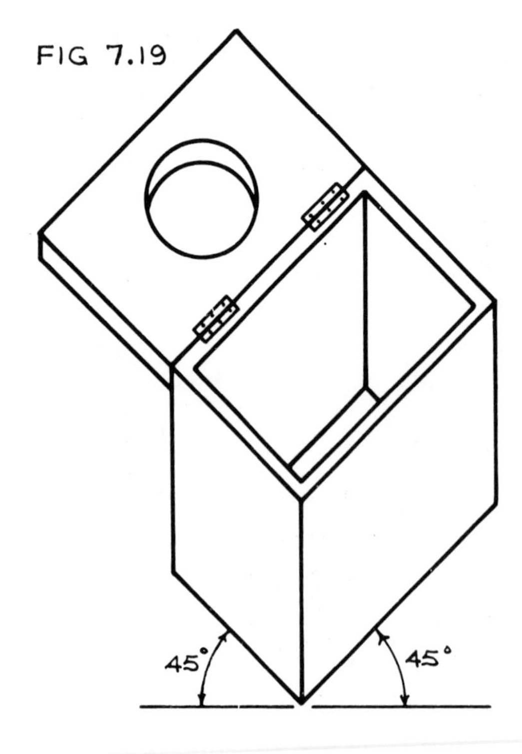

A BOX WITH LID OPEN

FIG 7.20

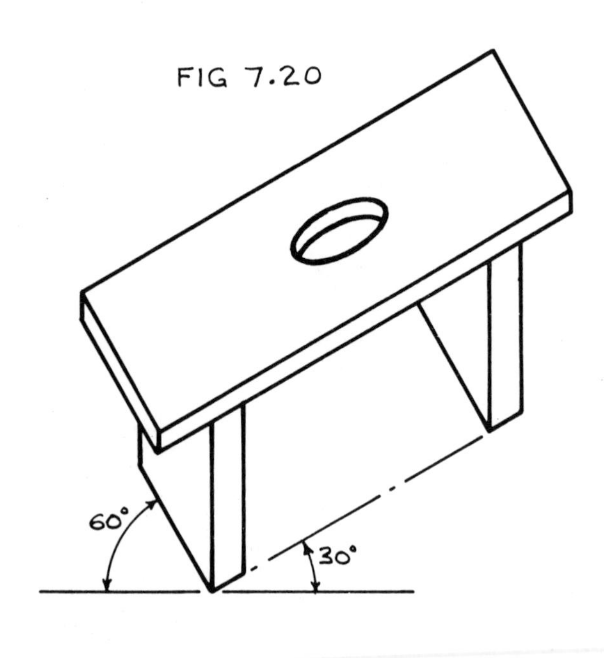

A WOODEN STOOL

Axonometric projection is usually applied to large objects and layouts of buildings, furniture, etc., whereas isometric projection is usually confined to portrayal of relatively small objects such as mechanical engineering components. A more realistic portrayal of a large object, such as a building, is achieved by use of perspective projection. In this latter system, which will be dealt with briefly next in this chapter, parallel edges appear to converge towards a point as in actual vision.

FIG 7.21

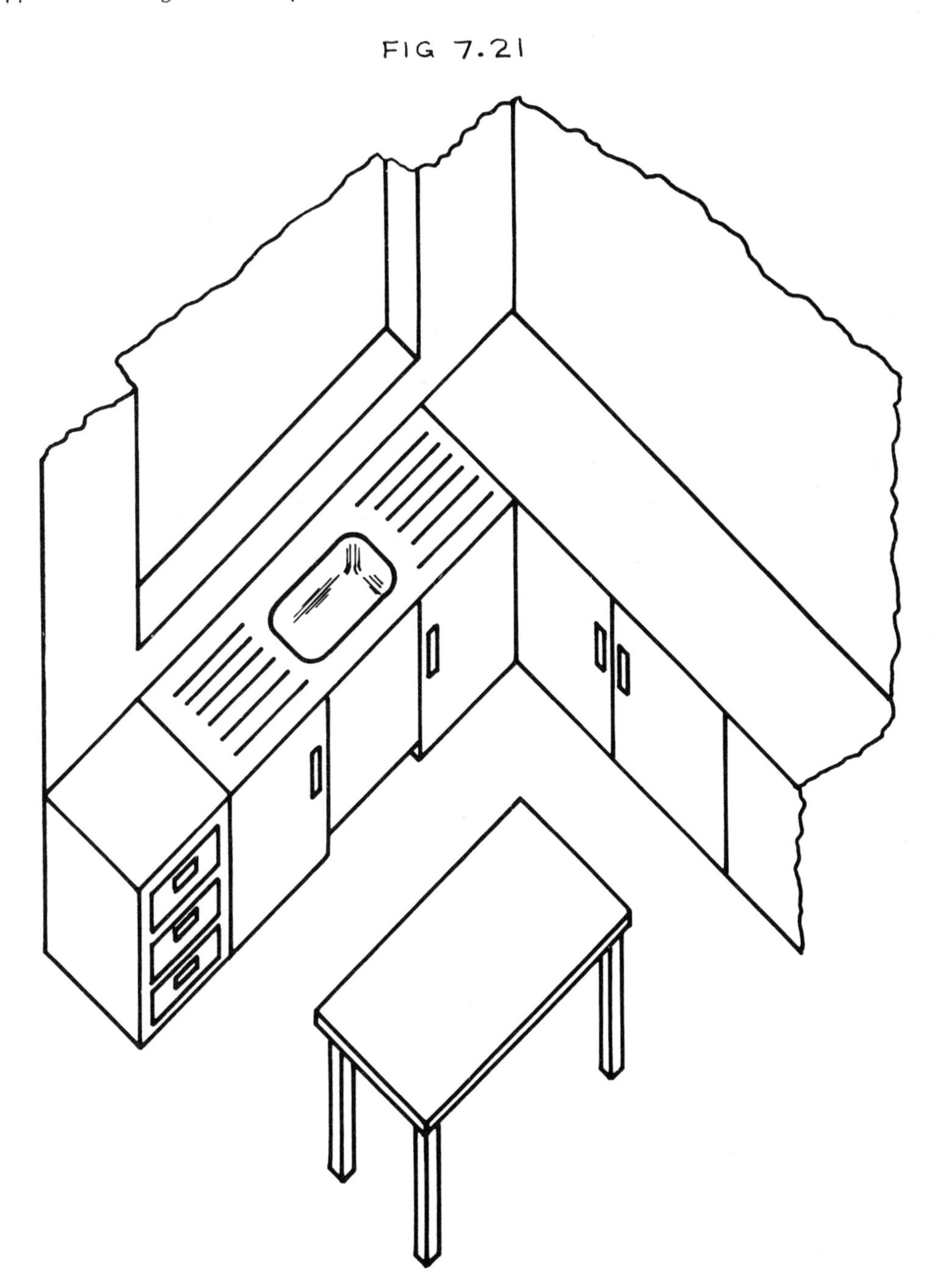

AN AXONOMETRIC PROJECTION SHOWING PART INTERIOR LAYOUT OF A KITCHEN

PERSPECTIVE PROJECTION

Perspective drawing is used by architects to illustrate an object such as a building, more realistically. The principles employed may also be of use in artistic freehand sketching. Briefly, perspective drawing makes use of the fact that the parallel edges or lines of any object appear to converge towards a point. The larger the object the more noticeable is this converging effect. Also, a relatively small rectangular object, such as a box, may exhibit strong convergence of parallel edges when viewed closely in a particular direction. Photographs of buildings, etc., show the 'vanishing parallels' effect clearly.

Books have been written on the theory of the various methods of making perspective drawing and the student is referred to such works for further study of the subject in depth. Only the simplest basic principles and methods will be outlined here in this work. The methods to be referred to, though simple, are nevertheless extremely useful.

ONE-POINT PERSPECTIVE

One-point perspective is the simplest form, that is, where the parallel edges appear to meet at one vanishing point. For example, a long, straight stretch of railway track appears to converge to one distant point on the horizon; a hoarding advertisement of some years ago extolling the magical properties of a well known beverage, depicted a workman carrying a steel girder which, by vanishing parallels principle, appeared to be at least half a kilometre long!

A simple rectangular block shape may be depicted in one point perspective by drawing a front view to start with, figure 7.22, and then choosing a suitable vanishing point. This would be similar to an oblique parallel projection drawing, figure 7.23, except that the vanishing parallels would give the impression of size to the object. On the other hand, the vanishing parallels effect can convey the idea of closeness of point of vision to a rectangular block object and so result in a more natural looking drawing than that obtained by oblique projection. Figure 7.24 shows a one-point perspective view of a room interior.

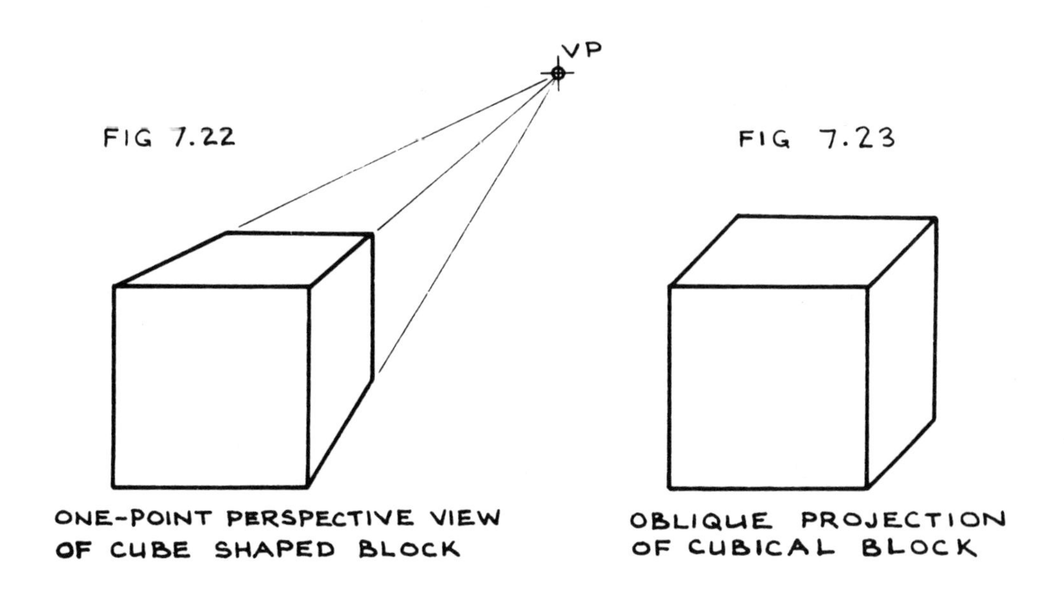

ONE-POINT PERSPECTIVE VIEW OF CUBE SHAPED BLOCK

OBLIQUE PROJECTION OF CUBICAL BLOCK

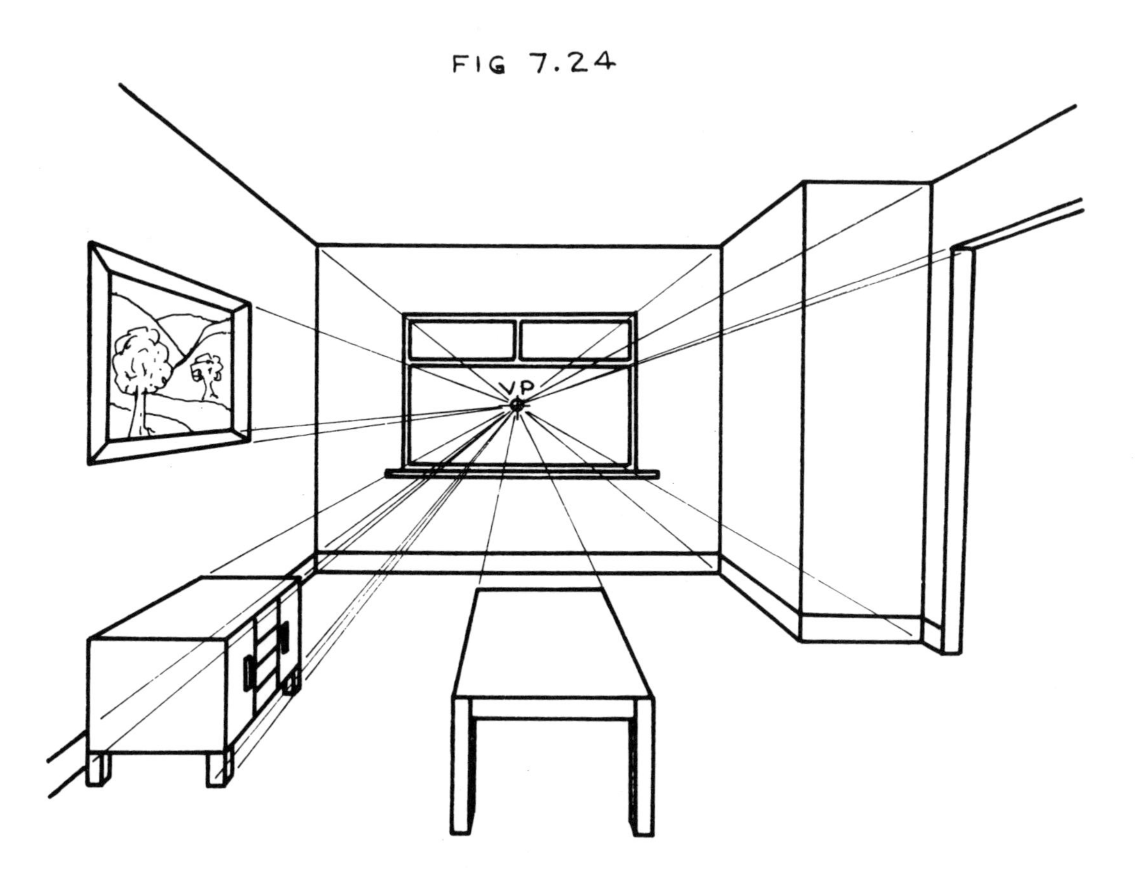

ONE-POINT PERSPECTIVE VIEW OF ROOM INTERIOR

TWO-POINT PERSPECTIVE

In two-point perspective, two vanishing points are used. Again, a simple rectangular object shown in two-point perspective may appear as shown in figure 7.25. The comparable isometric projection of the object is shown in figure 7.26. The perspective view again adds realism. The line joining the vanishing points, in figure 7.25, is the horizon line and corresponds with the eye level. Various perspective views may be obtained as illustrated in figure 7.27.

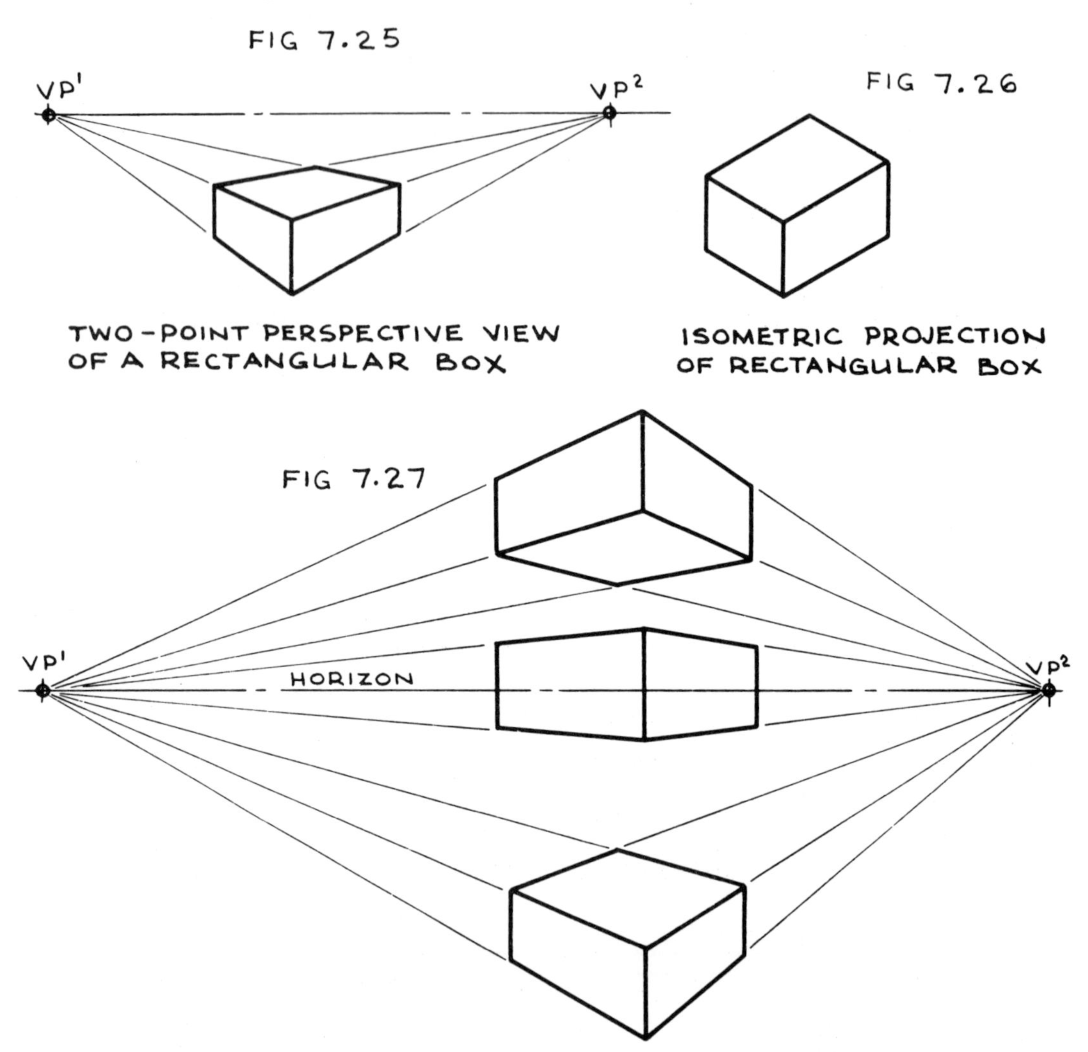

TWO-POINT PERSPECTIVE VIEW OF A RECTANGULAR BOX

ISOMETRIC PROJECTION OF RECTANGULAR BOX

VARIOUS TWO-POINT PERSPECTIVE VIEWS OF AN OBJECT

PROCEDURE FOR EXERCISE 9 :
Draw scale front view, side view and plan of bungalow.
Transfer plan on to tracing paper.
Set off angle of vision, approximately 57° in figure 7.34.
Place plan suitably to accommodate this angle and transfer tracing to drawing sheet in this position.
Choose picture plane, base or ground level and eye level lines. [See note referring to figure 7.28]
Complete by procedure set out for exercise 8 .

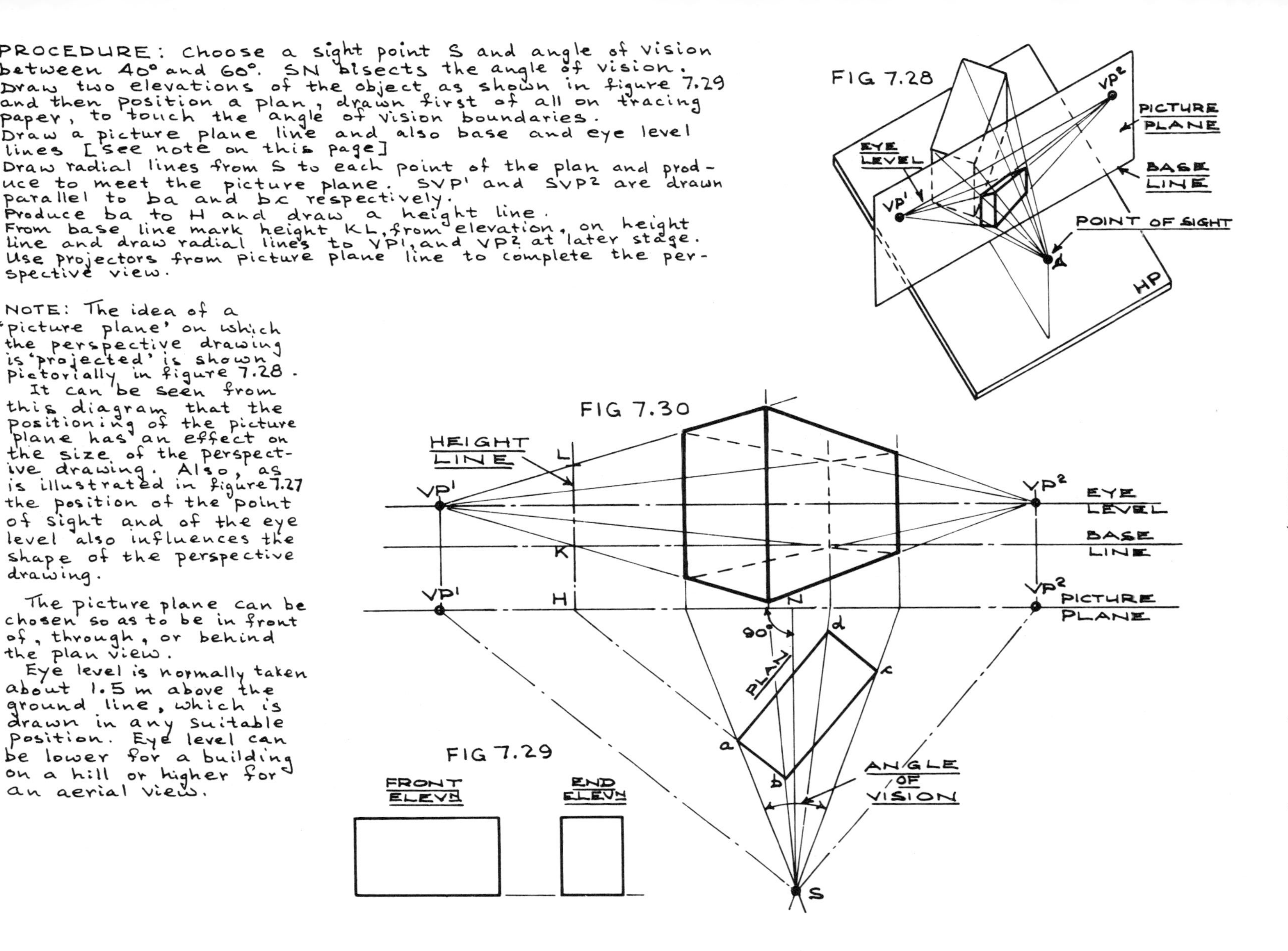
PROCEDURE: Choose a sight point S and angle of vision between 40° and 60°. SN bisects the angle of vision.
Draw two elevations of the object as shown in figure 7.29 and then position a plan, drawn first of all on tracing paper, to touch the angle of vision boundaries.
Draw a picture plane line and also base and eye level lines [see note on this page]
Draw radial lines from S to each point of the plan and produce to meet the picture plane. SVP¹ and SVP² are drawn parallel to ba and bc respectively.
Produce ba to H and draw a height line.
From base line mark height KL, from elevation, on height line and draw radial lines to VP¹, and VP² at later stage.
Use projectors from picture plane line to complete the perspective view.

NOTE: The idea of a 'picture plane' on which the perspective drawing is 'projected' is shown pictorially in figure 7.28.

It can be seen from this diagram that the positioning of the picture plane has an effect on the size of the perspective drawing. Also, as is illustrated in figure 7.27 the position of the point of sight and of the eye level also influences the shape of the perspective drawing.

The picture plane can be chosen so as to be in front of, through, or behind the plan view.

Eye level is normally taken about 1.5 m above the ground line, which is drawn in any suitable position. Eye level can be lower for a building on a hill or higher for an aerial view.

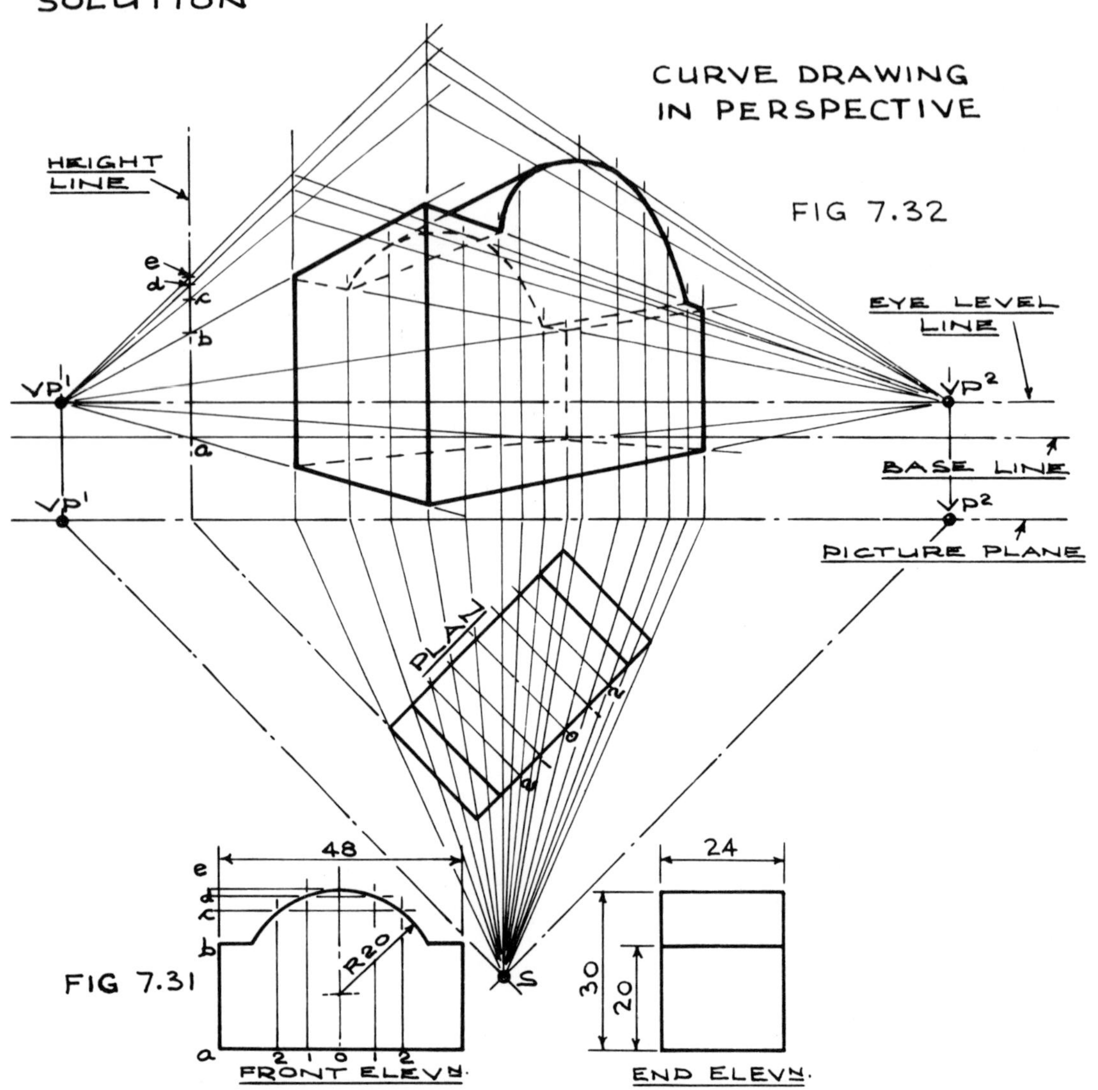

PROCEDURE FOR EXERCISE 8 :

The method is similar to the previous example of the rectangular box shaped object. The curve is dealt with as follows:

Draw ordinates in the front view and transfer spacings to plan. The ordinate heights are marked along the height line and radial lines from VP¹ through a, b, c, d and e are drawn.

Radial lines are drawn from sight point S through each point of the plan and produced to meet the picture plane.

Projectors from the picture plane points to meet radial lines through VP2 enable the perspective view to be completed.

NOTE: Exercise 9 procedure is given following figure 7.27

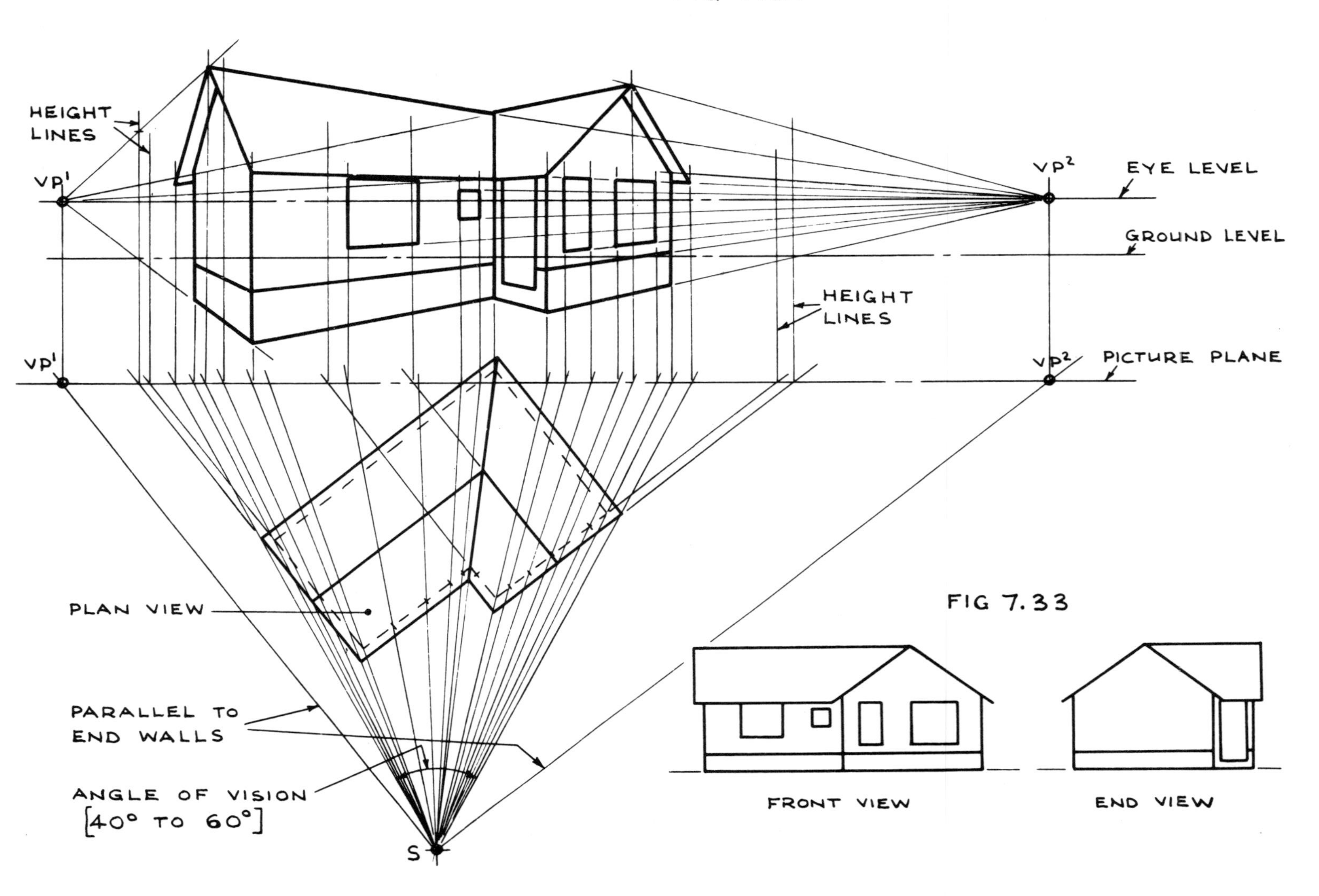

FIG 7.34

FIG 7.33

8
ENGINEERING DRAWING NOTES AND EXERCISES

FREEHAND SKETCHING

Freehand sketching is an important part of the training of the professional Engineer or Architect. Much preliminary work is greatly facilitated by free use of sketching. Again, it frequently happens that the engineer is at the job site without drafting facilities and it is necessary for him to be able to execute a freehand drawing on his sketch pad. This sketch must be complete with dimensions and all relevant information to enable him to prepare a finished drawing on return to the office.

The two types of sketching most used are:

(i) Orthographic views, sometimes in section.
(ii) Pictorial or isometric views.

Occasionally it may be found convenient to sketch in Oblique, Axonometric, or even Perspective projection.

Note. Basic principles of the foregoing types of projection are explained in other chapters of this book.

It is not necessary to be artistically gifted to produce good freehand sketches. If the principles of orthographic and isometric projection are clearly understood, then it is a matter of making these drawings without the aid of drafting equipment.

Some hints on sketching may be of value at this stage:

(i) All lines should be sketched lightly first of all in order to achieve reasonably correct proportions. Centre lines are important preliminary lines.

(ii) When drawing preliminary lines, straight or curved, make them broken and not full lines.

(iii) Take a long hold of the pencil for initial faint lining and never attempt to finish lining in curves against the natural sweep of hand or arm. Instead, turn the sketch pad round or, if working at a fixed board, move around the board.

Figures 8.1 to 8.10 illustrate the various points already referred to.

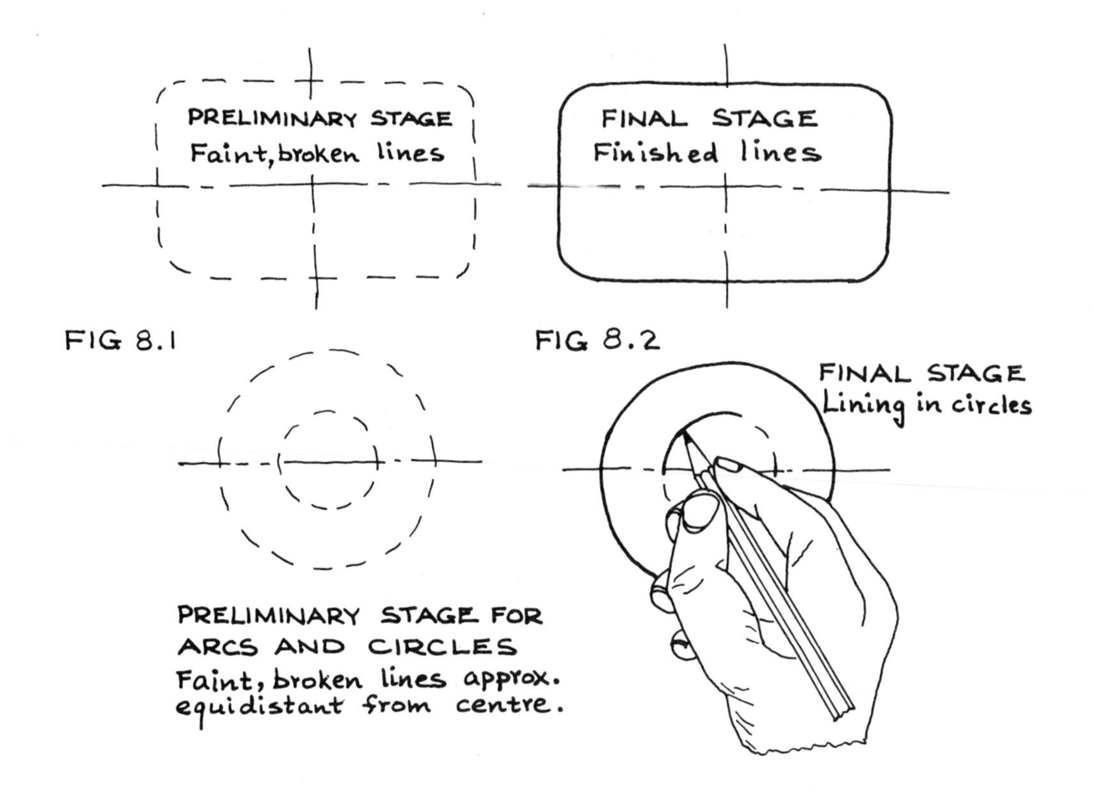

FIG 8.1

FIG 8.2

ISOMETRIC SKETCHING OF CIRCLES

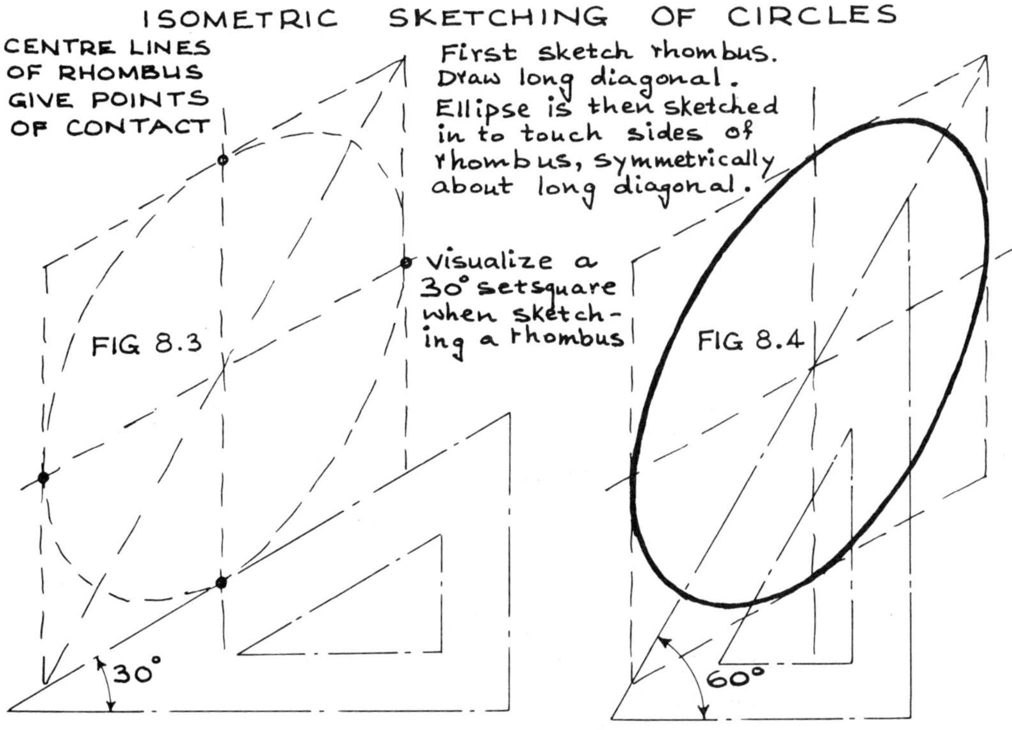

FIG 8.3

FIG 8.4

NOTE : SIDE OF RHOMBUS = DIAMETER OF CIRCLE

ISOMETRIC SKETCHING OF CIRCLES

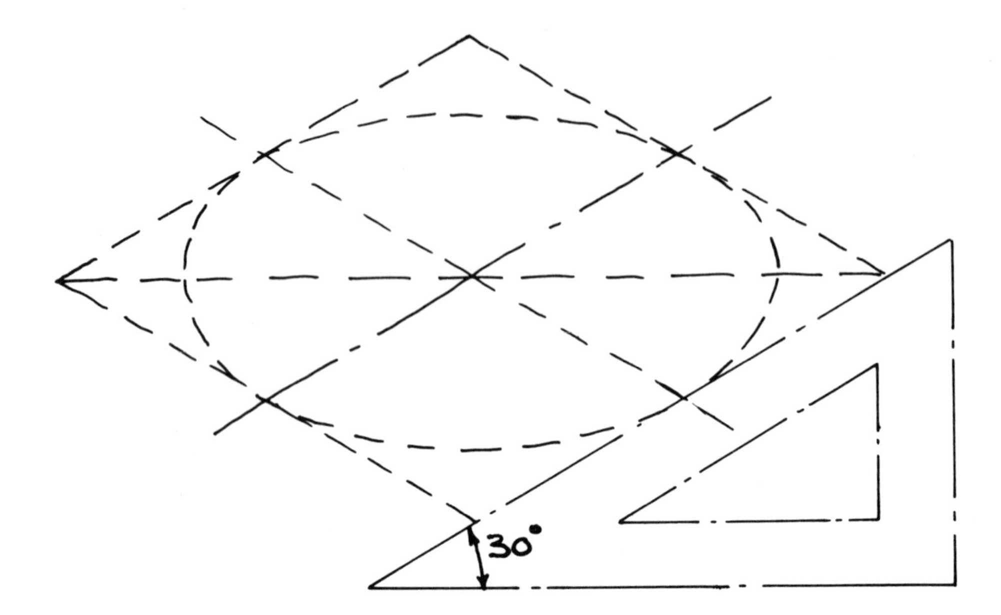

FIG 8.5

It may be found helpful to visualize a 30° set square when sketching a rhombus.

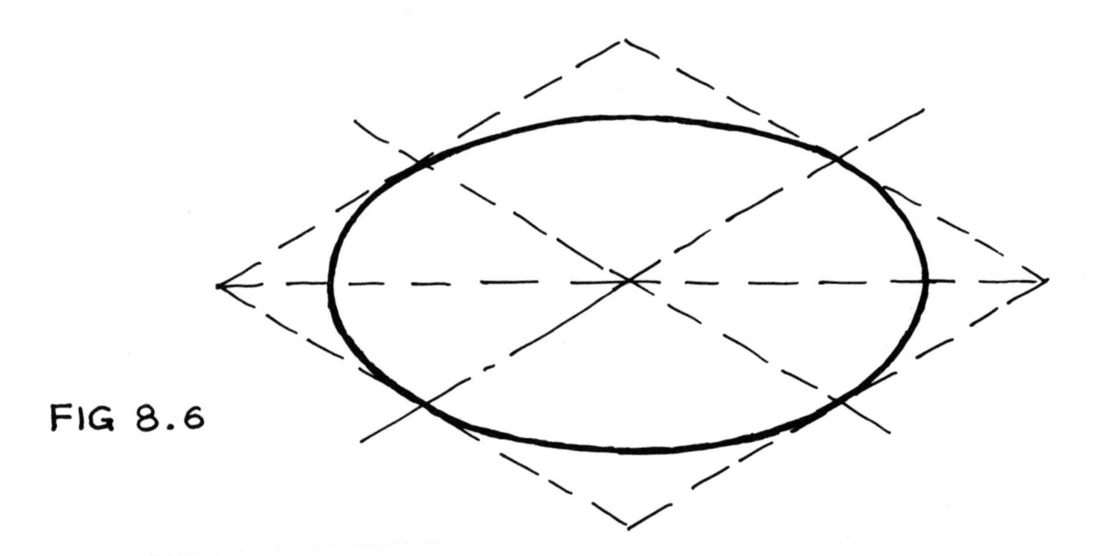

FIG 8.6

Sketched ellipse is symmetrical about the long diagonal, which is horizontal in this case.

FREEHAND SKETCH EXAMPLE

THIRD ANGLE PROJECTION (DIMENSIONS OMITTED)

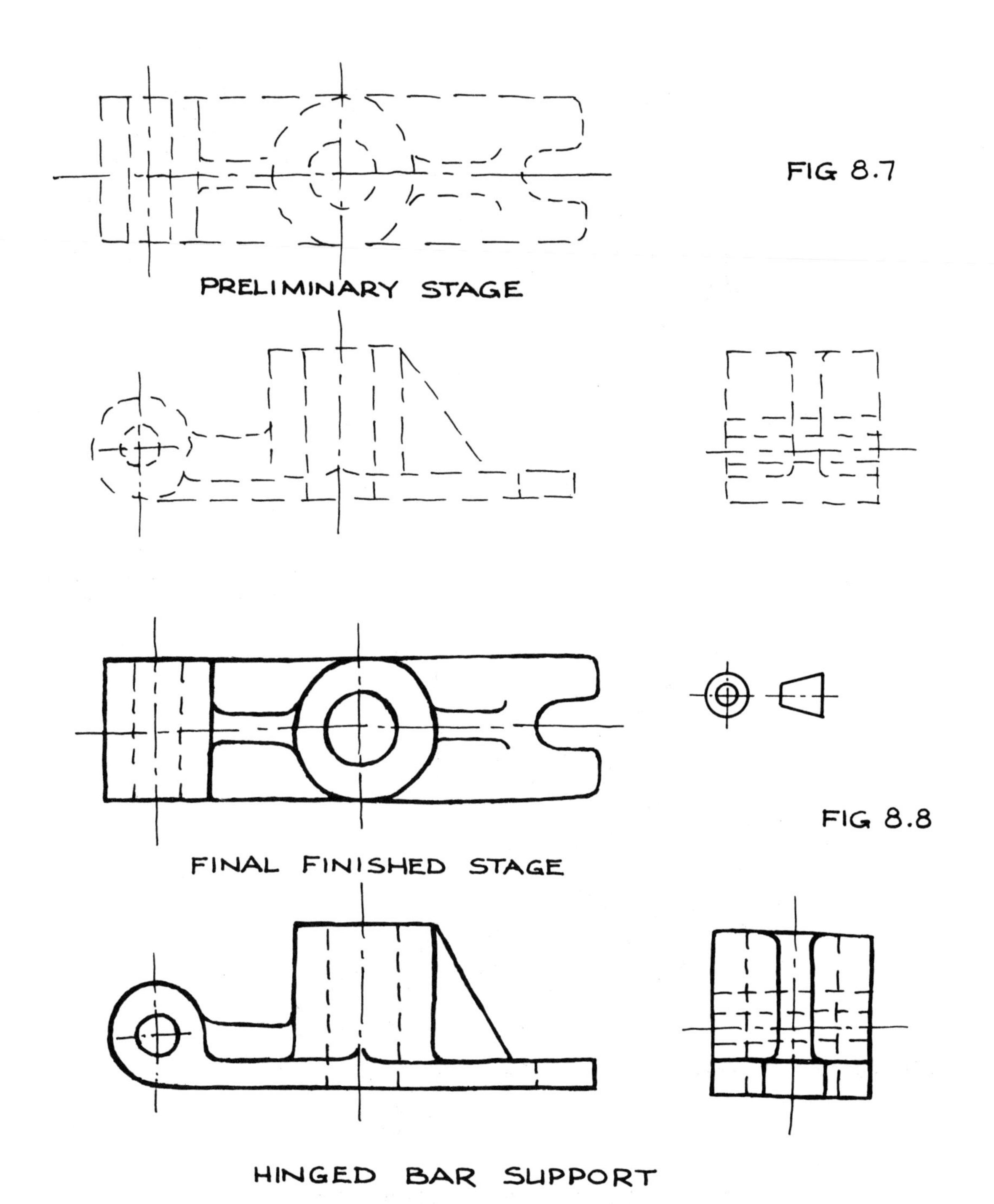

HINGED BAR SUPPORT

FREEHAND SKETCH EXAMPLE

ISOMETRIC DRAWING
(DIMENSIONS OMITTED)

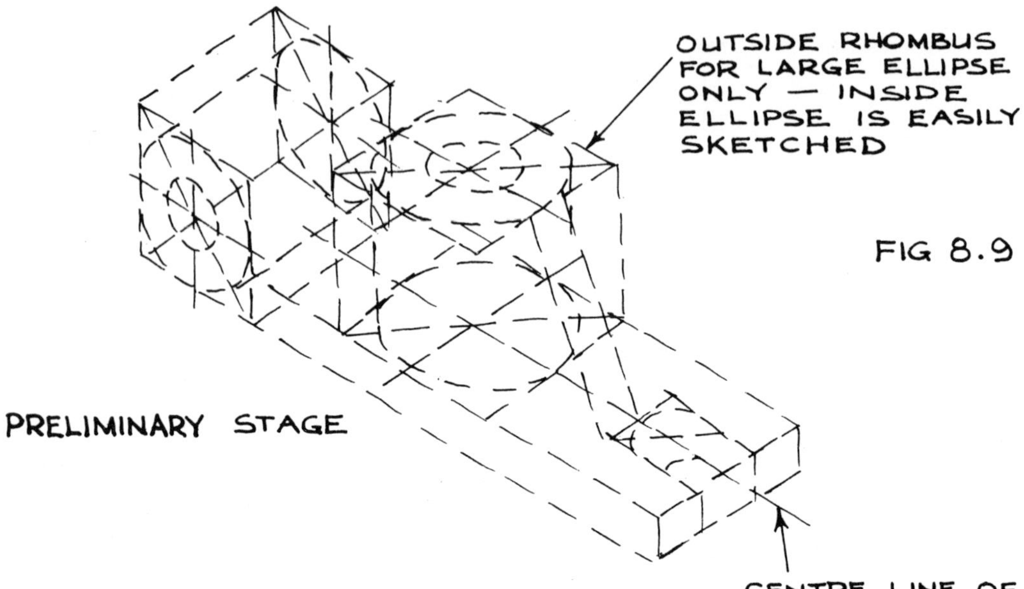

FIG 8.9

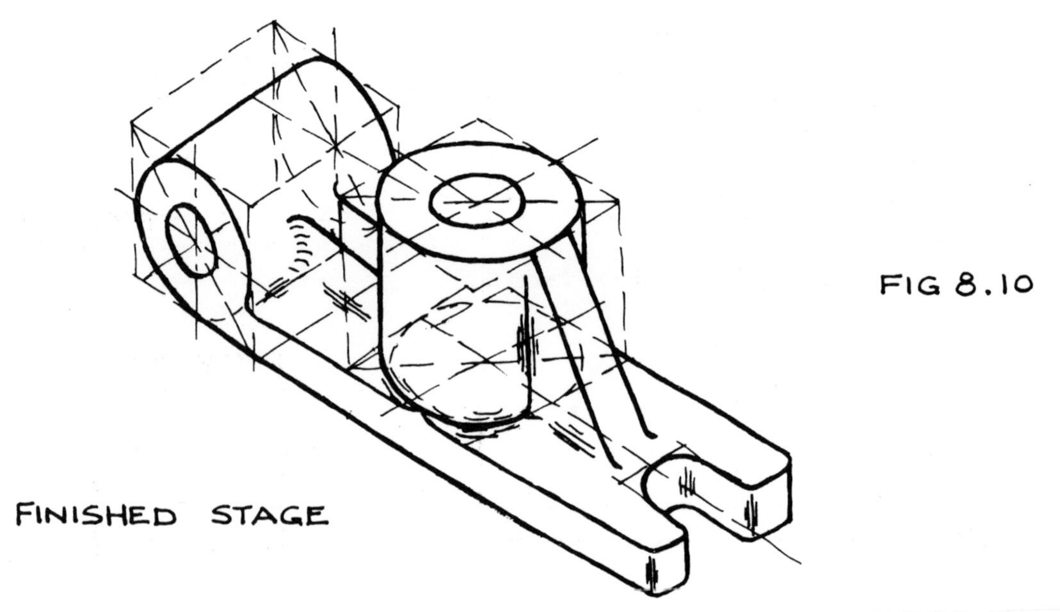

FIG 8.10

HINGED BAR SUPPORT

ORDER OF DRAWING PROCEDURE

The stages will be explained with reference to the light alloy casting of a FULCRUM BRACKET, which is here sketched, Fig. 8.11.

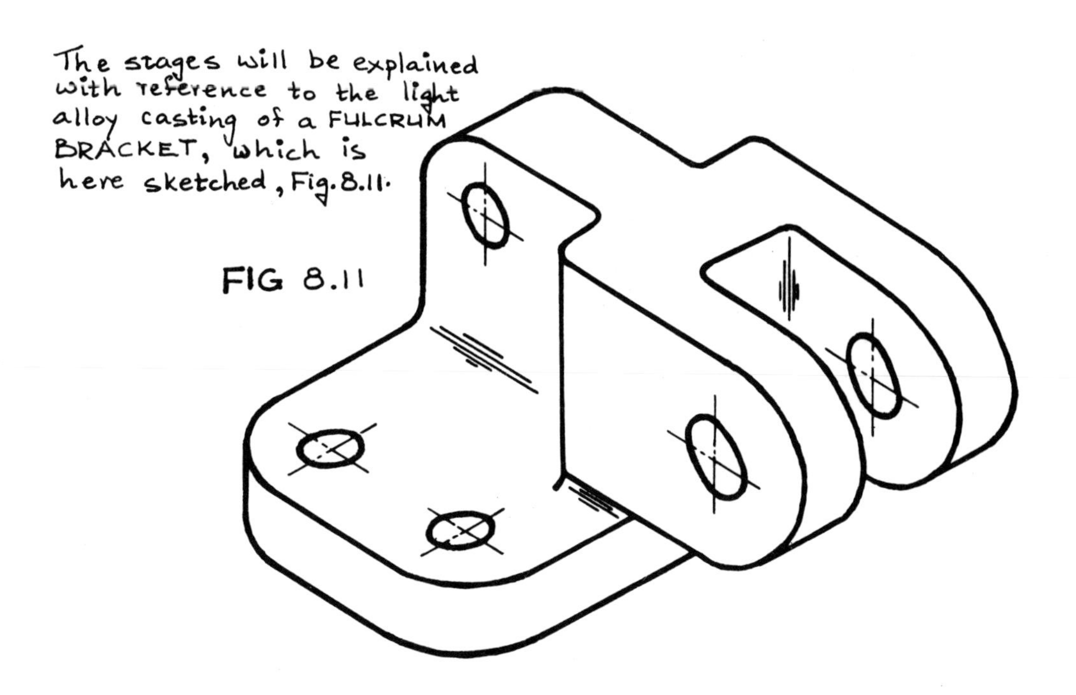

FIG 8.11

The first stage is to make a freehand sketch of suitable views as shown in Fig. 8.12. Views shown are in Third Angle projection.

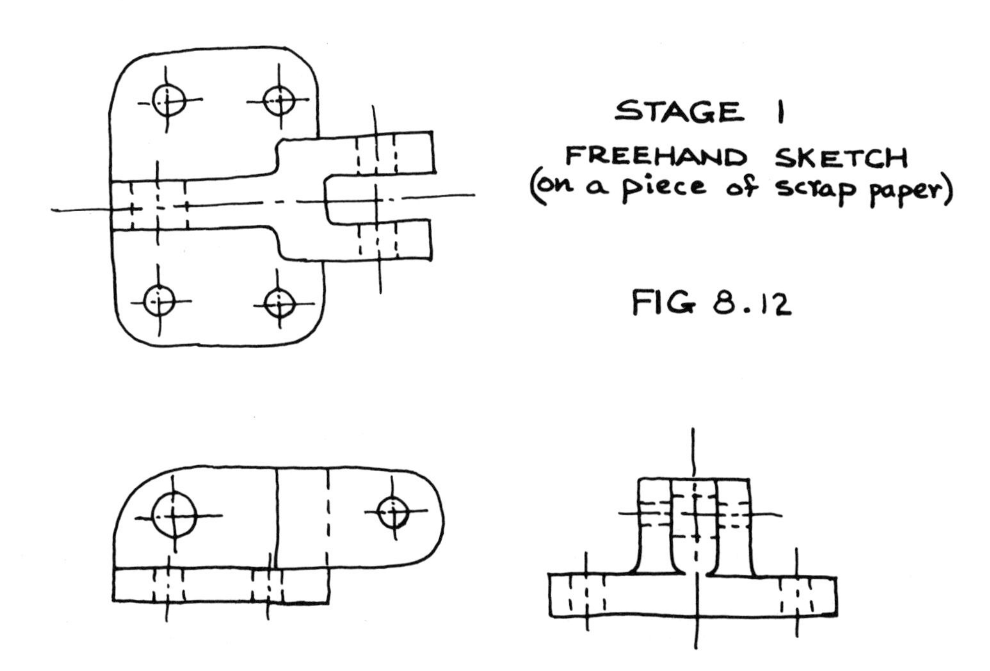

FIG 8.12

ORDER OF DRAWING PROCEDURE

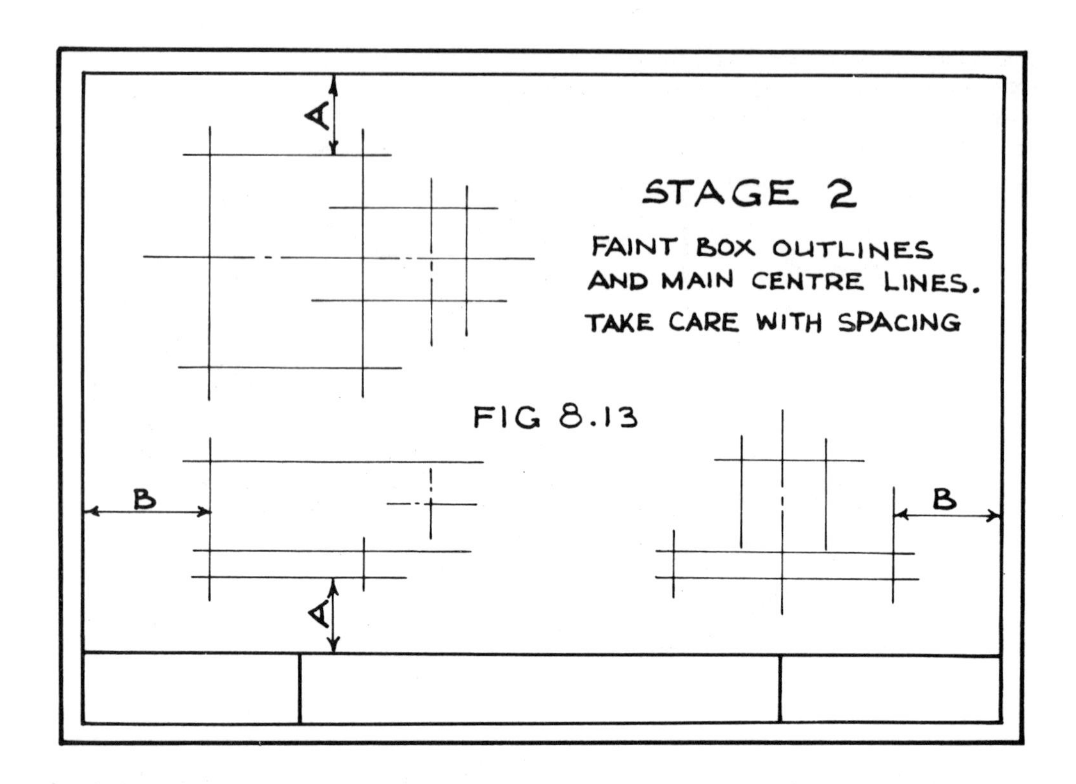

FIG 8.13

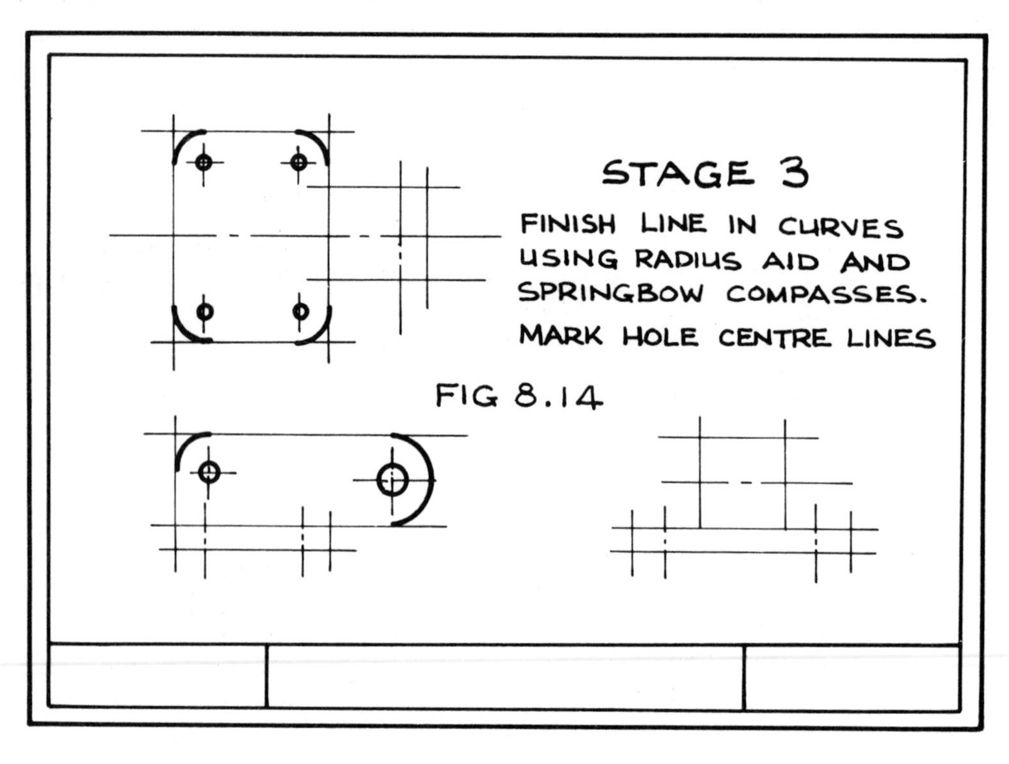

FIG 8.14

ORDER OF DRAWING PROCEDURE

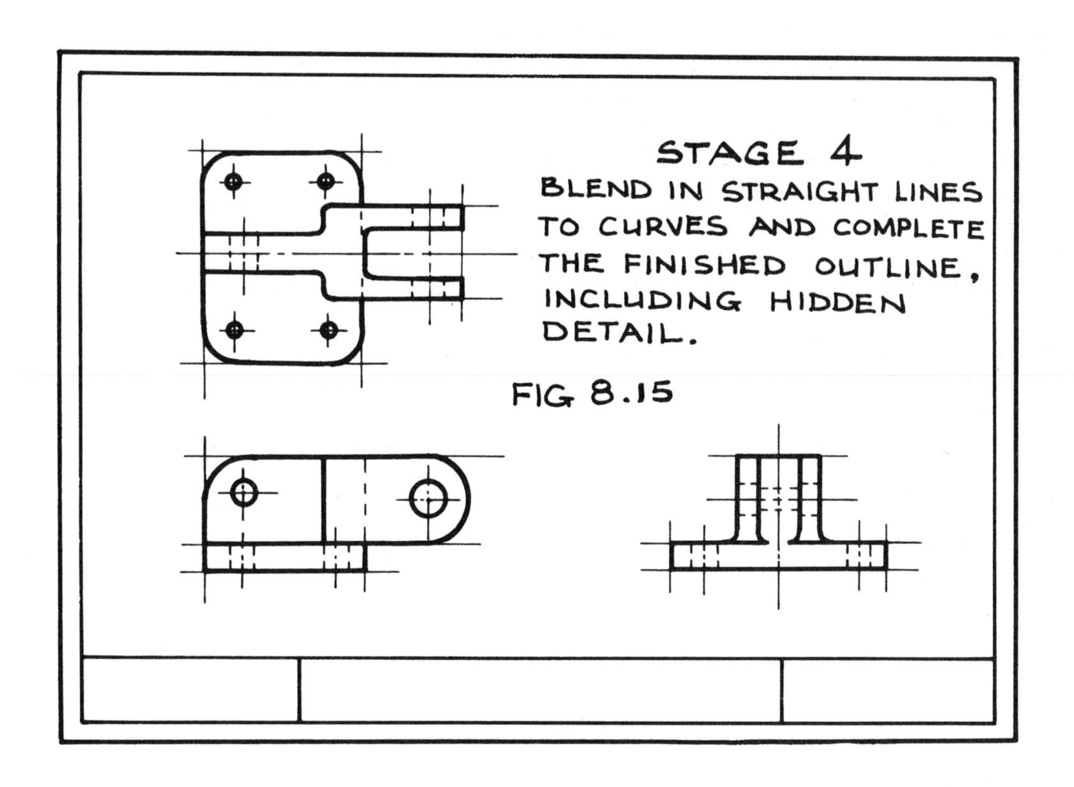

FIG 8.15

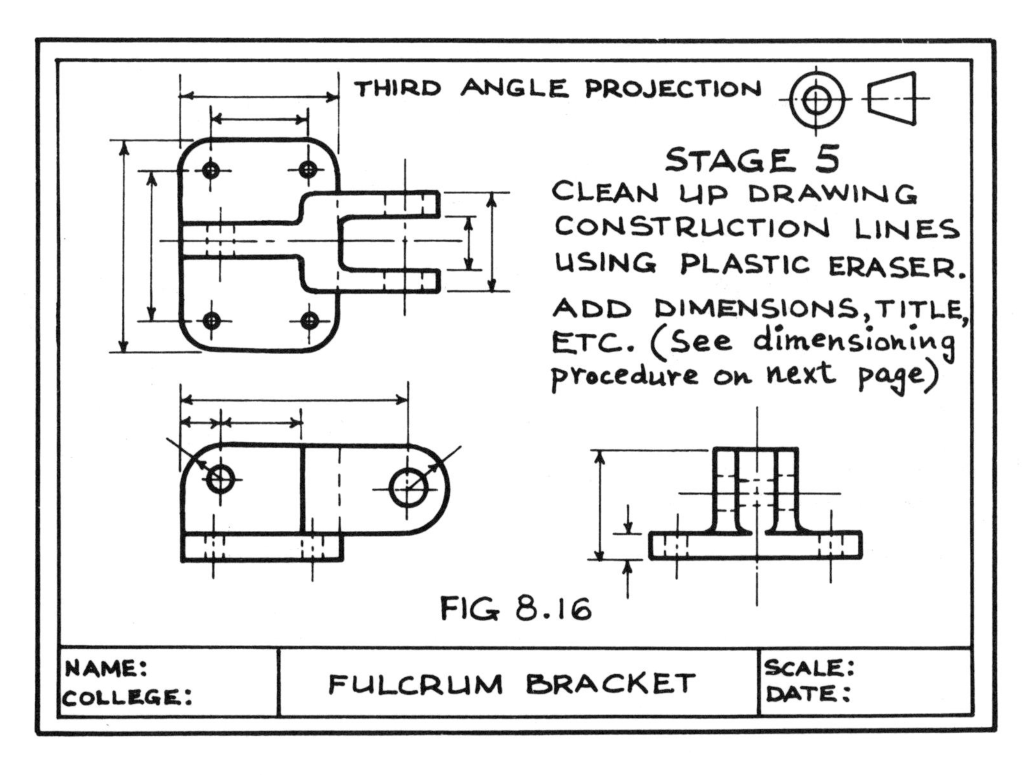

FIG 8.16

DIMENSIONING AN ENGINEERING DRAWING

The following general points should be noticed about the dimensioning of a drawing:

(1) Where possible, dimension lines are kept clear of the drawing.
(2) Dimension lines and projector lines are thin lines (2H pencil) and arrowheads are small.
(3) It is important to leave a space between projector lines and main outlines of the drawing.
(4) Dimension figures are kept clear of the dimension line.
(5) Overall dimensions are placed outside intermediate dimensions.
(6) If all intermediate dimensions are given in 'chain' form then the overall dimension, which is really redundant but useful for information, is placed in brackets.
(7) Dimension lines for circles or arcs are never placed to coincide with centre lines.
(8) Notice the different ways of dimensioning large circles, smaller circles and very small circles. Similarly for radii.
(9) Notice the dimensioning of small spaces.
(10) Figures of the dimensions are placed so as to read from the bottom or from the right-hand side of the drawing sheet.
(11) The symbols ϕ, R or □, placed before the figures, denote *diameter*, *radius* or *square*, respectively.

Note. The foregoing general rules for dimensioning are illustrated by figure 8.17, with the exception of the symbol for a square.

Figure 8.18 illustrates the staggering of dimension figures for greater clarity and particularly the use of the symbol for diameter.

FIG 8.17

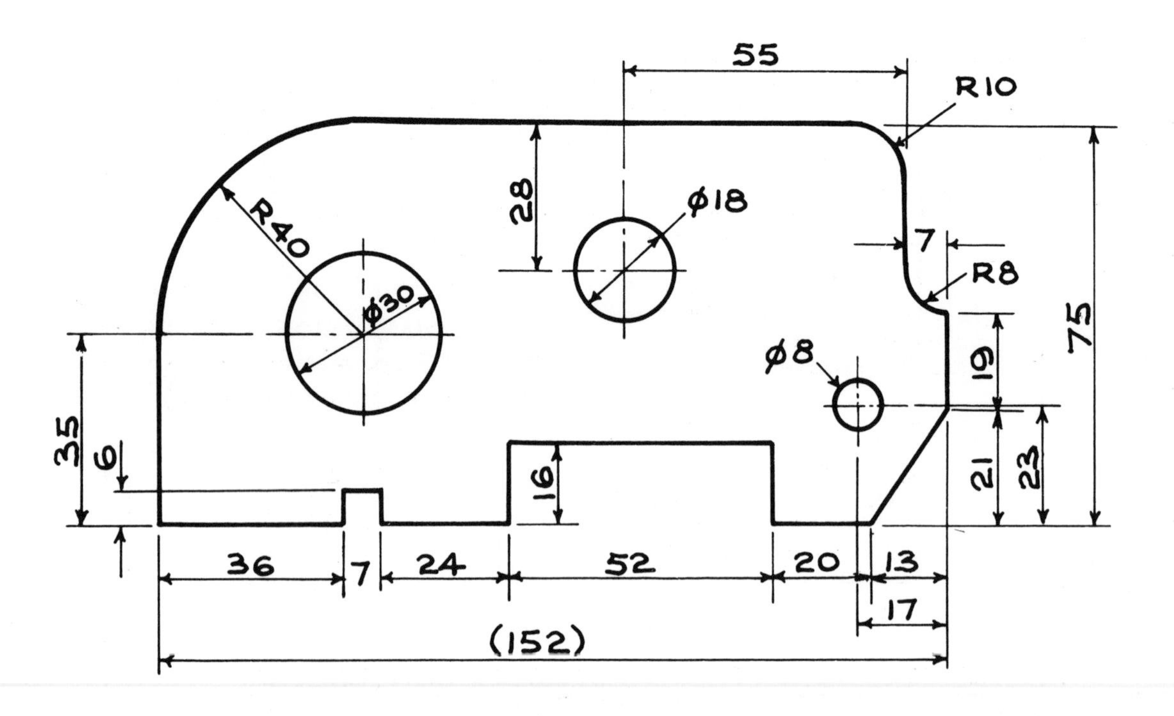

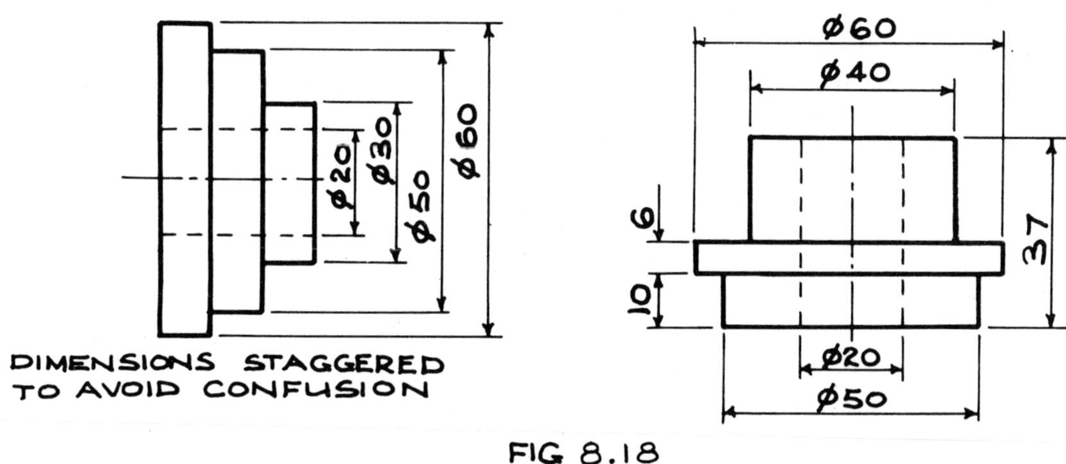

FIG 8.18

STAGGERED DIMENSIONS AND USE OF SYMBOL ø

Sometimes *datum dimensioning* is used in preference to the common method of *chain dimensioning*. The reason for this is that errors in manufacture from a chain dimensioned drawing are cumulative and hence for a component which has to be manufactured to very close tolerances, datum dimensioning is preferable.

Datum dimensioning is illustrated in figure 8.19.

Note. The student is referred to the British Standards Institution publication BS 308 1972, Part 2, for a detailed treatment of dimensioning conventions.

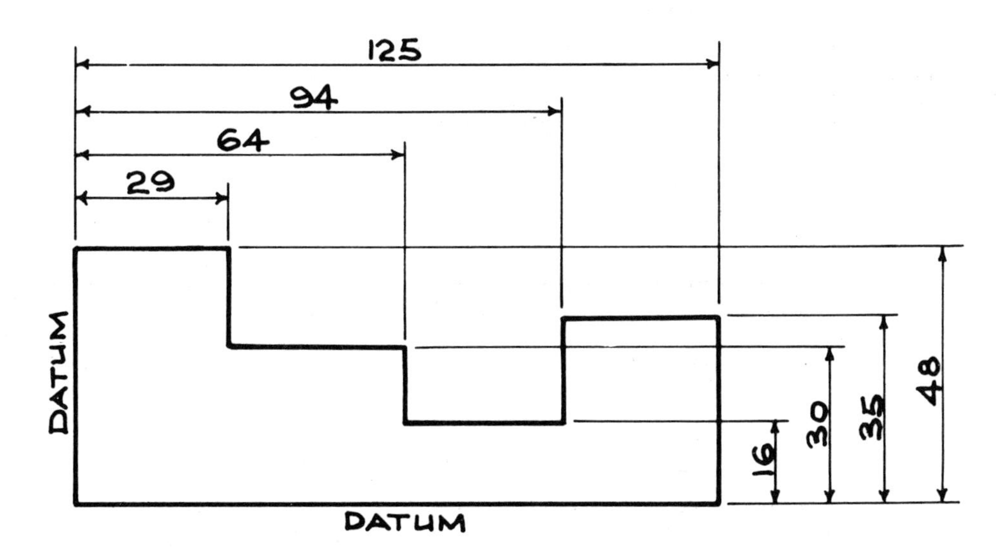

FIG 8.19

DATUM DIMENSIONING

SECTIONING

Sectioning is used to reveal the inside of a component or an assembly of components.

In the example shown, if one imagines a saw cut to divide the *slide block* in halves then when the right-hand half only is viewed in the direction of the arrow 'A' the result will be as shown in the sectioned side view of figure 8.21.

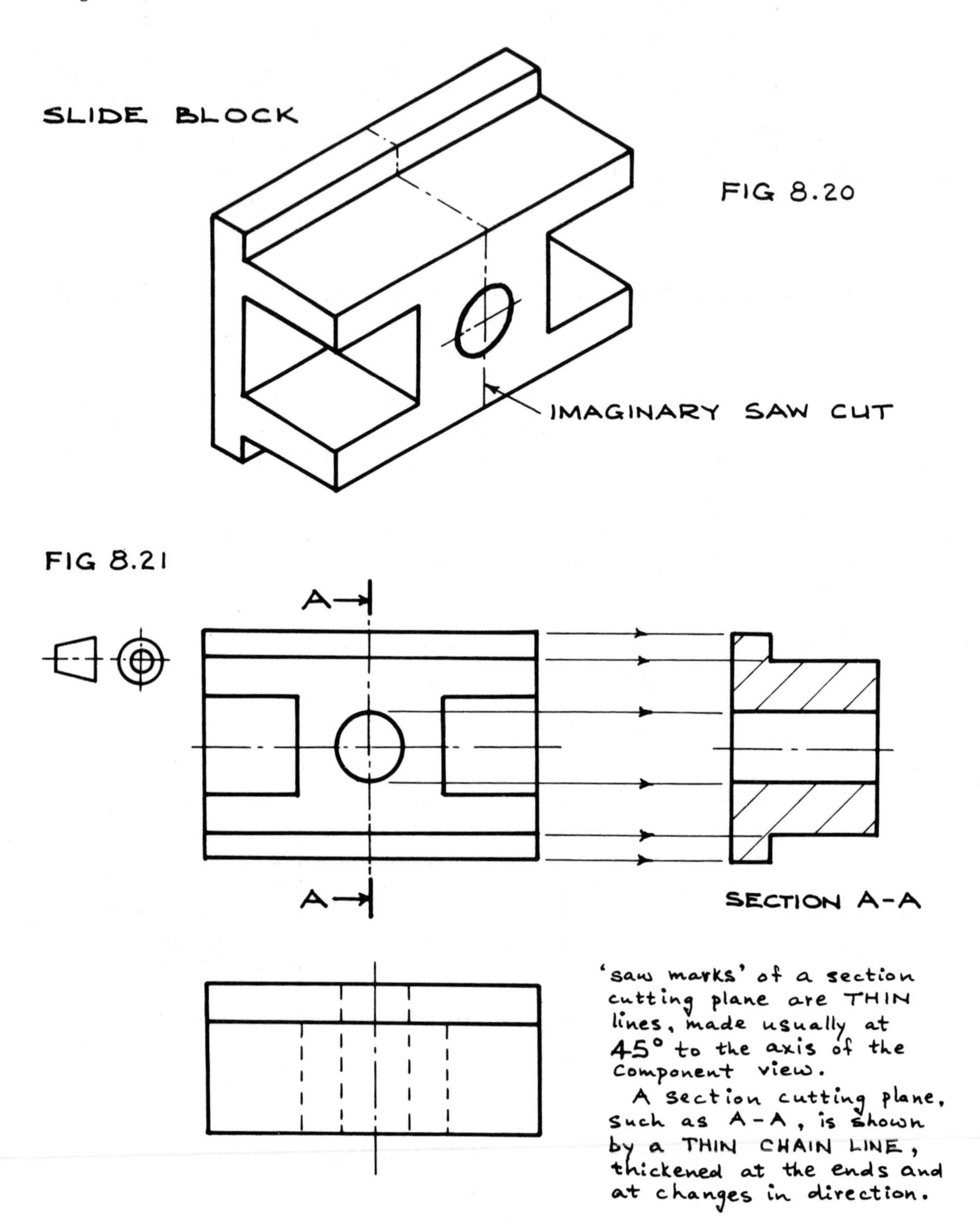

POINTS TO OBSERVE ON SECTIONING

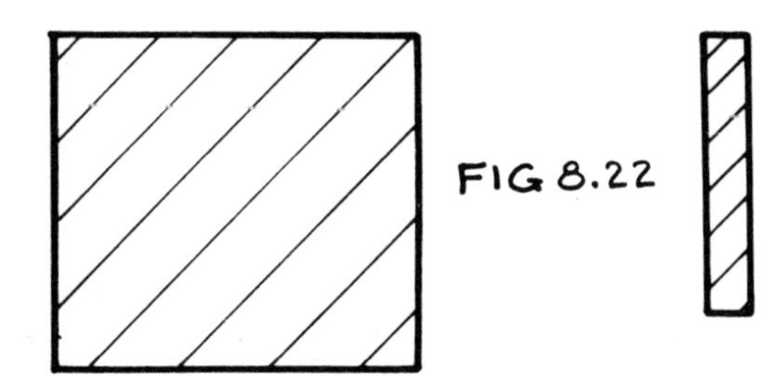

The spacing of hatching lines depends on the size of the area being shaded.

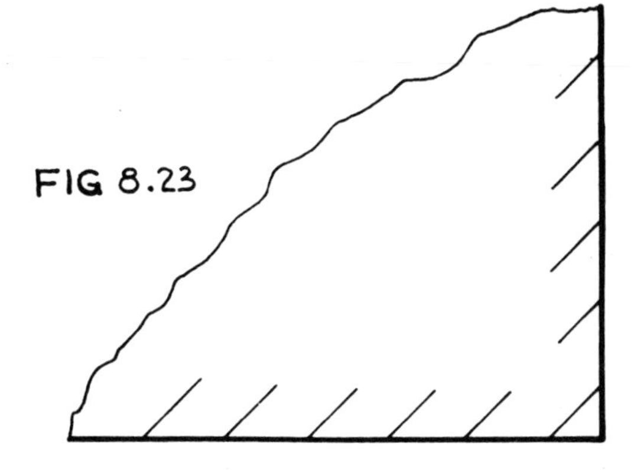

For a large area the hatching may be limited to a zone following the contour of an adjacent sectioned portion.

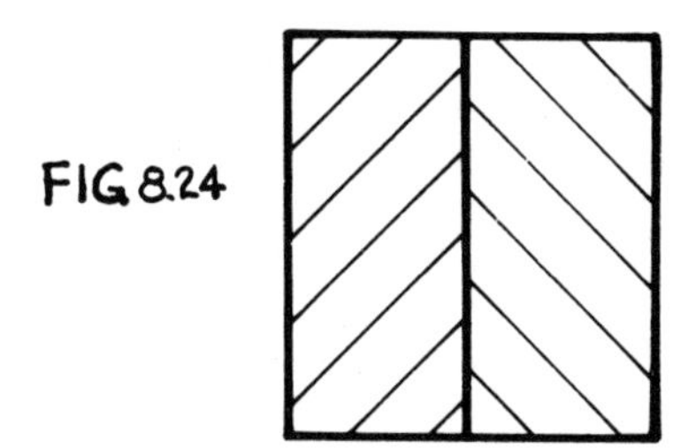

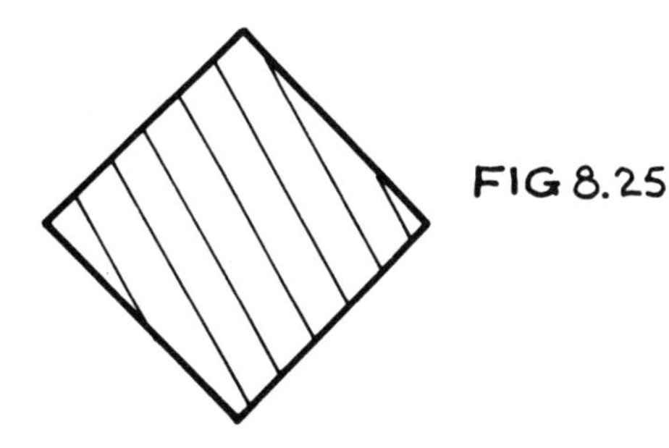

Adjacent parts are hatched in opposite directions.

For a component tilted at 45°, another angle, greater or less than 45°, may be used for hatching.

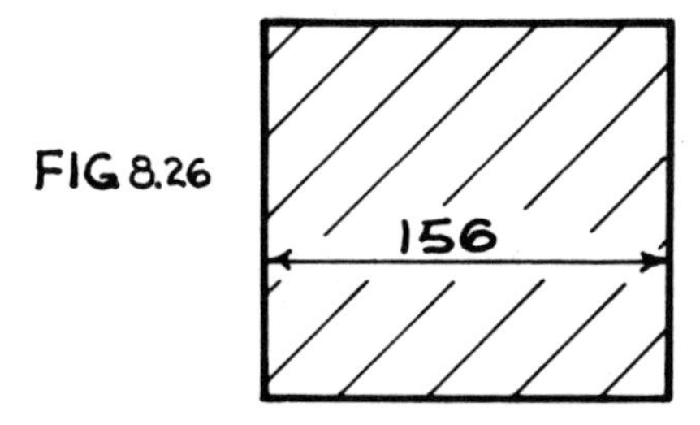

Where possible, dimension lines should be kept clear of the drawing, but if it is necessary to put a dimension on a sectioned part make sure that hatching is omitted in way of dimension line and figures.

POINTS TO OBSERVE ON SECTIONING

FIG 8.27

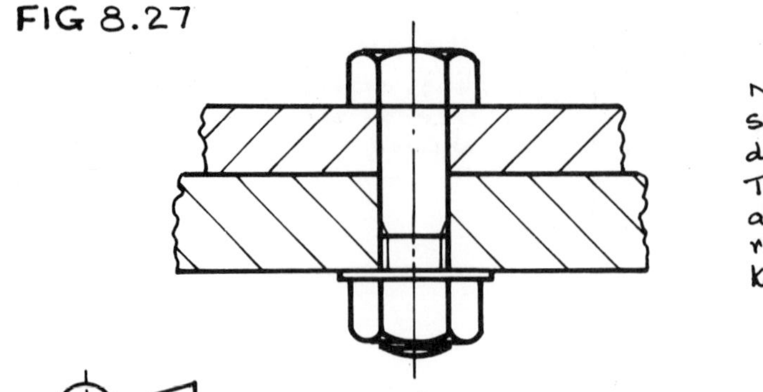

Nuts and bolts are never shown sectioned in a drawing.
The same convention applies to screws, studs, rivets, spindles, shafts, keys, cotters, etc.

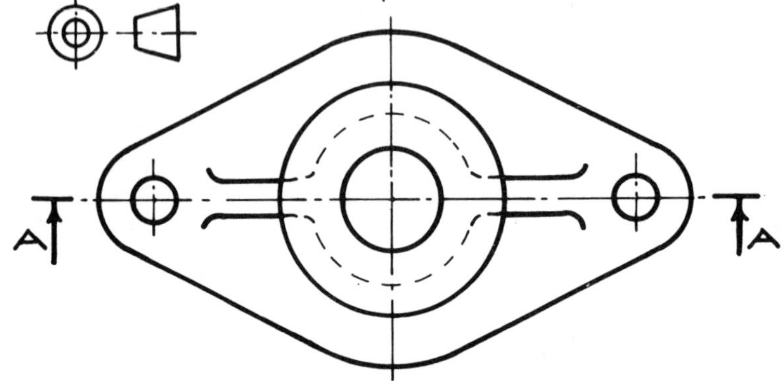

FIG 8.28

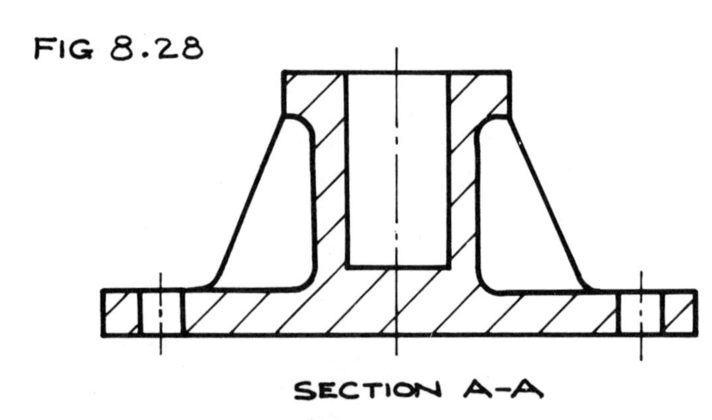

SECTION A-A

A longitudinal section of a web is never hatched even though, geometrically speaking, the section plane A–A splits the web along the centre line.

In cross-section a web is hatched just as the plane cuts it.

FIG 8.29

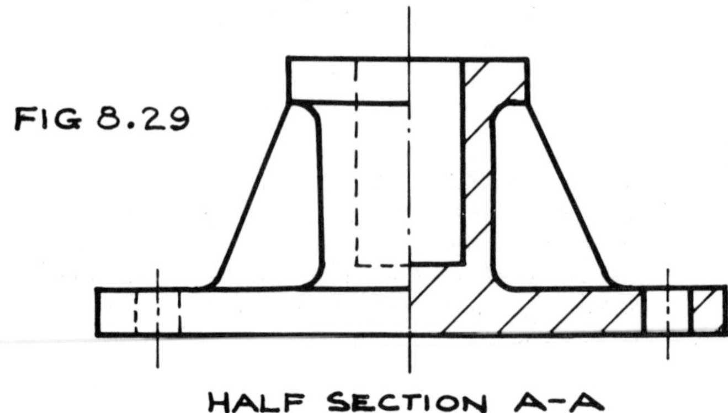

HALF SECTION A-A

Sometimes a half section is used in order to save draughting time or to save space on the drawing sheet.

Hidden detail is not normally shown in a sectioned view.

POINTS TO OBSERVE ON SECTIONING

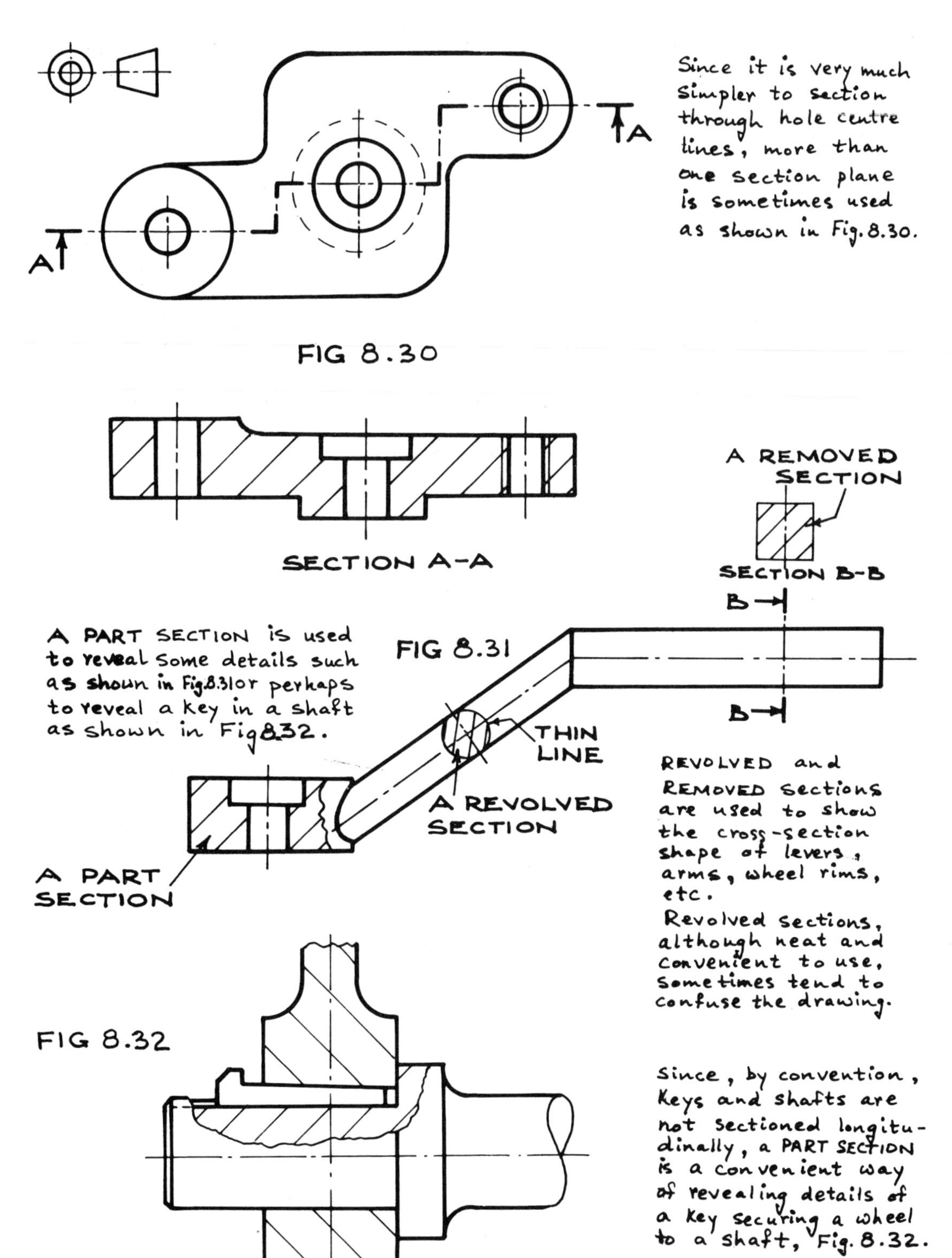

SECTIONING EXERCISES

A few simple engineering components used as sectioning exercises are shown in figures 8.33 to 8.37. The solutions to these sectioning exercises, numbered 1 to 5, are given on the pages after those figures.

Use squared paper to answer all the following exercises.

Exercise 1. Obtain a sectioned Front View of the Lathe Tool Holder shown in figure 8.33.

Exercise 2. The Bearing Cap shown in figure 8.34 is cut by vertical section planes A-A and B-B. Draw a sectioned Front View (Section A-A) and a sectioned Side View (Section B-B).

Exercise 3. Repeat the instructions for *Exercise 2* but with reference to the Tool Post Holder shown in figure 8.35.

Exercise 4. The Shaft Bracket shown in figure 8.36 is cut by vertical section plane A-A and horizontal section plane B-B. Obtain a sectioned Side View (Section A-A) and a sectioned Plan (Section B-B).

Exercise 5. Figure 8.37 shows a Slotted Bracket cut by vertical section plane A-A and horizontal section plane B-B. Obtain a sectioned Front View (Section A-A) and a sectioned Plan (Section B-B).

Exercise 6. Two views of a Channel Iron Bracket are shown in figure 8.38. Obtain a sectioned Front View on vertical plane A-A and a sectioned Side View on vertical plane B-B.

Exercise 7. Draw a sectioned Front View of the Gland, figure 8.39, cut by vertical plane A-A. Obtain also a sectioned Side View on vertical plane B-B.

Exercise 8. Two views of a Cast Iron Crank as shown in figure 8.40. Obtain a sectioned Front View by vertical plane A-A and a revolved section on vertical plane B-B.

Exercise 9. One half of a Flanged Coupling, figure 8.41, is cut by vertical plane A-A. Draw a half-sectioned Side View to the centre line of the coupling half.

Exercise 10. Follow the instructions given on the diagram of the Shaft Support Bracket, figure 8.42.

Exercise 11. Show a sectioned Side View of the Simple Thrust Bearing, figure 8.43, with the Shaft in position.

SECTIONING EXERCISES

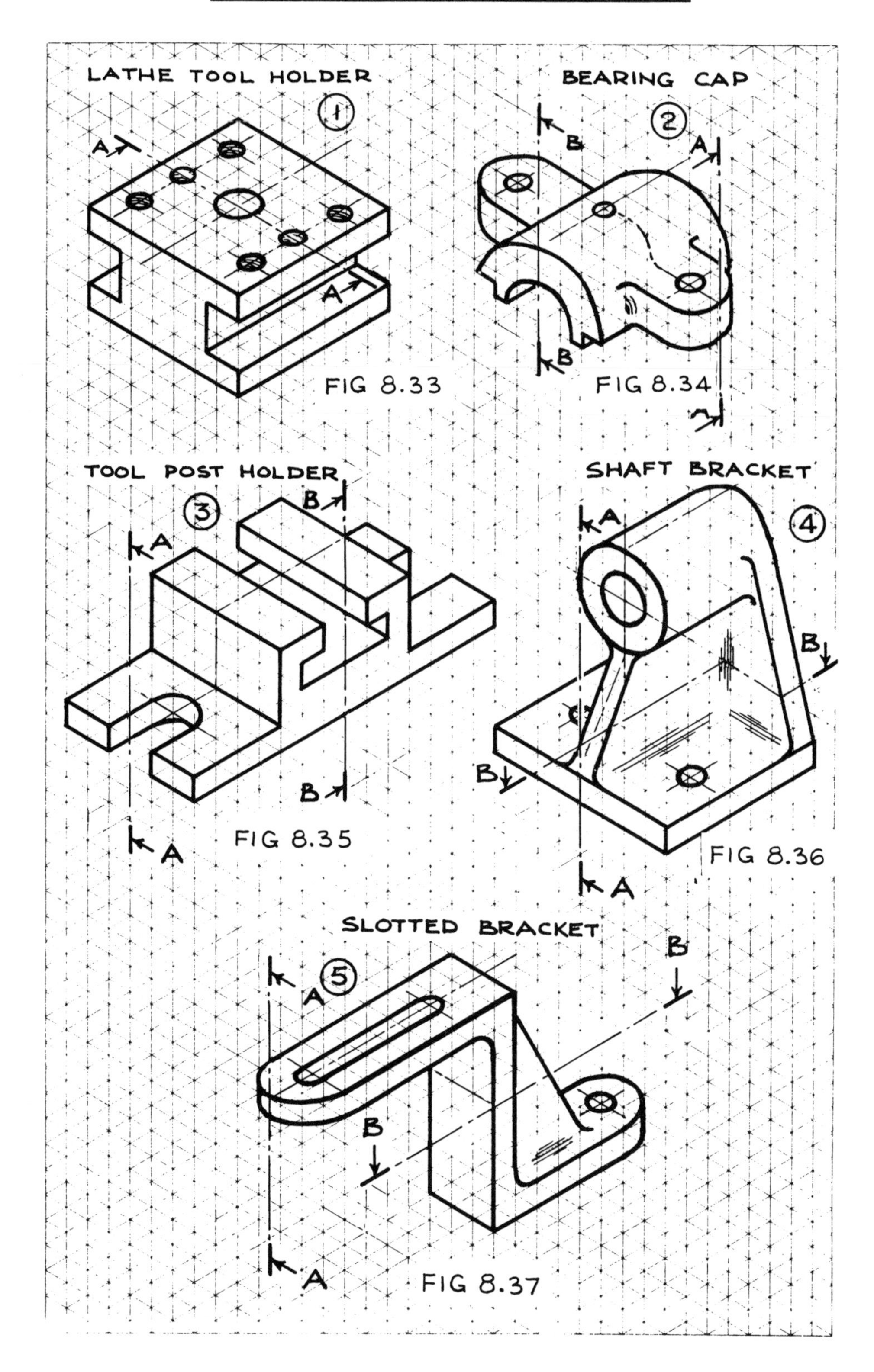

SOLUTIONS TO SECTIONING EXERCISES 1 TO 5

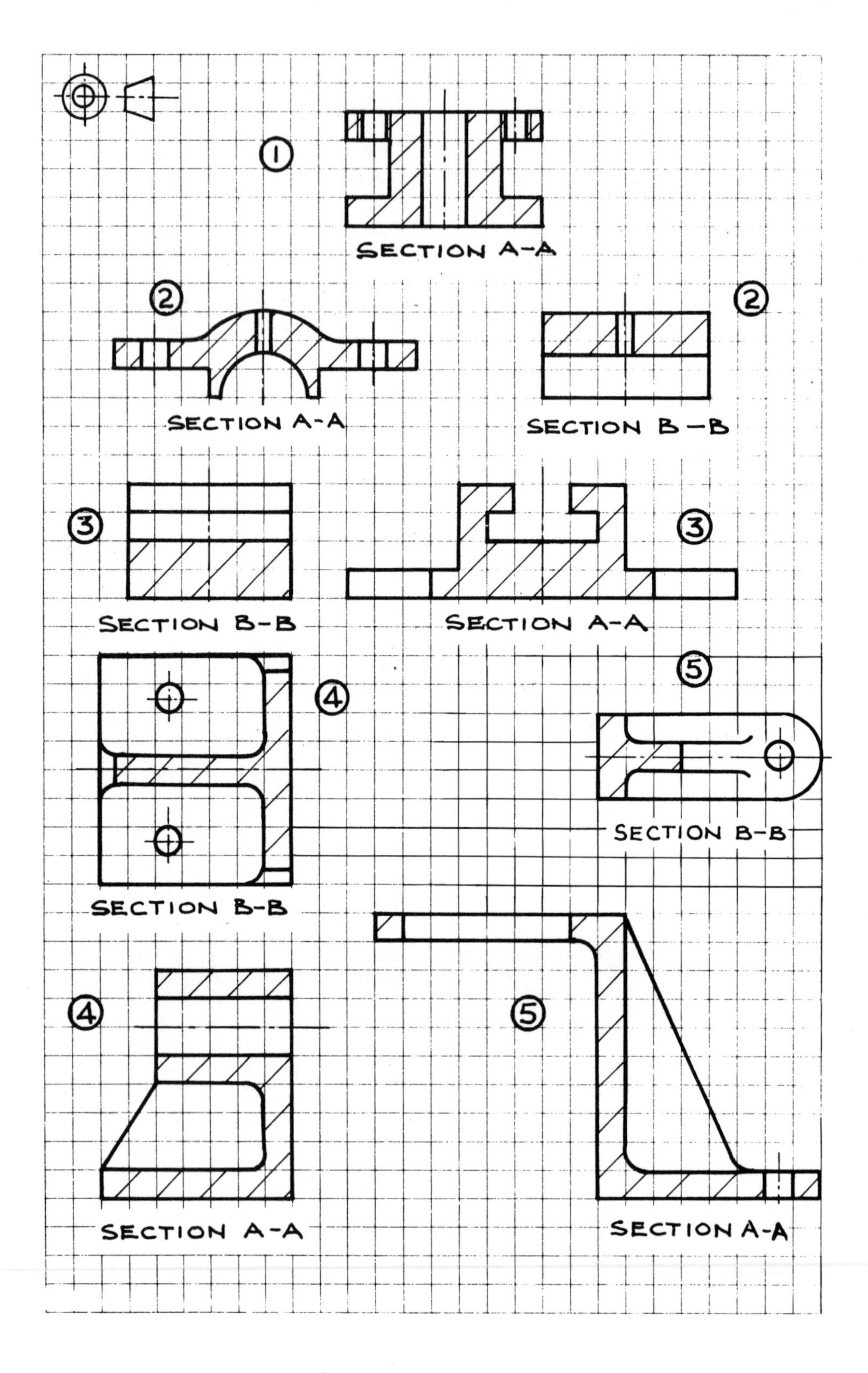

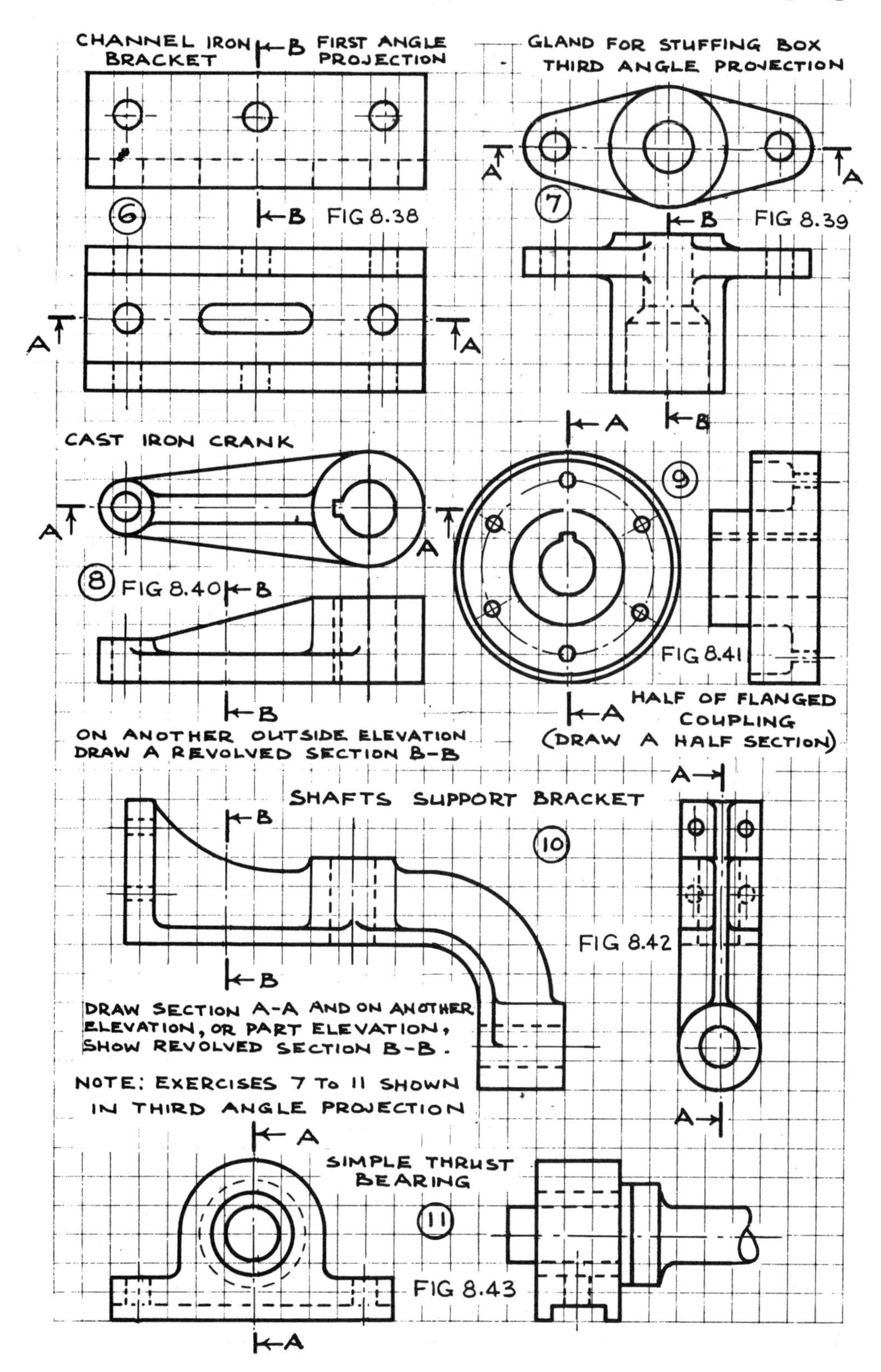
SQUARED PAPER SECTIONING EXERCISES
CHANNEL IRON BRACKET
FIRST ANGLE PROJECTION
6
FIG 8.38
GLAND FOR STUFFING BOX
THIRD ANGLE PROJECTION
7
FIG 8.39
CAST IRON CRANK
8
FIG 8.40
ON ANOTHER OUTSIDE ELEVATION DRAW A REVOLVED SECTION B-B
9
FIG 8.41
HALF OF FLANGED COUPLING
(DRAW A HALF SECTION)
SHAFTS SUPPORT BRACKET
10
FIG 8.42
DRAW SECTION A-A AND ON ANOTHER ELEVATION, OR PART ELEVATION, SHOW REVOLVED SECTION B-B.
NOTE: EXERCISES 7 To 11 SHOWN IN THIRD ANGLE PROJECTION
SIMPLE THRUST BEARING
11
FIG 8.43

ENGINEERING FASTENINGS

The most commonly used methods of fastening in Engineering are : (i) SCREWED FASTENINGS, (ii) RIVETING and (iii) WELDING.

The techniques of welding have advanced so much in recent years that this means is preferred to riveting in a great many instances.

Screwed fastening may be by Bolt and Nut, Fig 8.44 ; Setscrew, Fig 8.45, or Stud and Nut, Fig 8.46. Self-tapping type screws are used extensively in light engineering, particularly in the motor industry.

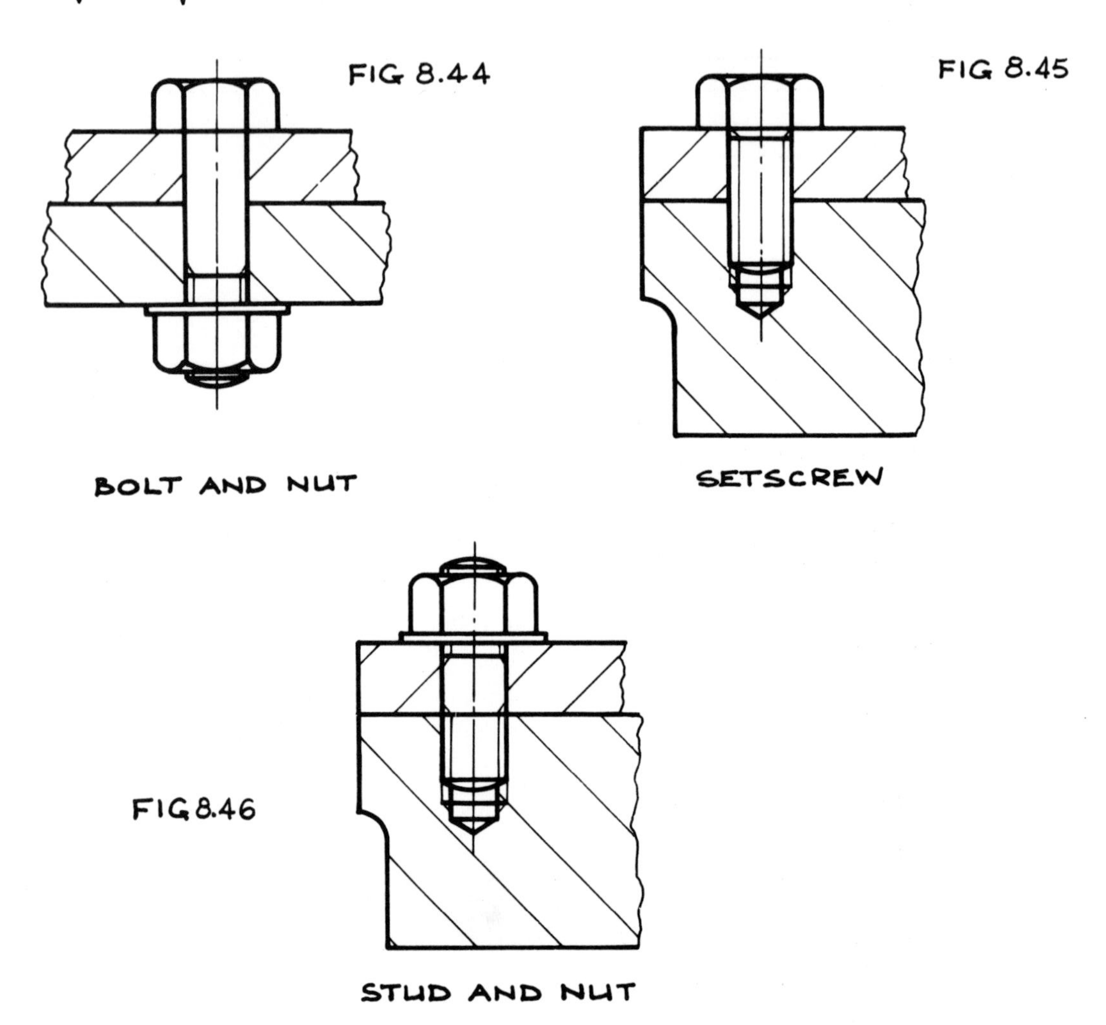

The most common type of head for a bolt or screw is hexagonal, but other shapes are frequently used, as shown on the following page.

FITTED BOLTS are sometimes used in precision engineering work, in which case the bolt is made a fairly tight fit in a specially finished hole. Generally however, bolts are assembled into clearance holes, i.e. holes drilled a little larger than the bolt size and this facilitates assembly. It is not necessary to show this clearance on a drawing.

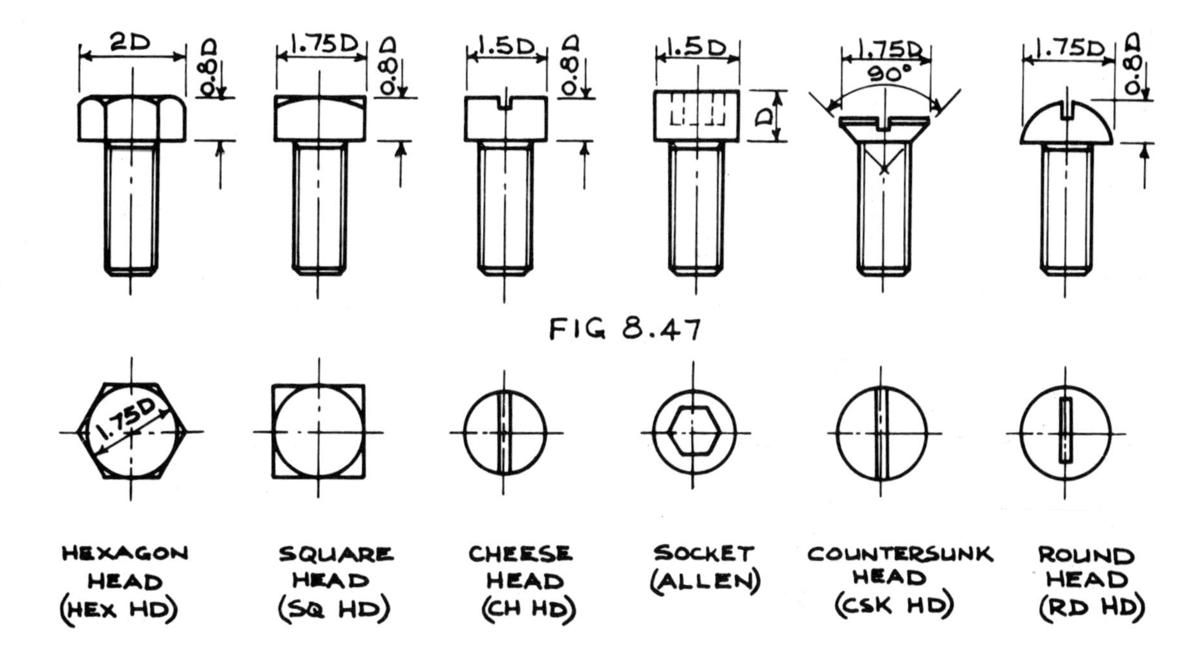

FIG 8.47

Approximate drawing proportions, based on the screw diameter, are given for the common screws shown above. The abbreviated reference is also shown in brackets.

Since hexagonal head fastenings are by far the most common, special instructions for their representation will be given on the following pages.

DOWELS are steel pins, sometimes ground silver steel bars, which are used to locate parts accurately and prevent relative movement between parts. An example of the use of dowels is in press tool work where the die plate has to be located accurately on its base plate without any chance of movement when the press is operating. The conventional way of representing a dowel on a drawing is shown by Fig 8.48.

FIG 8.48

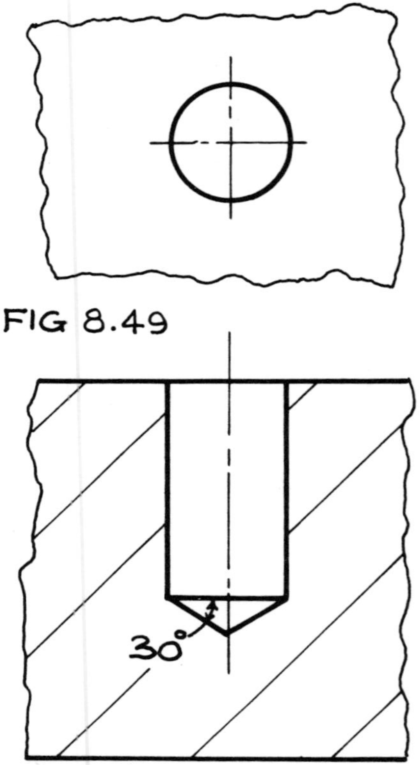

FIG 8.49

A DRILLED HOLE

Notice that the bottom of a drilled hole is finished by 30° set square.

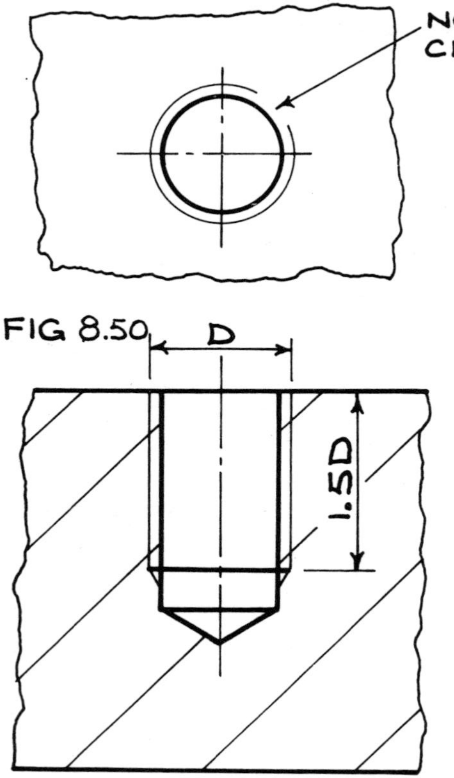

FIG 8.50

A TAPPED HOLE

Tapping operation is done by hand or by machine.

The hole is not threaded to the full depth.

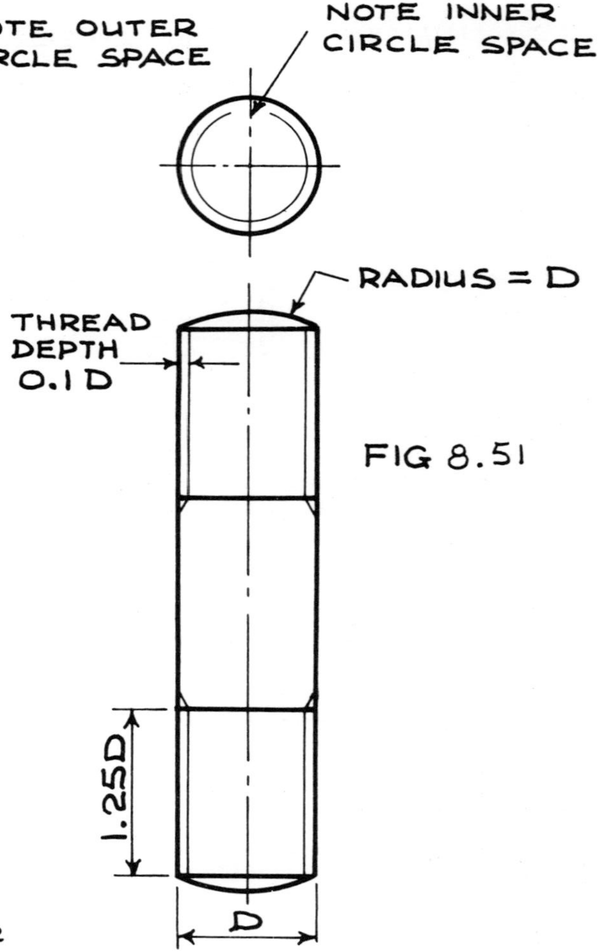

FIG 8.51

A BAR THREADED AT BOTH ENDS (A STUD)

D = OUTSIDE DIAMETER OF THREAD

STUD ASSEMBLY

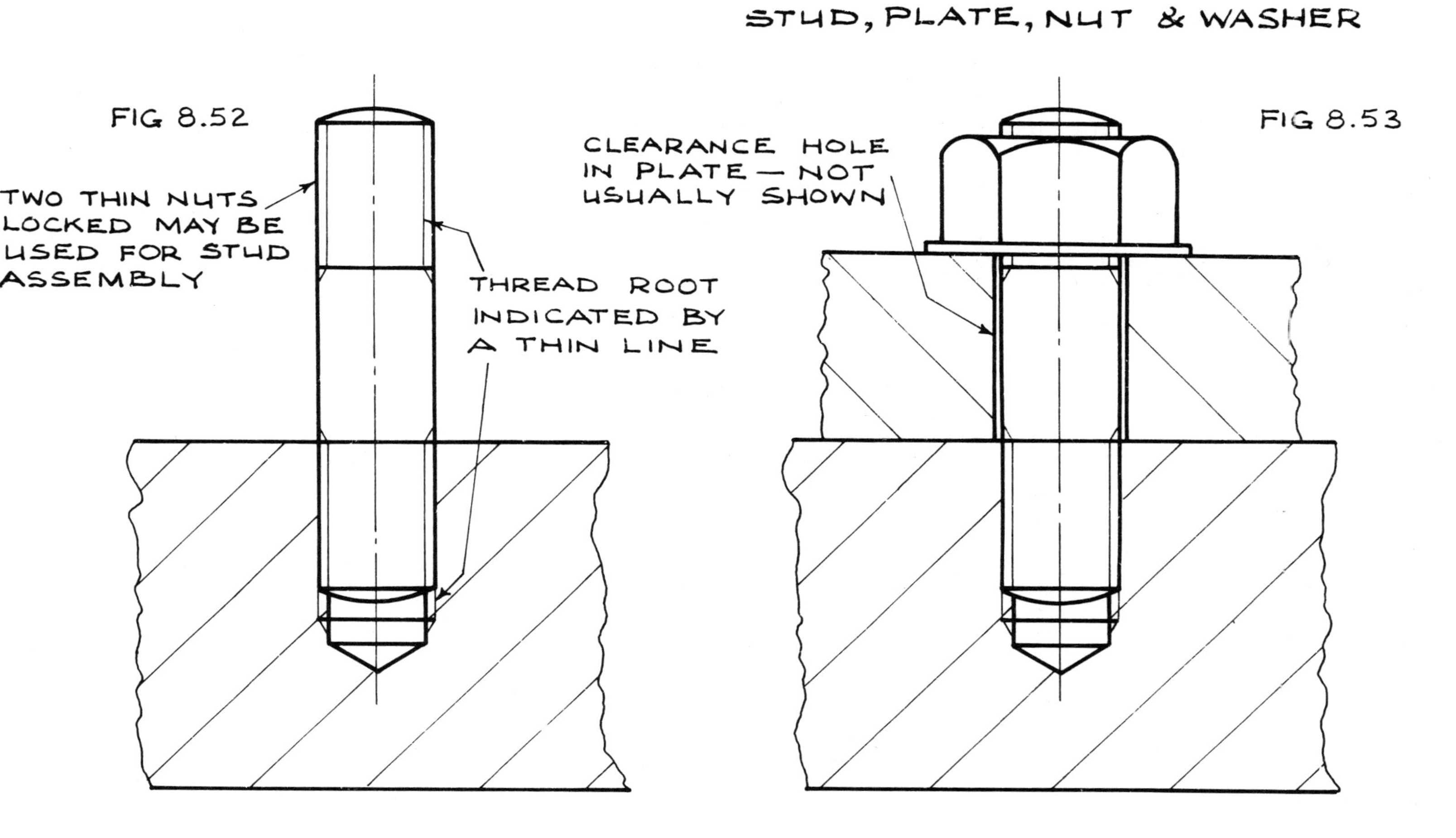
COMPLETE ASSEMBLY
OF
STUD, PLATE, NUT & WASHER
FIG 8.52
TWO THIN NUTS LOCKED MAY BE USED FOR STUD ASSEMBLY
THREAD ROOT INDICATED BY A THIN LINE
CLEARANCE HOLE IN PLATE — NOT USUALLY SHOWN
FIG 8.53
NOTE: STUDS AND SETSCREWS MUST BE TIGHT BEFORE REACHING THE END OF THE THREADED PART OF THE HOLE.

REPRESENTATION OF A NUT

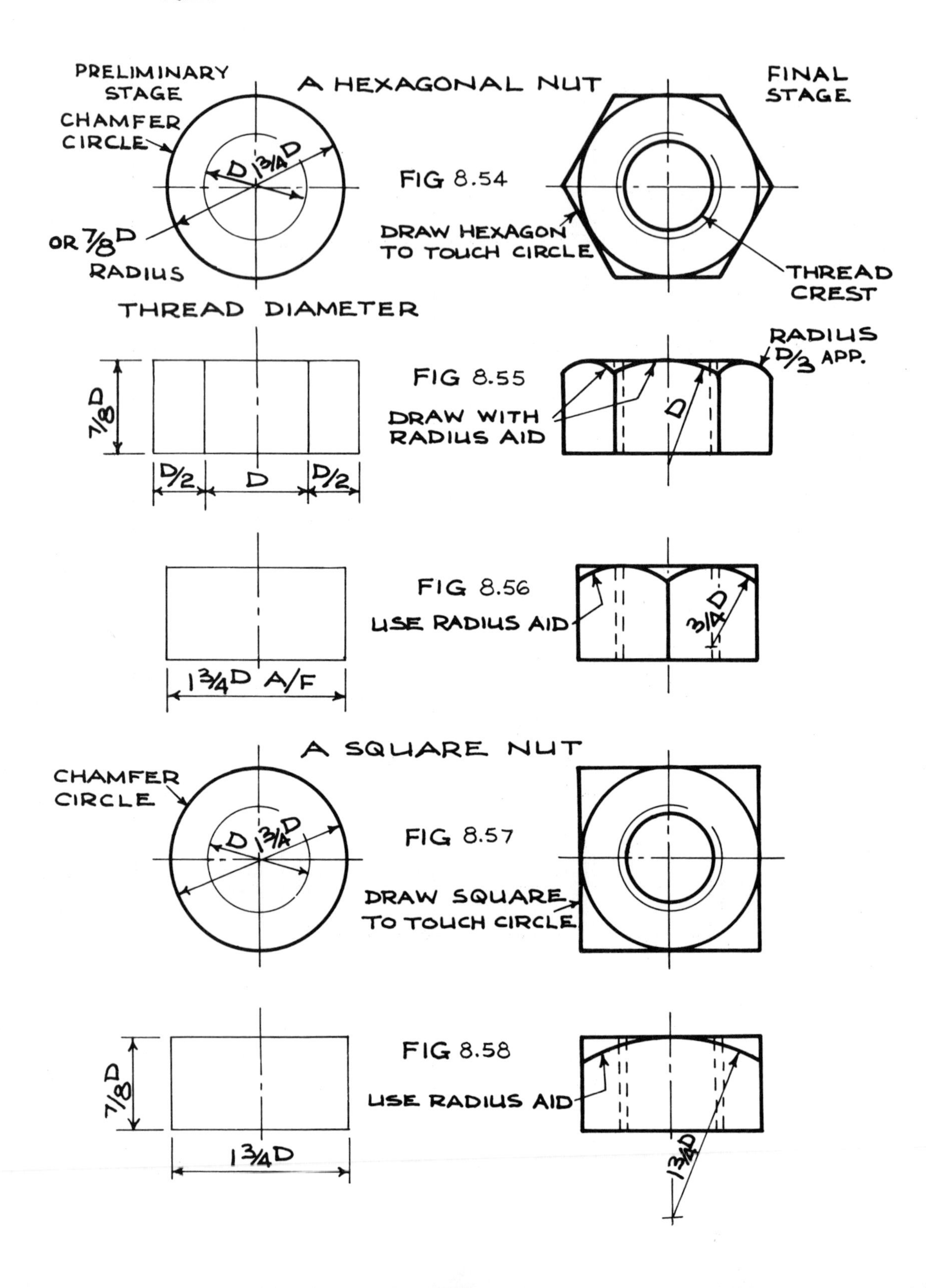

REPRESENTATION OF A BOLT

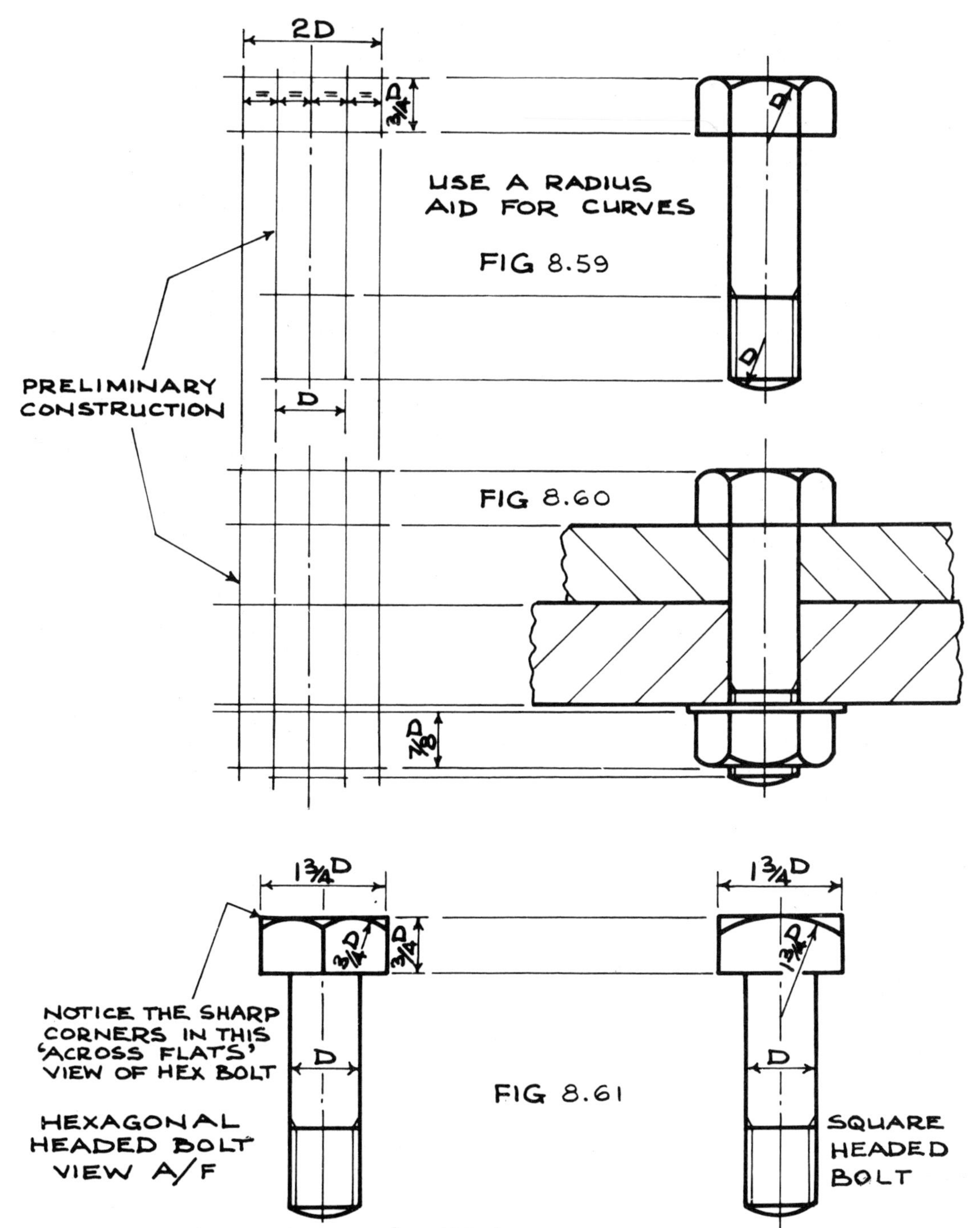

NOTE: As well as simplifying the proportions for the representation of nuts and bolts, the foregoing methods have the additional advantage of being reasonably accurate. For example, the chamfer circle diameter of 1 3/4 D leads to a hexagon face approximately equal to the bolt diameter, which is very convenient for drawing.

LOCKING DEVICES

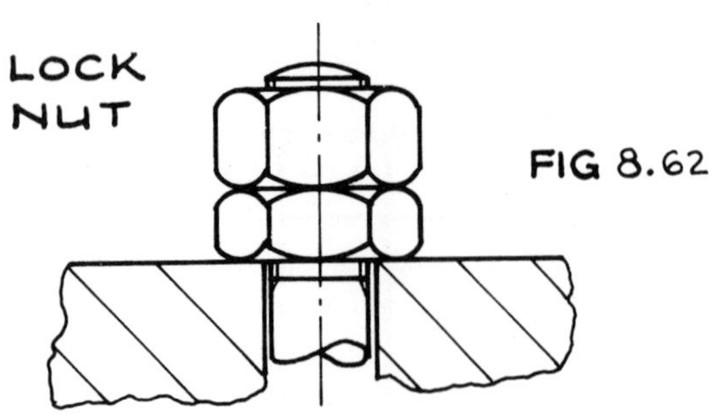

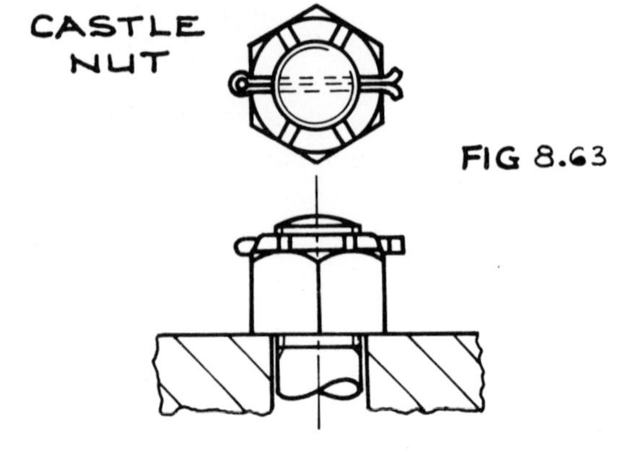

Two spanners are used and locking is effected by slackening back the thin lock nut by a small amount whilst the outer nut is held by the second spanner.

SIMMOND'S NUT

NYLON COLLAR INSERT

FIG 8.65

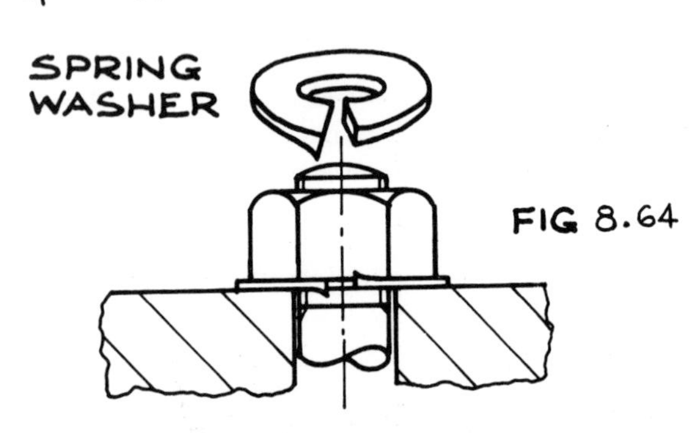

Vibration is what causes a nut to slacken. The spring washer or the nylon insert of the Simmond's type nut is an attempt to damp the vibrations of the nut. A patch of nylon on the threads of the bolt is a recent nut locking development.

A split pin with a castle or a slotted nut is perhaps the most secure means of nut locking. A very effective, but permanent, means sometimes used in aircraft work, is simply to spread the end of the bolt using a hammer and locking punch.

Two or more set screws close together may be secured by means of a soft tab washer folded back against the hexagon flats or perhaps by drilling a small hole through each bolt head and linking the screws by a wire loop.

SCREW THREAD FORMS

The first British Screw Thread to be standardised was the Whitworth thread (B.S.W.), designed by Sir Joseph Whitworth about 130 years ago. It is still used extensively in general engineering along with a finer version, the British Standard Fine (B.S.F.).

The British Standard Whitworth thread form applied to pipe work is known as the British Standard Pipe thread (B.S.P.) and in pipe work the thread size depends upon the internal diameter of the pipe. For example, a 1" B.S.P. thread would be used on pipes having a 1" diameter bore and outside diameter approximately 1.309". B.S.P. tapered threads are used for high pressure pipe work, to give greater leakage resistance.

The British Association Thread (B.A.), with dimensions in millimetres, has also been used extensively for fine screws in instrument work.

However, in view of the fact that other countries have been developing their own standards down the years, so that now there are literally scores of different screw threads in world wide use, it has been decided in this country to adopt the International Standards Organisation Metric, known as ISO Metric screw threads, to replace B.S.W. and B.S.F. thread forms. This change over to ISO threads obviously cannot be completed in a short time since many valuable machine tools with years of useful life remaining, motor vehicles, ships, etc., still require B.S.W. and B.S.F. screw threads for replacement of spare parts. The ISO Metric thread form is identical to the Unified thread which was agreed between G.B., Canada and U.S.A. in 1948. Coarse and fine threads of the Unified classification are designated UNC and UNF respectively.

Other special purpose thread forms in general use are: (i) the *square* thread, used for valve spindles, lathe cross slides, power transmission, etc., (ii) the *acme* thread, which has largely superseded the square thread since it is more easily machined and is more suitable, for example, for use with a split nut in a lathe lead screw, and (iii) the *buttress* thread, which is used to transmit power in one direction only, for example, a vice or a press screw.

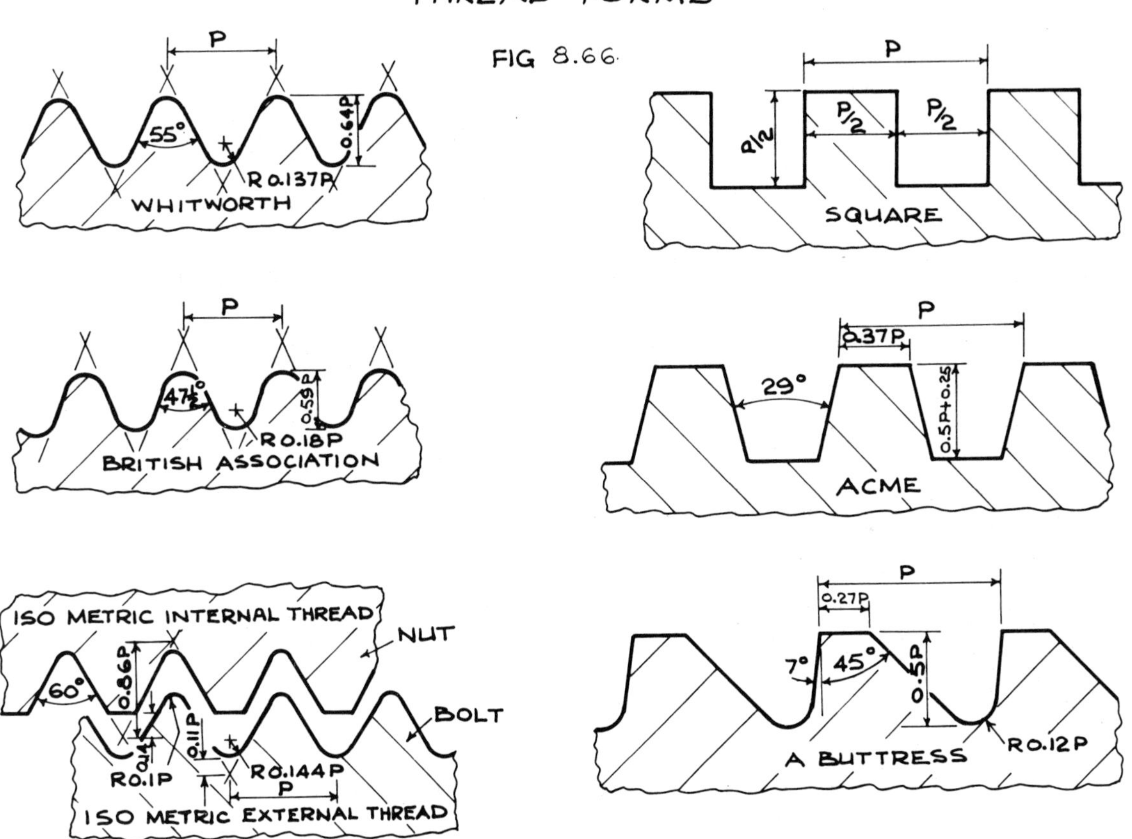

The table, figure 8.67, gives useful information about screw threads at present in common use.

The International Standards Organisation Metric thread (ISO Metric), as already noted has the same form as Unified, but the dimensions are in millimetres. In metric countries, the general practice has been to use coarse threads on all standard fasteners, fine threads being used only for special purposes. Thus, ISO Metric coarse threads will, in general, replace B.S.W. or B.S.F. threads.

The letter M followed by the diameter and the pitch in millimetres is the method of designating an ISO Metric thread. Thus, M20 x 2.5 is a coarse ISO thread of 20 mm diameter and pitch 2.5 mm and M20 x 1.5 is a fine ISO thread of 20 mm diameter and pitch 1.5 mm.

Figure 8.68 table gives equivalent inch sizes of the recommended ISO Metric thread sizes, useful for the transition period of years.

SCREW THREAD DATA

FIG 8.67

THREAD SIZE. DIAMETER IN INCHES	THREADS PER INCH					BSP MAJOR DIAMETER IN INCHES
	BSW	BSF	UNC	UNF	BSP	
$\frac{1}{8}$ (0.125)	40				28	0.383
$\frac{3}{16}$ (0.1875)	24					
$\frac{1}{4}$ (0.250)	20	26	20	28	19	0.518
$\frac{5}{16}$ (0.3125)	18	22	18	24		
$\frac{3}{8}$ (0.375)	16	20	16	24	19	0.656
$\frac{7}{16}$ (0.4375)	14	18	14	20		
$\frac{1}{2}$ (0.500)	12	16	13	20	14	0.825
$\frac{9}{16}$ (0.5625)	12	16	12	18		
$\frac{5}{8}$ (0.625)	11	14	11	18	14	0.902
$\frac{3}{4}$ (0.750)	10	12	10	16	14	1.041
$\frac{7}{8}$ (0.875)	9	11	9	14	14	1.189
1 (1.000)	8	10	8	12	11	1.309
$1\frac{1}{4}$ (1.250)	7	9	7	12	11	1.650
$1\frac{1}{2}$ (1.500)	6	8	6	12	11	1.882
$1\frac{3}{4}$ (1.750)	5	7	5		11	2.116
2 (2.000)	4.5	7	4.5		11	2.347

FIG 8.68

ISO METRIC THREAD SIZE	APPROX. INCH EQUIVALENT SIZE	APPROX. BA EQUIVALENT SIZE
1.6		10
2		9 OR 8
2.5		7
3	$\frac{1}{8}$	5
4	$\frac{1}{8}$ OR $\frac{3}{16}$	3
5	$\frac{3}{16}$	2 OR 1
6	$\frac{1}{4}$	0
8	$\frac{5}{16}$	
10	$\frac{3}{8}$	
12	$\frac{7}{16}$ OR $\frac{1}{2}$	
16	$\frac{5}{8}$	
20	$\frac{13}{16}$	
24	$\frac{15}{16}$	
30	$1\frac{3}{16}$	
36	$1\frac{7}{16}$	
42	$1\frac{5}{8}$ OR $1\frac{11}{16}$	

Note: Metric thread sizes given in this table are 'preferred' sizes. There are other intermediate sizes but their general use is not recommended.

RIVETED JOINTS

Rivets are used to make a *permanent* fastening, taking *shearing* loads mainly. A bolt and nut can take *tensile* as well as *shearing* loads.

In ship-building, boiler-making and other types of heavy engineering, the steel rivets are heated to red heat before insertion in the clearance holes, previously drilled or punched, in the steel plates. A heavy dolly is then held against the head of the rivet by the riveter's mate and a new head is formed by a suitable shaped pneumatic hammer as the rivet is compressed to fill the hole. Alternatively the red hot rivet may be squeezed to shape by an air-operated machine.

In light engineering, such as Aircraft construction, similar techniques are employed on a smaller scale. The light alloy rivets are not used hot, but are softened by normalising just prior to assembly. Tubular rivets are used sometimes for greater strength combined with lightness. In cases where the assembly job is only accessible from one side, break-stem or *pop* rivets may be used.

Various types of riveted joints are illustrated on the following pages.

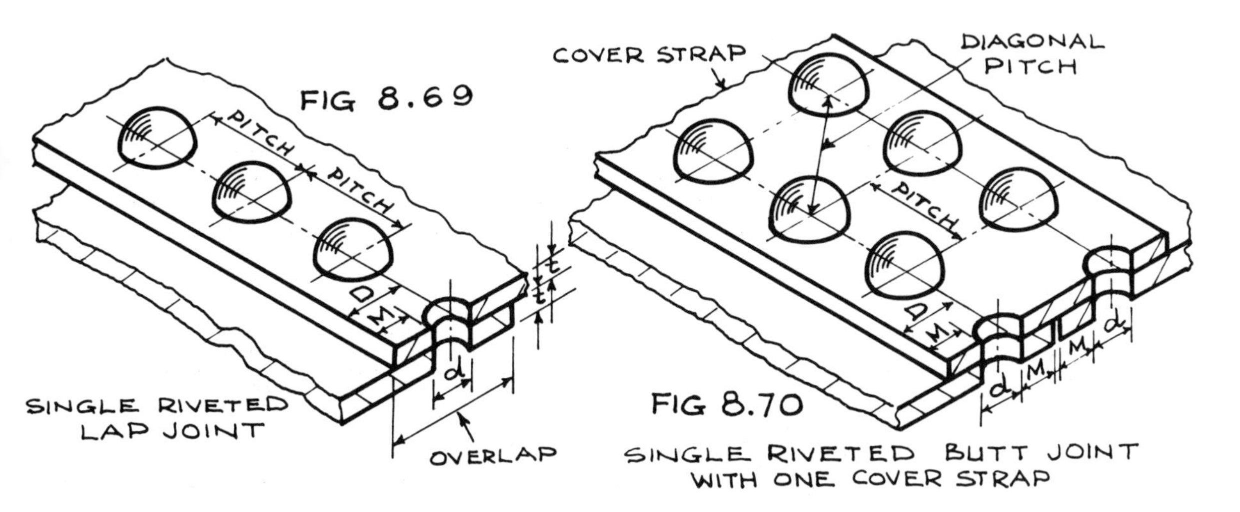

SINGLE RIVETED LAP JOINT

SINGLE RIVETED BUTT JOINT WITH ONE COVER STRAP

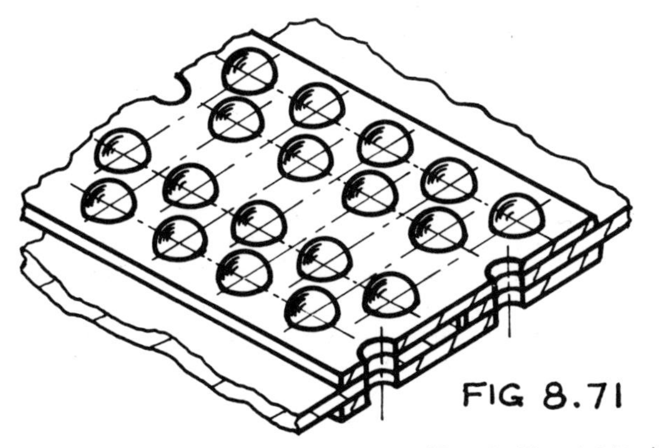

DOUBLE RIVETED BUTT JOINT WITH DOUBLE COVER STRAPS AND ZIG-ZAG OR DIAGONAL RIVETING

The diameter of the rivet (d) may be calculated using Unwin's Rule: $d = 6\sqrt{t}$, where t is the plate thickness in millimetres. The margin M is usually not less than $1\frac{1}{2}d$ and the coverstrap thickness varies from 0.875t to t.
The pitch P of the rivets is at least 2d, likewise the distance between rows of rivets. The pitch is obtained exactly from design calculations outlined in a later chapter.

RIVETED JOINTS

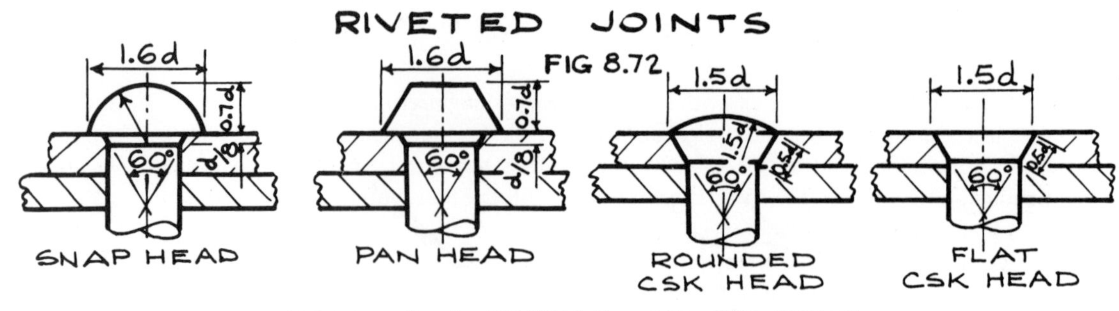

COMMON TYPES OF RIVETS

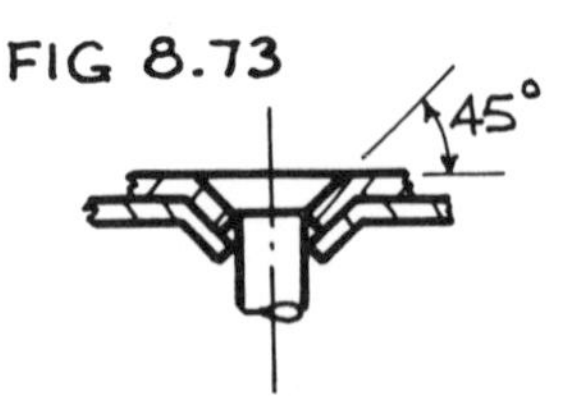

For riveting thin plates a flat countersunk head rivet of 45° or even 30° angle is used. The plate holes are prepared by 'dimpling', on a light press, and this increases the resistance to tearing of the thin plates.

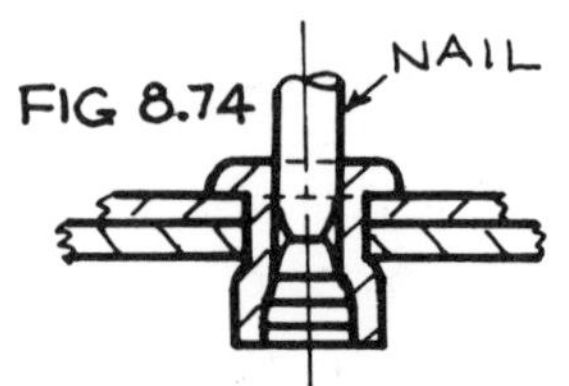

When the plate is accessible from one side only 'pop' rivets may be used. A tool, consisting of a lever system, grips the nail or mandrel which is pulled until it breaks, forming the rivet head and leaving the head of the nail to seal the hole.

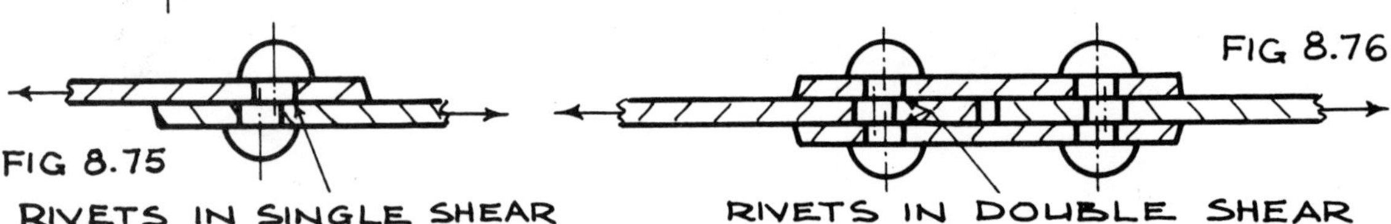

RIVETS IN SINGLE SHEAR

RIVETS IN DOUBLE SHEAR

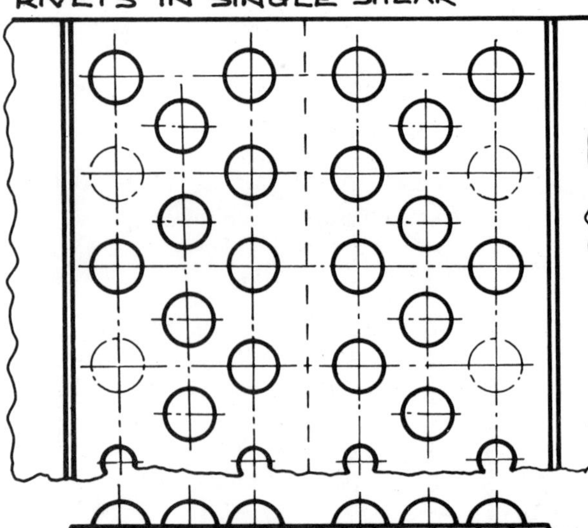

FIG 8.77

In Fig. 8.75 the rivets are said to be in SINGLE SHEAR, because under load the plates tend to slide one along the other thus producing a shearing force on the rivets.

In the case of the Butt joint shown by Fig. 8.76 there is a tendency for the plates to slide from between the cover straps, thus producing a DOUBLE SHEARING effect upon the rivets. In theory, a rivet is twice as strong in Double shear.

Fig. 8.77 shows a Triple Riveted Butt joint drawn in orthographic projection. It is worth noting that the omission of alternate rivets in the outer rows increases the efficiency of the joint.

STRUCTURES RIVETED JOINTS

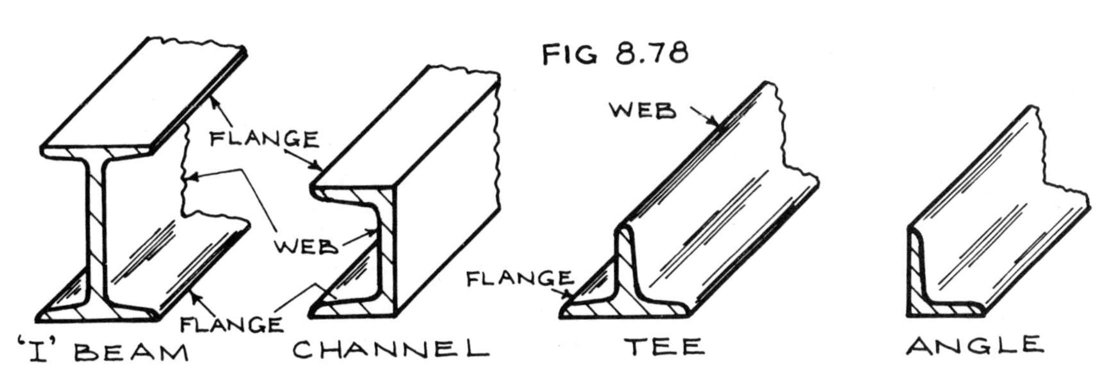

ROLLED STEEL SECTIONS USED IN STRUCTURAL ENGINEERING

COLUMNS, BEAMS, ROOF TRUSSES, ETC. ARE MADE BY RIVETING ROLLED STEEL SECTIONS. Fig. 8.79 SHOWS AN OUTLINE DRAWING OF A ROOF FRAME AND Fig. 8.80 SHOWS DETAILS OF JOINT A.

ROOF TRUSS

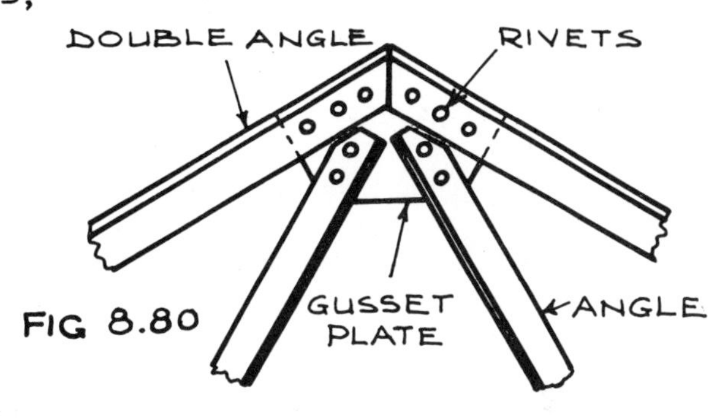

ENLARGED VIEW OF JOINT A

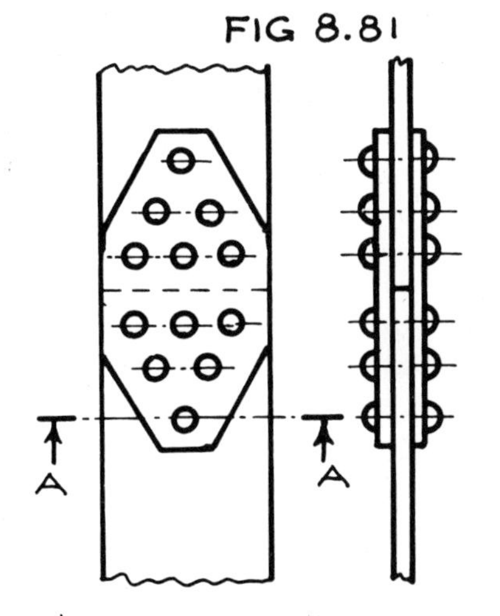

LOZENGE JOINT

This joint may be used on a tension member. The pattern of rivets ensures that the bar is not weakened by more than one rivet hole at A-A, the most likely place of failure.

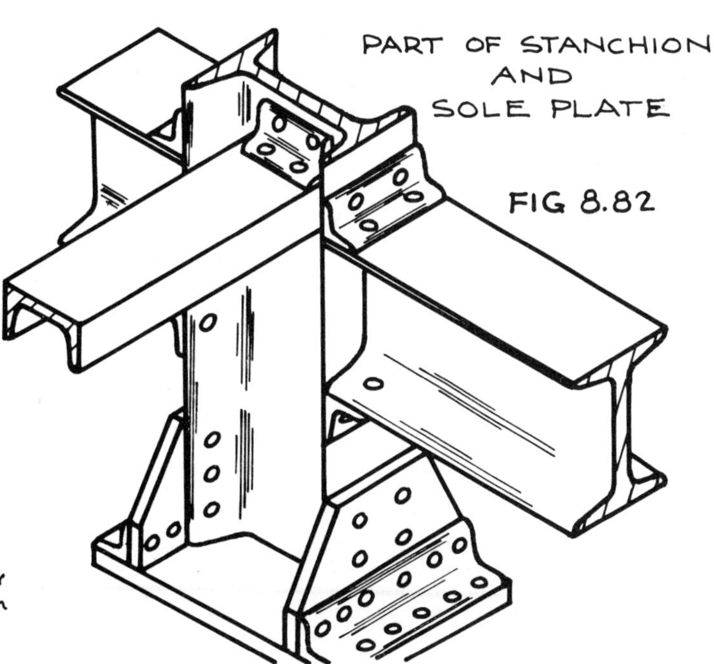

WELDING

The simplest type of welding is to be seen in the operation performed by the blacksmith on the anvil, when he hammers two pieces of white hot metal together to form a joint.

In *electric* welding, which is being used increasingly in industry today, there are two main divisions of the process. One method, the Electrical Resistance type, is carried out by pressing in contact the parts to be welded whilst passing a very heavy low voltage current through them. The electrical resistance at the contact points is much higher than the rest of the circuit and the temperature of the joint rises rapidly. When welding temperature is reached, mechanical pressure applied by machine completes the welded joint. *Butt* welding, *spot* welding – used extensively in the motor industry to replace riveting – and *seam* welding come into this class of welding.

The second method, and perhaps the more important, is the *electric-arc* process. The low resistance of the electrode in contact with the metal workpiece produces intense heat because of the heavy current. Thus the metal melts and vaporises and when the electrode (filler rod) is separated slightly from the metal a conducting arc is produced and the electrode and the metal to be welded are melted together. Temperatures approaching 4000°C are reached in electric-arc welding. This process is used extensively in ship-building where it replaces riveting to a fairly large extent and also in 'fabrication' work where large machine bed-plates, turbine and generator casings, etc., are made up from steel plates welded together instead of being produced as iron castings.

In *gas* welding the most commonly used gases are *oxygen* and *acetylene*, supplied in gas cylinder pressure vessels. The gases from the cylinders are passed through pressure reducing valves to the blowpipe where they are burned in approximately equal proportions, giving a flame temperature of around 3000°C. Gas welding is simply accomplished by applying the torch to a suitable metal filler rod and thereby melting it into the joint as the torch is passed slowly along the joint. This torch is also used extensively for metal cutting and the portability of the equipment is an advantage, particularly for use by demolition gangs, etc.

Two hands are required for gas welding whereas the welding rod in the electric-arc process is clamped in a holder and the operation may thus be carried out using one hand. Electric-arc welding is quicker and cheaper than gas welding for plates of more than 6 mm thick. Gas welding is unsuitable, unless with pre-heating of the work, for plates more than 20 mm thick.

Some modern developments of special welding techniques are worthy of brief mention at this stage. *Friction* welding is now being used very successfully for fabricating back axles for motor vehicles. A machine which resembles a heavy duty lathe is used for friction welding. One of the parts to be welded, such as a flange or perhaps a differential casting, is securely held in a chuck which is rotated at high speed. The tubular shaft is brought into contact with the rotating piece and as axial force is applied friction causes the two parts to heat up until welding temperature is reached. At this stage a brake is applied and the two parts are allowed to rotate together thus cooling and completing a very efficient welded joint. *Explosive* welding is another technique which has been successfully applied to the joining of plates of different metals in sandwich form and also to welding tubes into end plates in atomic reactor work. *Electron beam* welding of very fine components has been successfully carried out, but once again very special equipment and technical knowledge is necessary.

JOINTS USED IN WELDING

Joints used in welding may be divided into two classes: (i) BUTT JOINTS and (ii) FILLET JOINTS. An example of each type is given in Figs. 8.83 and 8.84

FIG 8.83

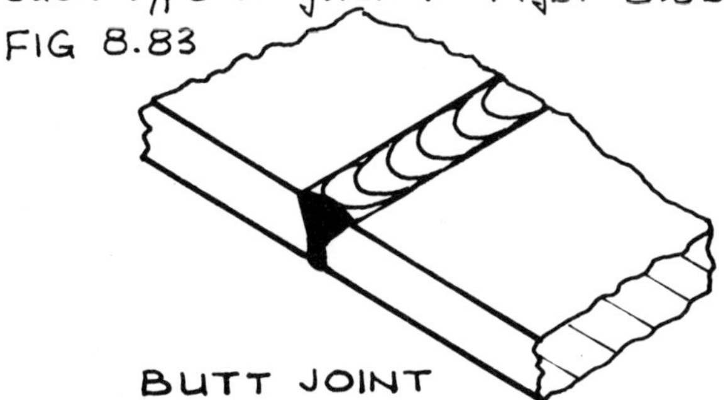

FIG 8.84

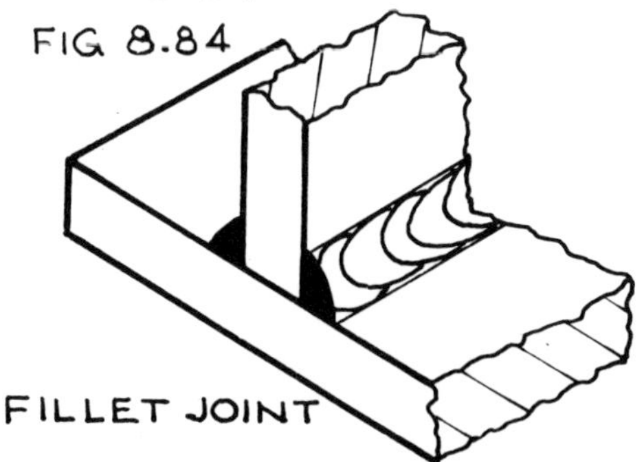

BUTT JOINT WITH BUTT WELD
Plate thickness less than 5 mm

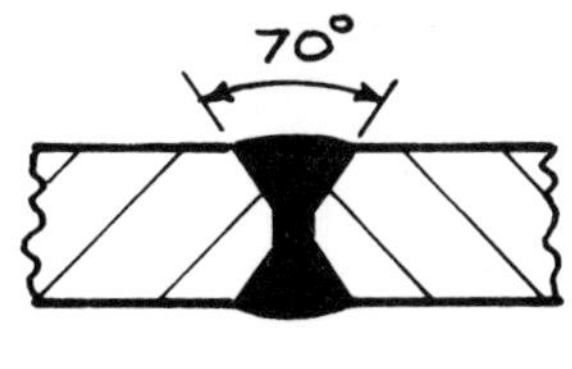

SINGLE-VEE BUTT WITH BUTT WELD
Plate thickness 5 to 15 mm

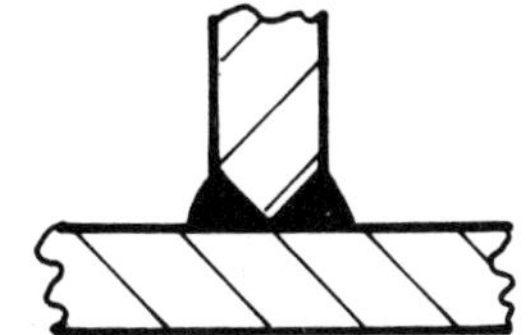

DOUBLE-VEE BUTT WITH BUTT WELD
Plate thickness over 15 mm

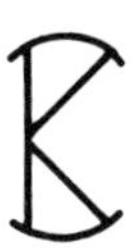

TEE JOINT WITH BUTT WELDS
Plate thickness over 12 mm

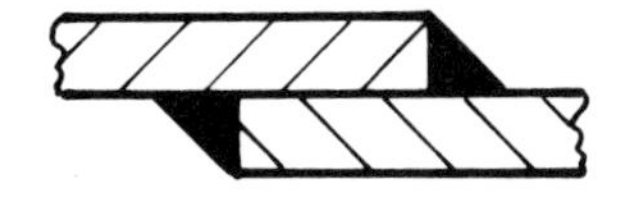

LAP JOINT WITH FILLET WELDS
Plate thickness over 5 mm

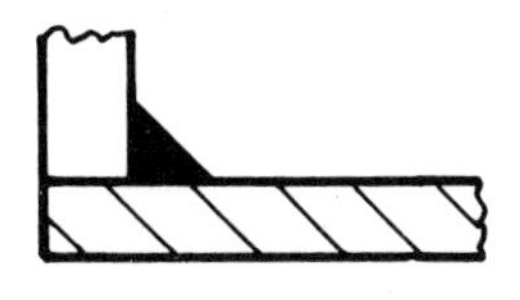

CORNER JOINT WITH FILLET WELD
Plate thickness over 5 mm

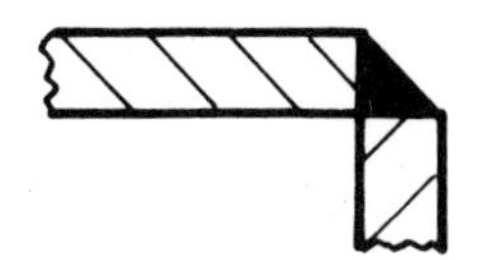

CORNER JOINT WITH FILLET WELD
Plate thickness over 5 mm

FABRICATION

The production of large cast iron pieces such as machine bed-plates, transformer casings, turbine casings, etc., presents some problems. The practice now is to make such components from mild steel plate welded and perhaps bolted together. In fact even for smaller components, which are not required in large enough quantities to justify expense of pattern making prior to casting, fabrication methods are adopted.

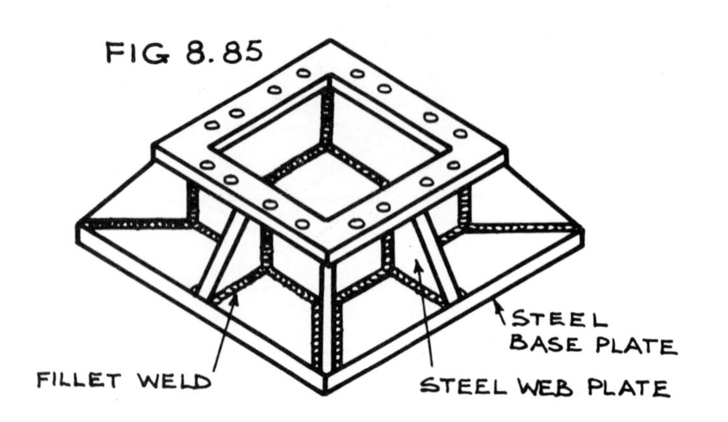

FIG 8.85

Fig. 8.85 shows a typical fabricated base plate for a large machine, made from mild steel plate parts which have been flame cut or sheared to shape and welded together.

USE OF WELD SYMBOLS ON A DRAWING

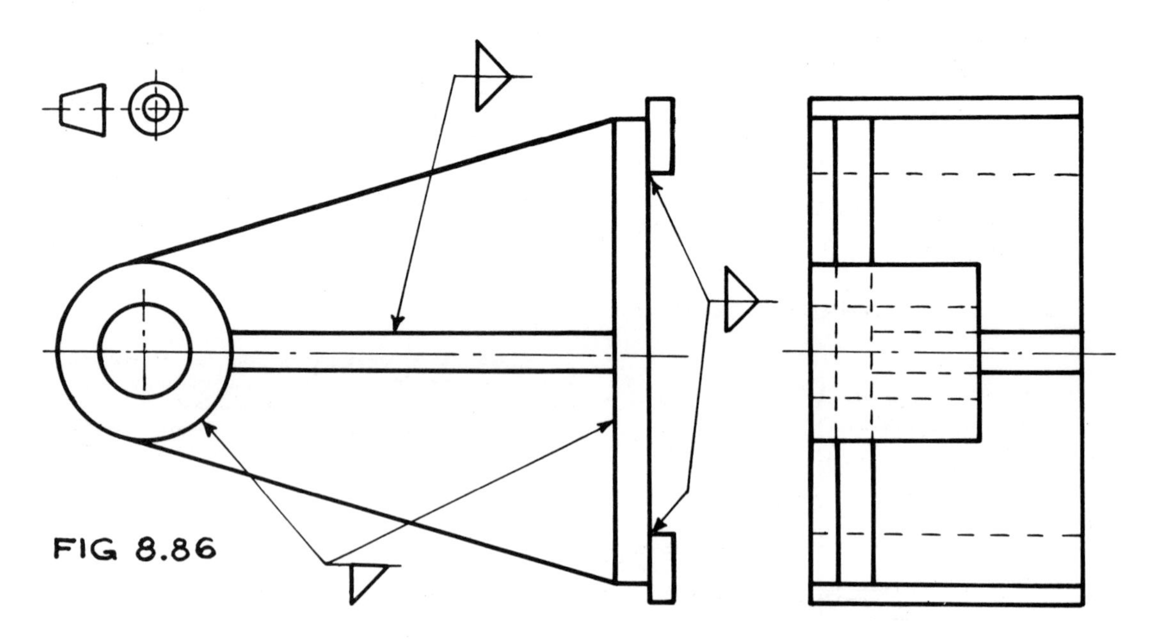

FIG 8.86

NOTE: Dimensions, machining instructions and other information have been omitted from the drawing.

BEARINGS

BEARINGS are used to support rotating parts and to minimise friction.

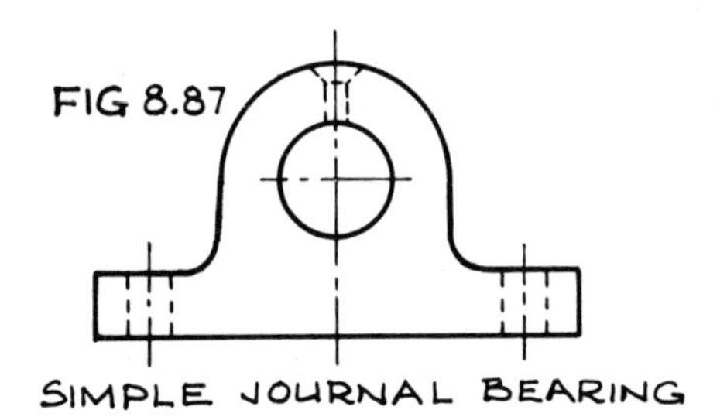

SIMPLE JOURNAL BEARING

A simple JOURNAL bearing may be made by boring a cast iron block to suit the shaft. Cast iron is a reasonably good bearing metal because of the presence of free graphite in the iron which acts as a lubricant. However, this type of bearing, which is suitable for slow speed shafting only, is rarely used in modern engineering practice.

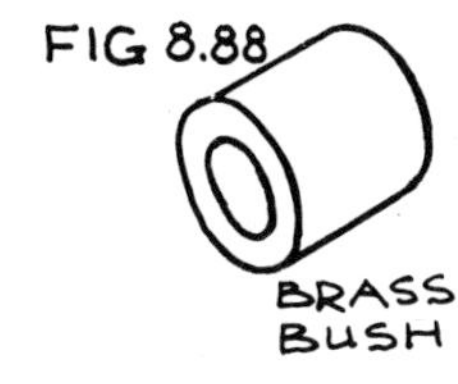

A more efficient bearing is one with a brass bush pressed in. The bush may be renewed when the hole has worn excessively. This type of bearing is used in car engines, on camshafts, oil-pump drives, distributor drive, the rockers, etc.

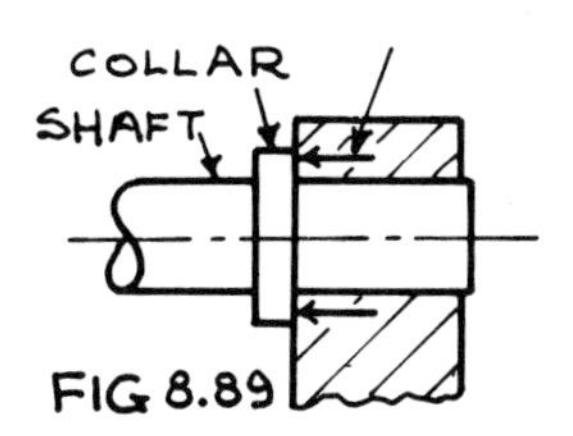

SIMPLE THRUST BEARING

The split bearing, shown by the 'exploded' drawing of figure 8.92 is an even more practical design which facilitates adjustment for wear.
Motor car 'big end' bearings use shell bearings, made in halves, of steel lined with bearing alloy.

When a shaft has to transmit a force along its axis a THRUST bearing is used. The shaft has a collar fixed on or the shaft end is machined to shape. THRUST WASHERS are sometimes used along with shell bearings. These are steel, lined with a bearing metal and help to resist end thrust.

BRASS BUSH FOR THRUST BEARING

To support a vertical shaft, a FOOTSTEP bearing, or its modern equivalent is necessary. The shaft end and the bronze pad are machined to the same spherical radius to minimise side thrust on the brass bush.

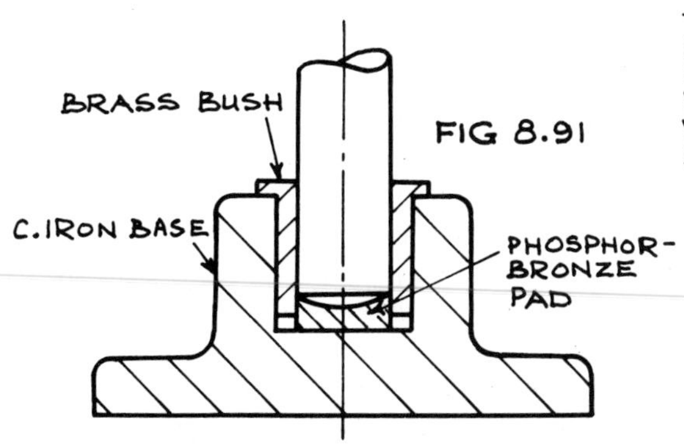

FOOTSTEP BEARING

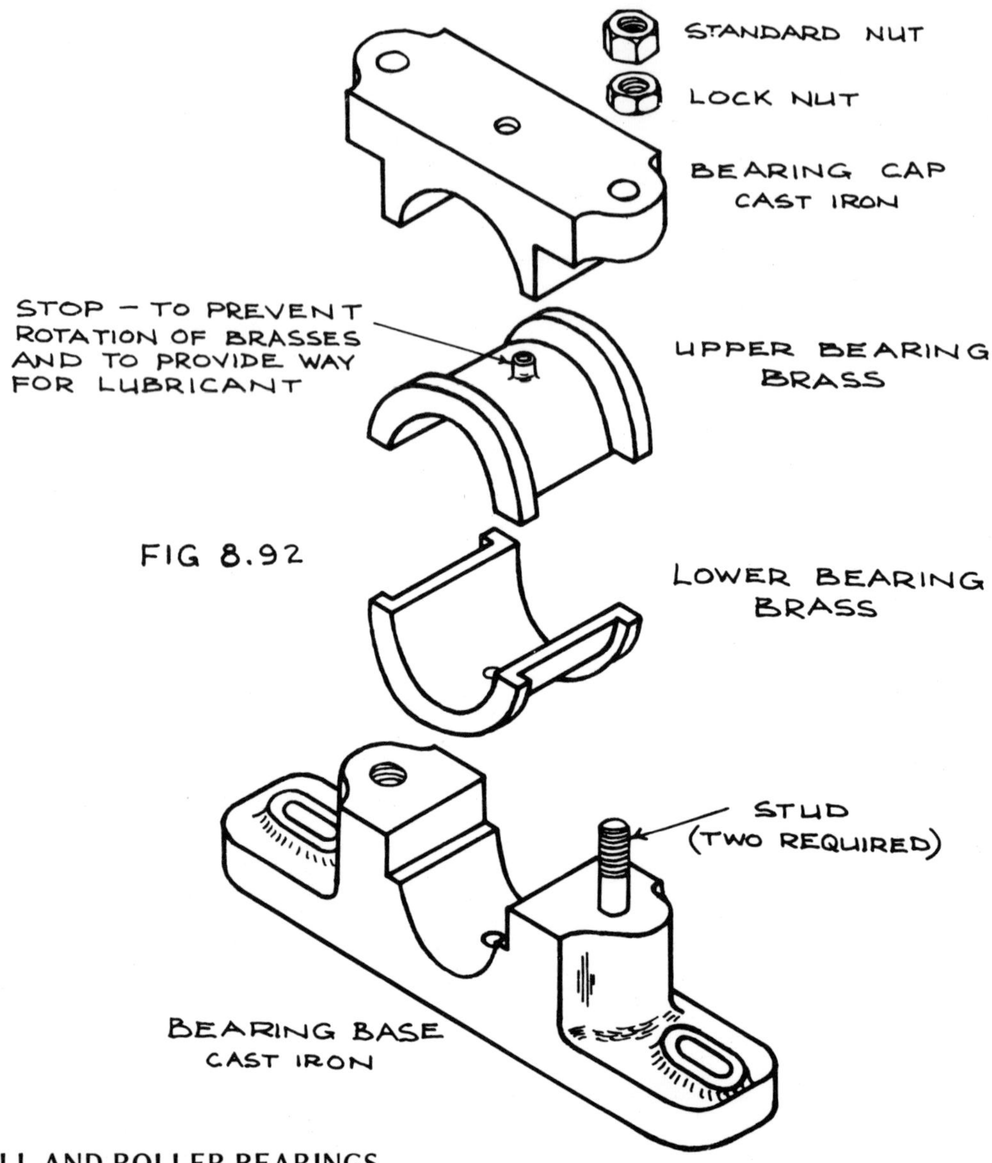

BALL AND ROLLER BEARINGS

These bearings have been developed extensively in recent years and have now largely replaced brass bush type bearings for many applications. Friction and load characteristics are very much better for ball and roller type bearings.

The outer race is a tight fit in the bearing housing which may be a wheel centre or a fixed bracket. The inner race is a tight fit on the Shaft. Both races are hardened and ground to finish. The balls are also hardened and finish ground to size and are separated around the race by means of a cage. The cage may be of brass, light alloy or other metal.

Double row ball bearings are sometimes employed for greater load carrying capacity. In the double row, self-aligning ball bearing shown in figure 8.96 the surface of the outer race is spherically ground so that the balls can run smoothly even though the shaft axis may be out of alignment by a few degrees.

Dust or grit is the enemy of all ball or roller bearings. For this reason a dust cover, often in the form of a felt ring, is fitted to act as grease retainer and grit excluder.

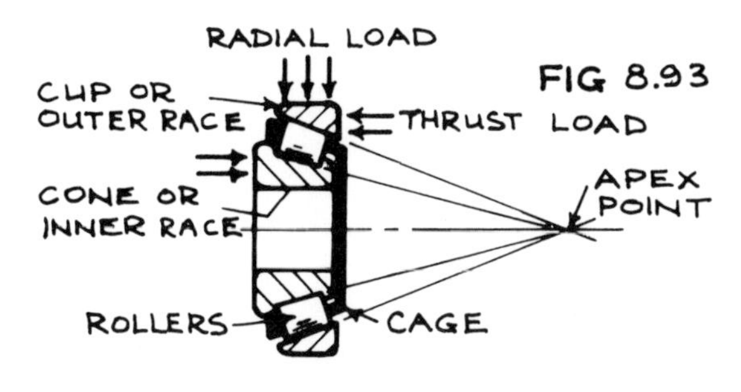

TAPERED ROLLER BEARING

The tapered roller bearing is capable of taking thrust as well as radial loads. The rollers are really conical, having apex points coinciding on the axis of the races.

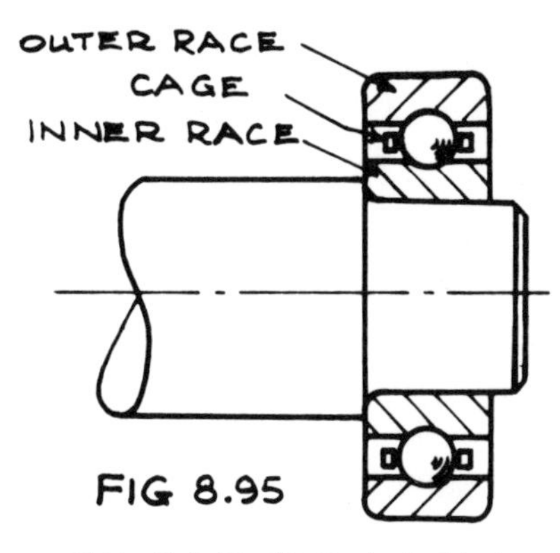

FIG 8.95

SINGLE ROW BALL BEARING

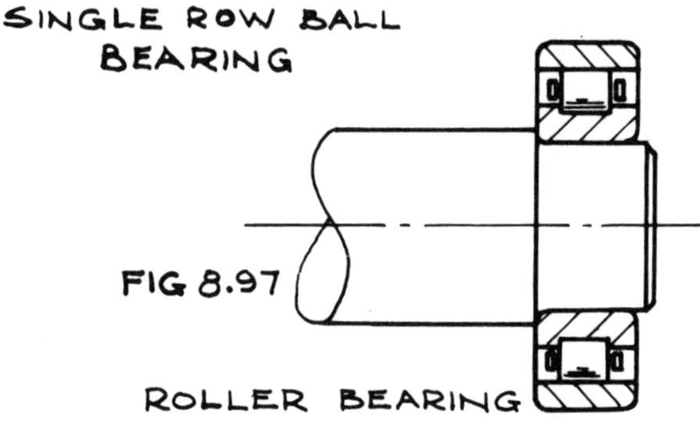

ROLLER BEARING

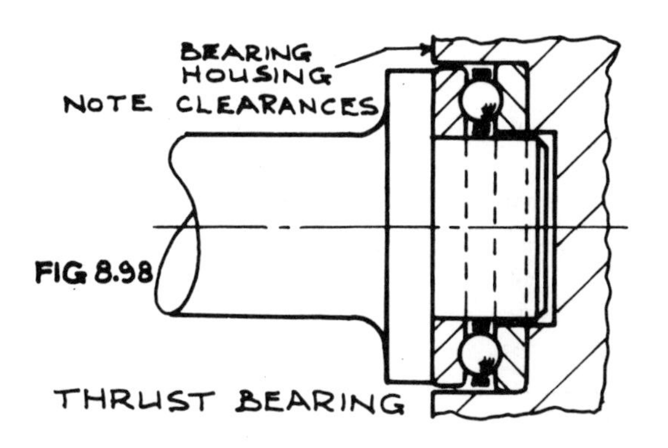

THRUST BEARING

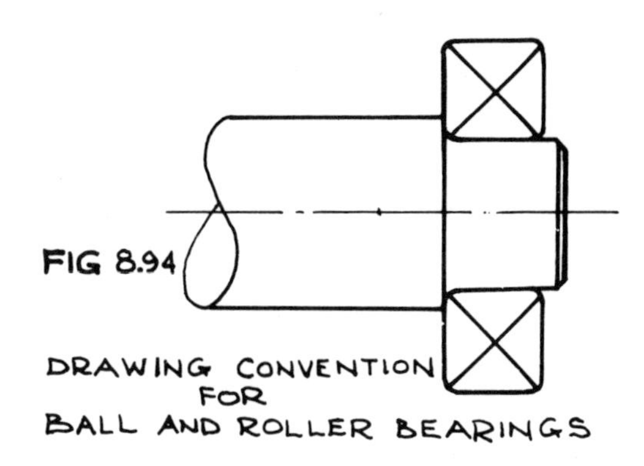

DRAWING CONVENTION FOR BALL AND ROLLER BEARINGS

On a drawing all types of ball and roller bearings are indicated by the convention shown, Fig. 8.94 A note is added to specify the exact type of bearing to be used. A ball bearing PILLOW BLOCK and other typical bearing mountings are shown on the following pages.

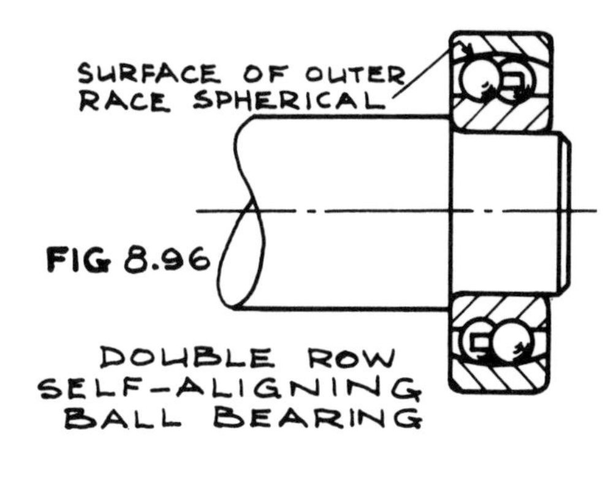

DOUBLE ROW SELF-ALIGNING BALL BEARING

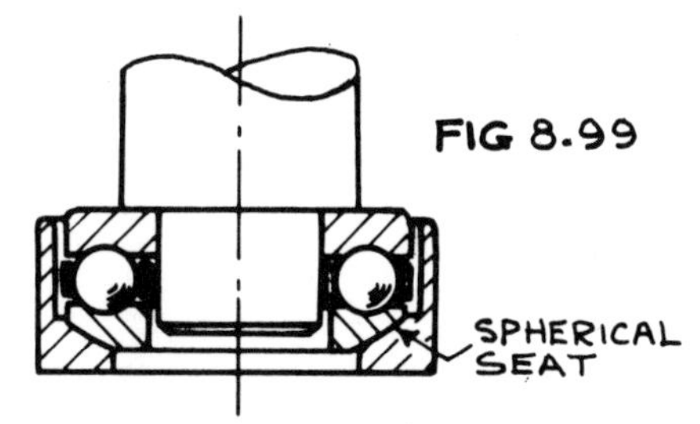

SELF-ALIGNING THRUST BEARING WITH SPHERICAL SEATING

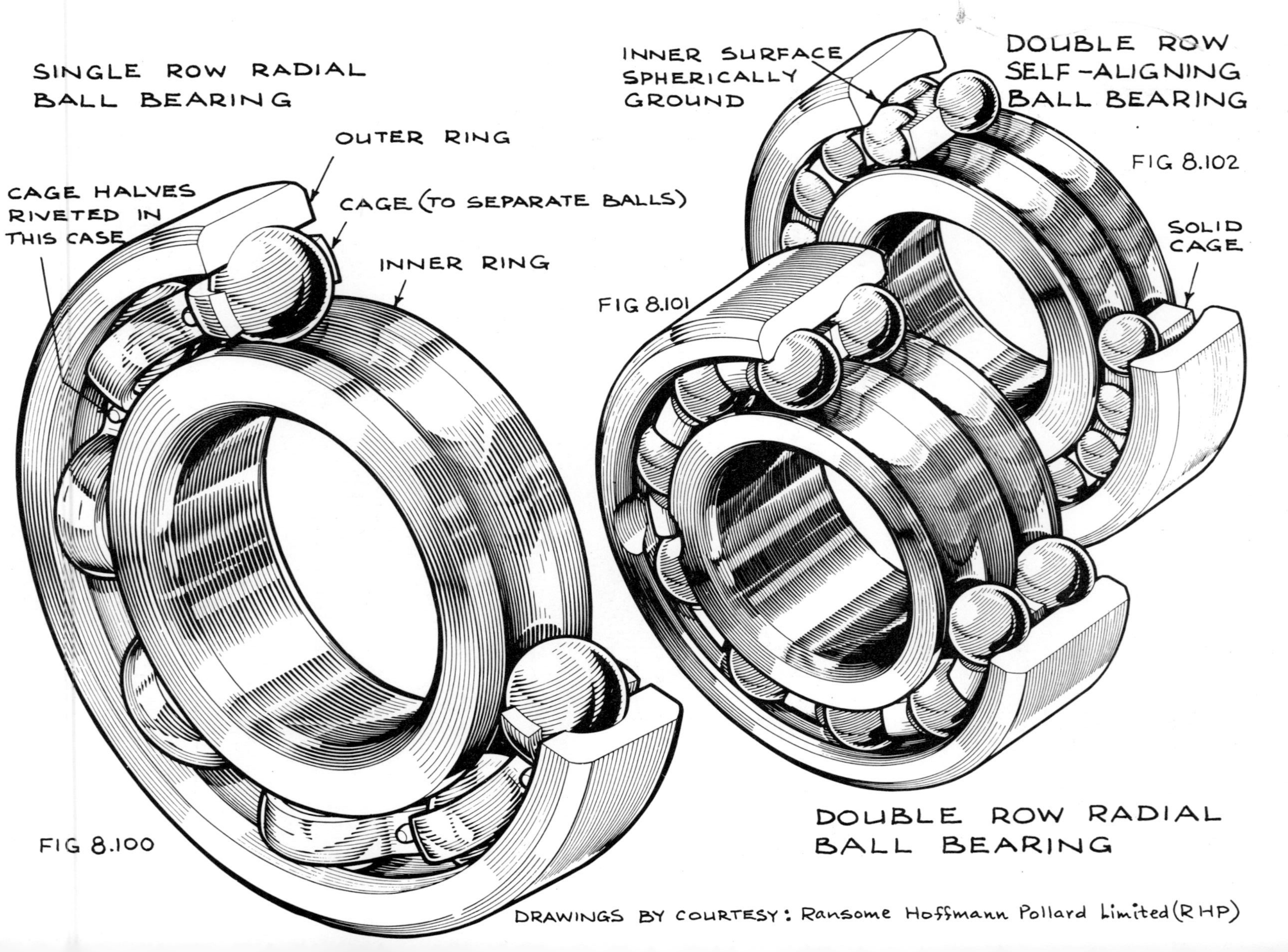
SINGLE ROW RADIAL
BALL BEARING
OUTER RING
CAGE HALVES
RIVETED IN
THIS CASE
CAGE (TO SEPARATE BALLS)
INNER RING
FIG 8.100
INNER SURFACE
SPHERICALLY
GROUND
DOUBLE ROW
SELF-ALIGNING
BALL BEARING
FIG 8.102
SOLID
CAGE
FIG 8.101
DOUBLE ROW RADIAL
BALL BEARING
DRAWINGS BY COURTESY: Ransome Hoffmann Pollard Limited (RHP)

MOUNTED BALL BEARINGS

FIG 8.103

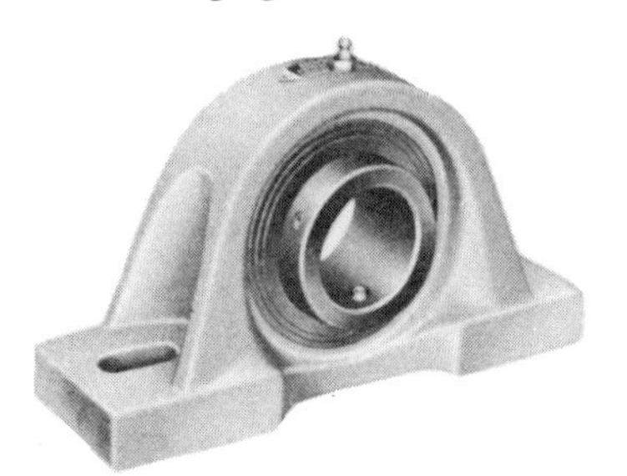

PILLOW BLOCK

Photographs by courtesy of :
MORSE CHAIN, DIVISION OF BORG-WARNER CORPN.

LOCKING PIN AND DIMPLE SYSTEM. Prevents rotation of outer race and locates lubricating hole in the outer race for free flow of grease.

FULL SELF ALIGNMENT

DEEP-GROOVED, ZONE HARDENED, BALL PATH. Darkened area is hardened leaving extended sections of inner race soft.

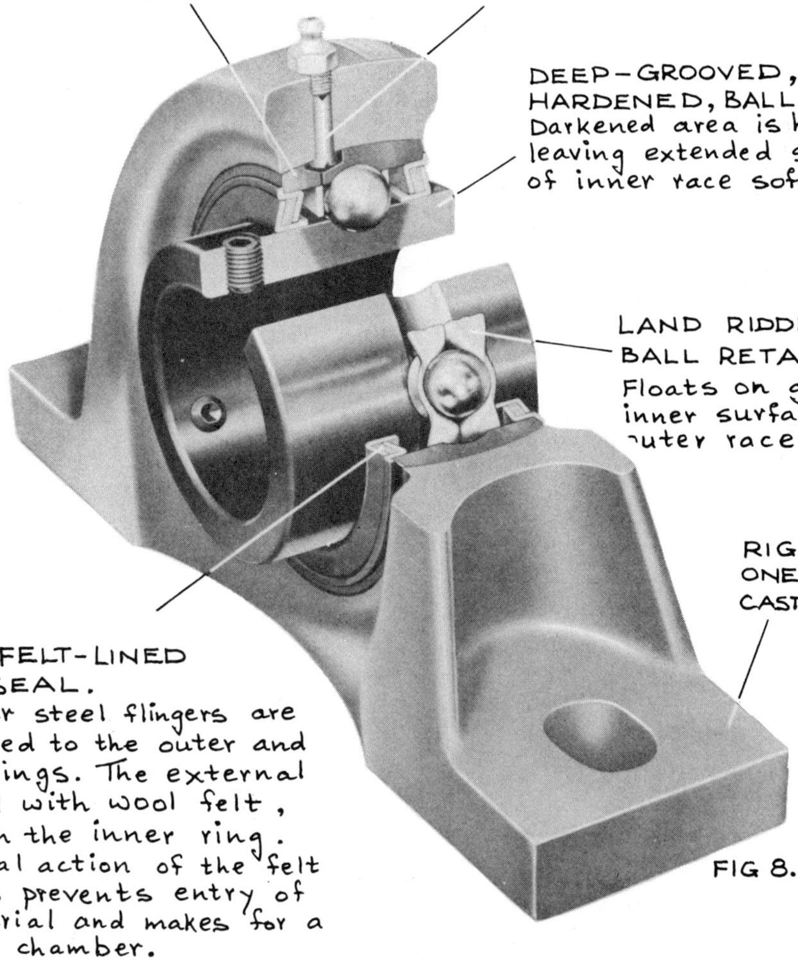

LAND RIDDEN BALL RETAINER. Floats on ground inner surface of outer race.

RIGID ONE-PIECE CAST HOUSING

PATENTED FELT-LINED FLINGER SEAL.
Two annular steel flingers are pressed fitted to the outer and inner race rings. The external flinger, lined with wool felt, rotates with the inner ring. The centrifugal action of the felt lined flingers prevents entry of foreign material and makes for a tight grease chamber.

FIG 8.104

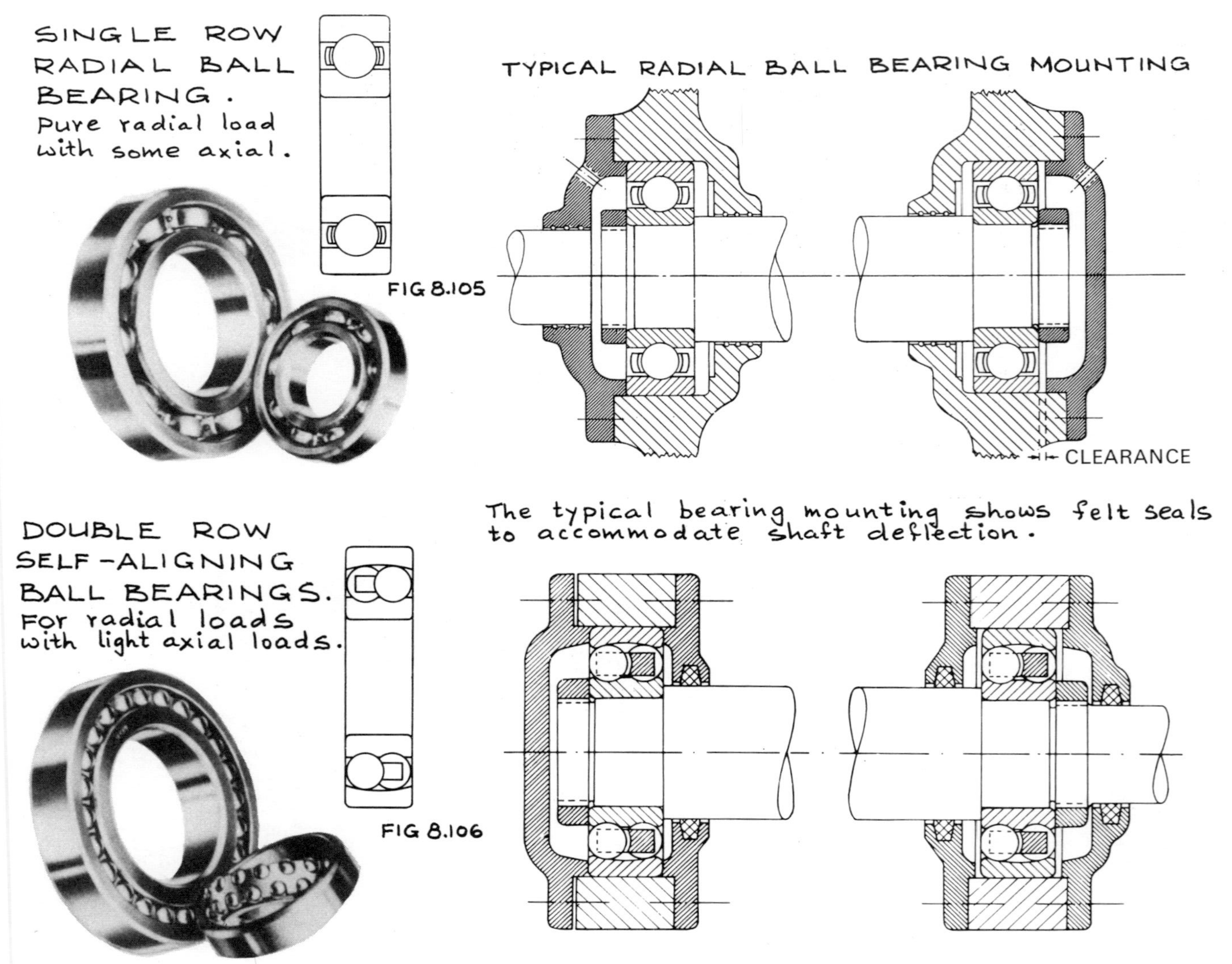

FIG 8.105

FIG 8.106

Photographs and drawings by courtesy of: Ransome Hoffmann Pollard Limited (RHP)

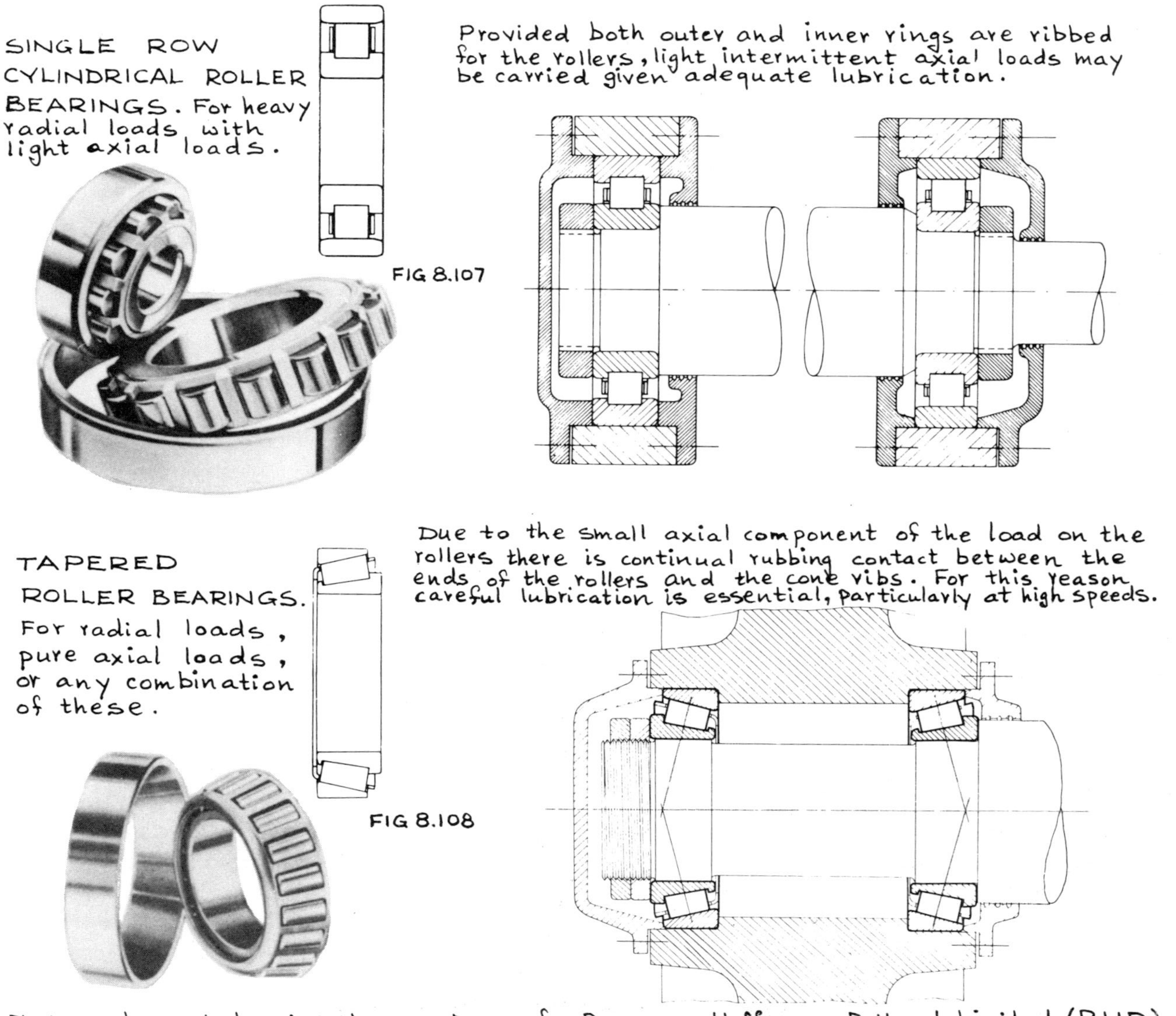

FIG 8.107

FIG 8.108

Photographs and drawings by courtesy of : Ransome Hoffmann Pollard Limited (RHP)

BEARING LUBRICATION

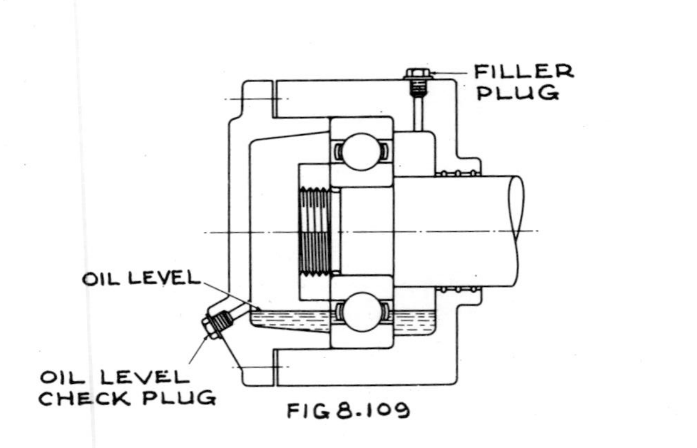

FIG 8.109

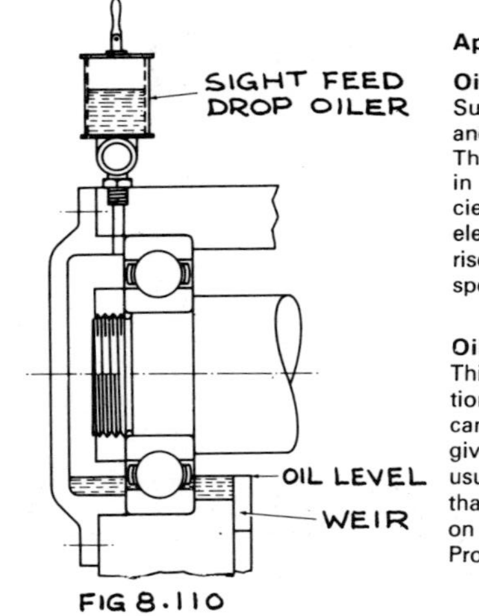

FIG 8.110

Application of oil

Oil bath lubrication (fig. 8.109)
Suitable for most horizontal shaft applications involving normal loads and speeds.
The optimum oil level is at the centre line of the lowest rolling element in the bearing. The surface area and volume of oil should be sufficiently large to maintain an adequate depth for the cage and rolling elements to dip into when running. A greater depth of oil could give rise to excessive oil churning and high temperatures – particularly if speeds are high.

Oil drip feed (fig. 8.110)
This method is suitable for most horizontal and vertical shaft applications. A metering device is used to supply oil to the bearing and this can be regulated to suit the operating conditions. Attention must be given to drainage to prevent over-filling the bearing housing, and it is usual on horizontal mountings to arrange the drain hole, or a weir, so that an oil bath is maintained at the correct level to provide lubrication on starting up.
Provision must be made to recirculate the oil or allow it to run to waste.

Oil mist lubrication(fig. 8.111)
This system lubricates and cools the bearing, and also helps to prevent foreign matter getting into the bearing housing. It is frequently used on applications such as machine tools where the same system can lubricate slideways, etc. The design should be in accordance with the recommendations of the lubrication system manufacturer or RHP engineers, to ensure that all working parts of the bearing are properly lubricated.
The compressed air used must be clean and dry and the bearings must be continually covered by a thin film of oil. It is advisable to switch on the lubrication system before the bearings start to rotate.

Automatic re-circulating system – vertical shaft(fig. 8.112)
There are several methods of providing a self-contained re-circulating oil system, using rotating flingers or screw thread devices. A typical system using a rotating flinger is illustrated. It is important to ensure that the oil level when the machine is stationary is below the lip on the bearing end cover.

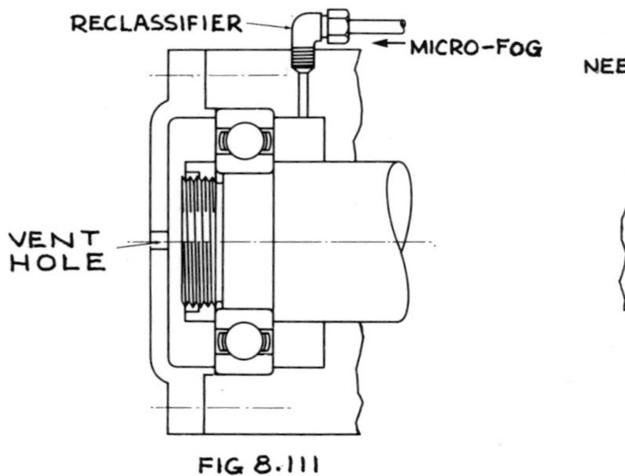

FIG 8.111

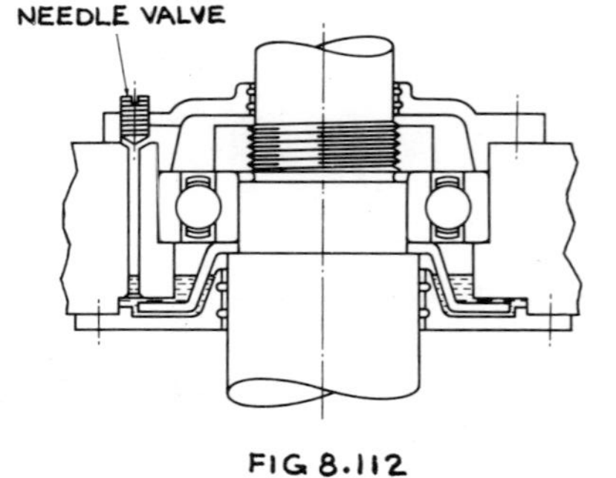

FIG 8.112

Drawings and information by courtesy of : Ransome Hoffmann Pollard Limited (RHP)

KEYS

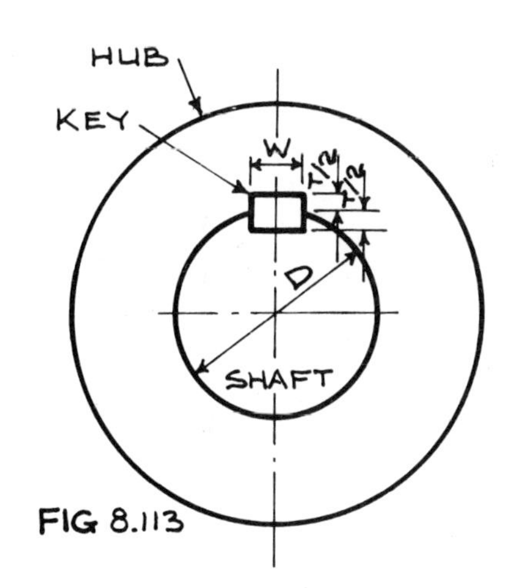

FIG 8.113

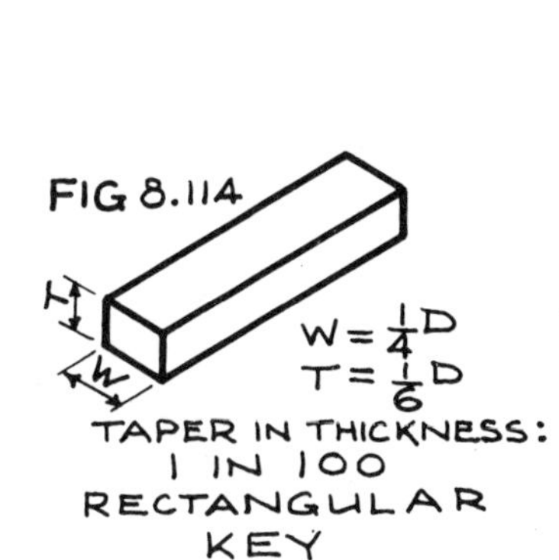

FIG 8.114

RECTANGULAR KEY

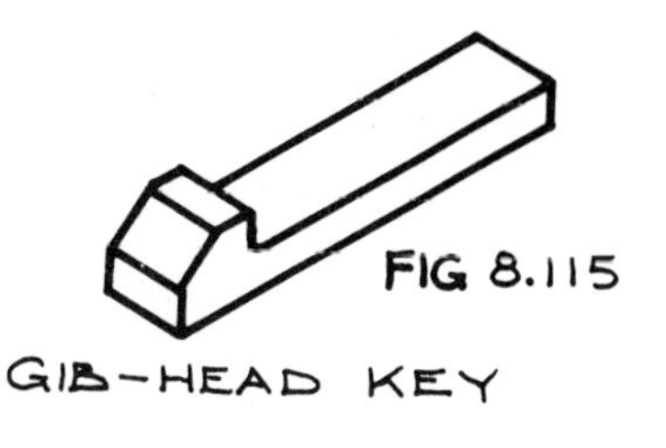

FIG 8.115

GIB-HEAD KEY

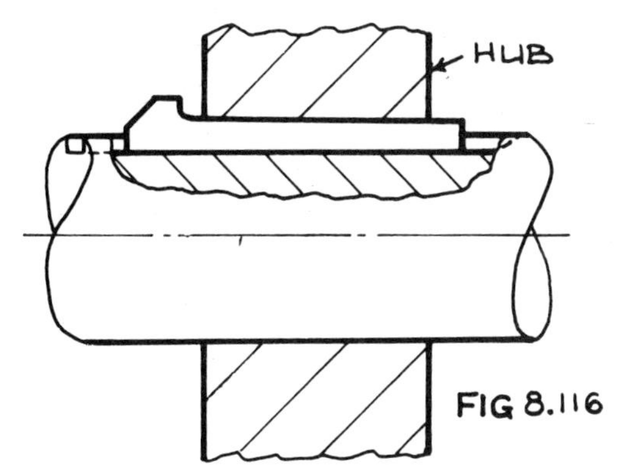

FIG 8.116

PART SECTION SHOWING GIB-HEAD KEY IN PLACE

A KEY is used to secure a wheel to a shaft or coupling-half to shaft, etc. and prevent relative movement between these parts. Keys are usually of steel and the rectangular key is the most common. It is driven into a keyway which is half in the shaft and half in the hub.

The GIB-HEAD KEY is shaped to facilitate withdrawal, when it is not possible for the key to be driven out.

A FEATHER KEY usually has parallel faces and sufficient clearance to permit axial movement between wheel and shaft. It is partly sunk into the shaft and secured by countersunk screws.

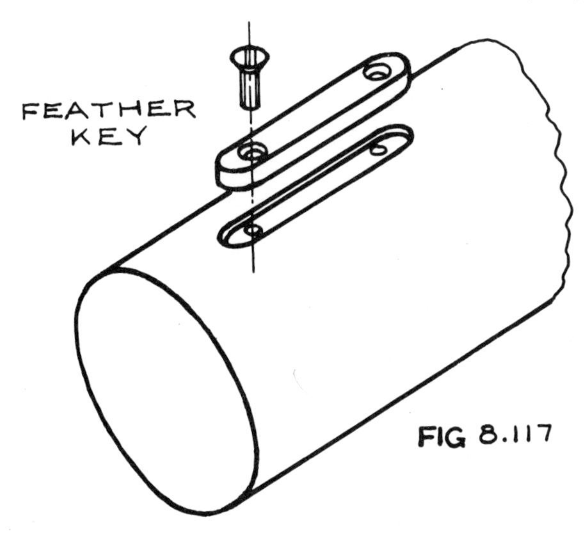

FIG 8.117

KEYS

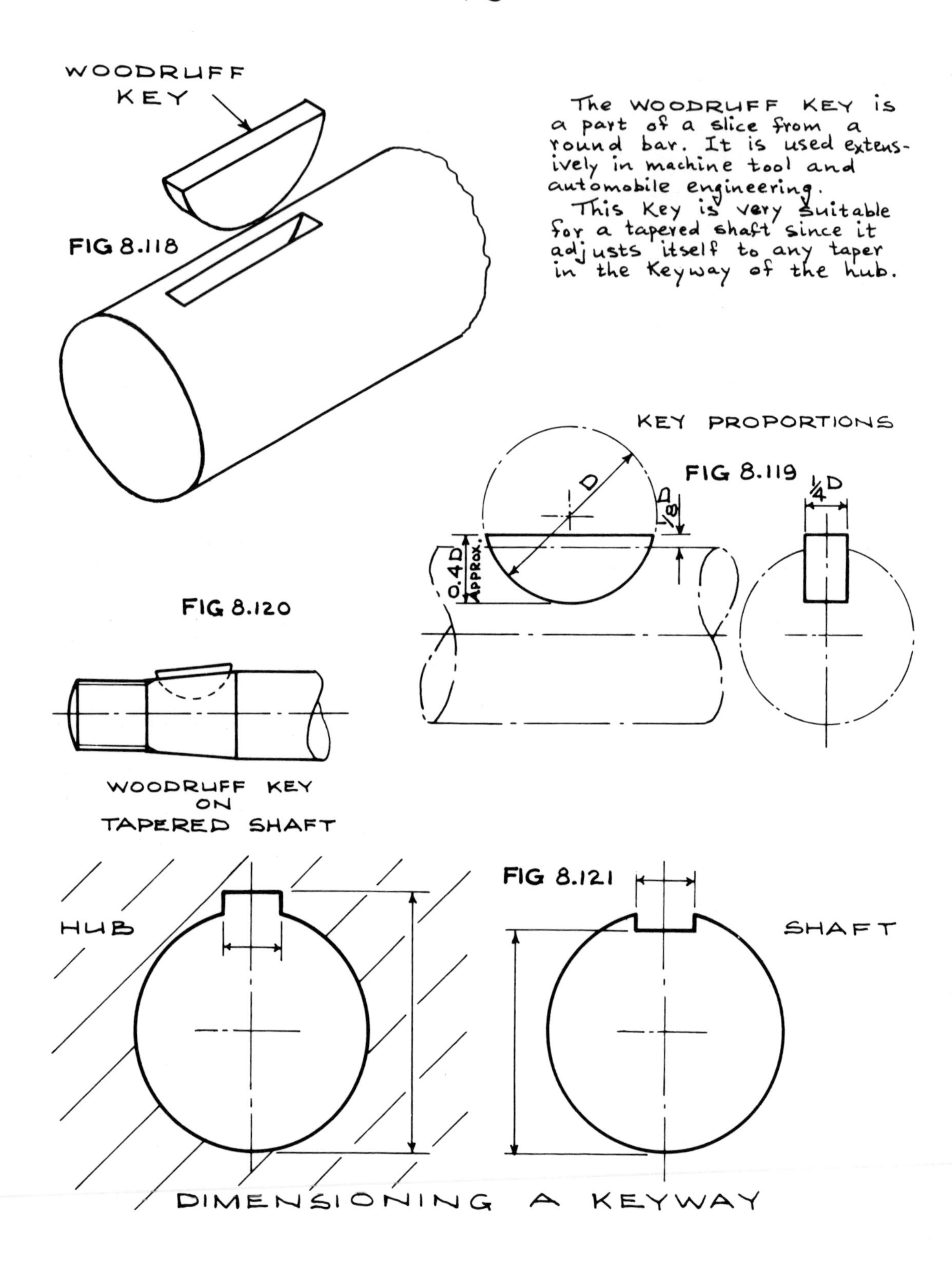
WOODRUFF KEY
FIG 8.118
The WOODRUFF KEY is a part of a slice from a round bar. It is used extensively in machine tool and automobile engineering.
This Key is very suitable for a tapered shaft since it adjusts itself to any taper in the Keyway of the hub.
KEY PROPORTIONS
FIG 8.119
D
1/8 D
1/4 D
0.4 D APPROX.
FIG 8.120
WOODRUFF KEY ON TAPERED SHAFT
FIG 8.121
HUB
SHAFT
DIMENSIONING A KEYWAY

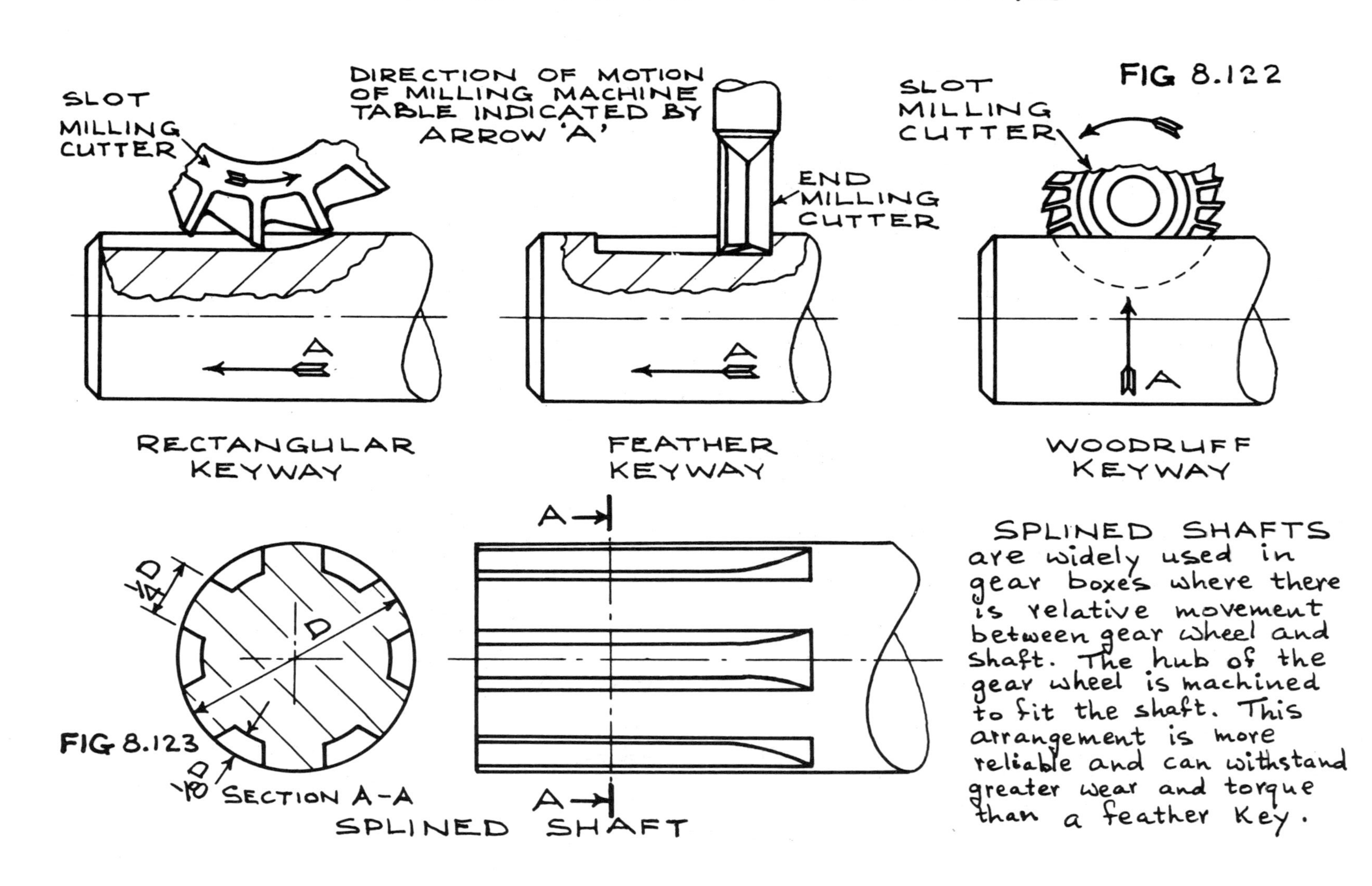
MACHINING OF KEYWAYS
SLOT MILLING CUTTER
DIRECTION OF MOTION OF MILLING MACHINE TABLE INDICATED BY ARROW 'A'
END MILLING CUTTER
SLOT MILLING CUTTER
FIG 8.122
A
RECTANGULAR KEYWAY
FEATHER KEYWAY
WOODRUFF KEYWAY
D
D/4
D/10
FIG 8.123
SECTION A-A
SPLINED SHAFT
SPLINED SHAFTS are widely used in gear boxes where there is relative movement between gear wheel and shaft. The hub of the gear wheel is machined to fit the shaft. This arrangement is more reliable and can withstand greater wear and torque than a feather key.

COUPLINGS

MUFF COUPLING

SQUARE

FIG 8.124

Two lengths of shafting are joined by a shaft coupling. There are many forms of coupling in use and three common types are here illustrated.

The SPLIT MUFF coupling is used in cases where removal of shaft is not practicable. The shafts almost touch end to end and the halves of the coupling are fitted over the shaft and bolted together. There is a keyway in one half of the coupling which fits the key on each shaft end. Square neck bolts are used to facilitate assembly.

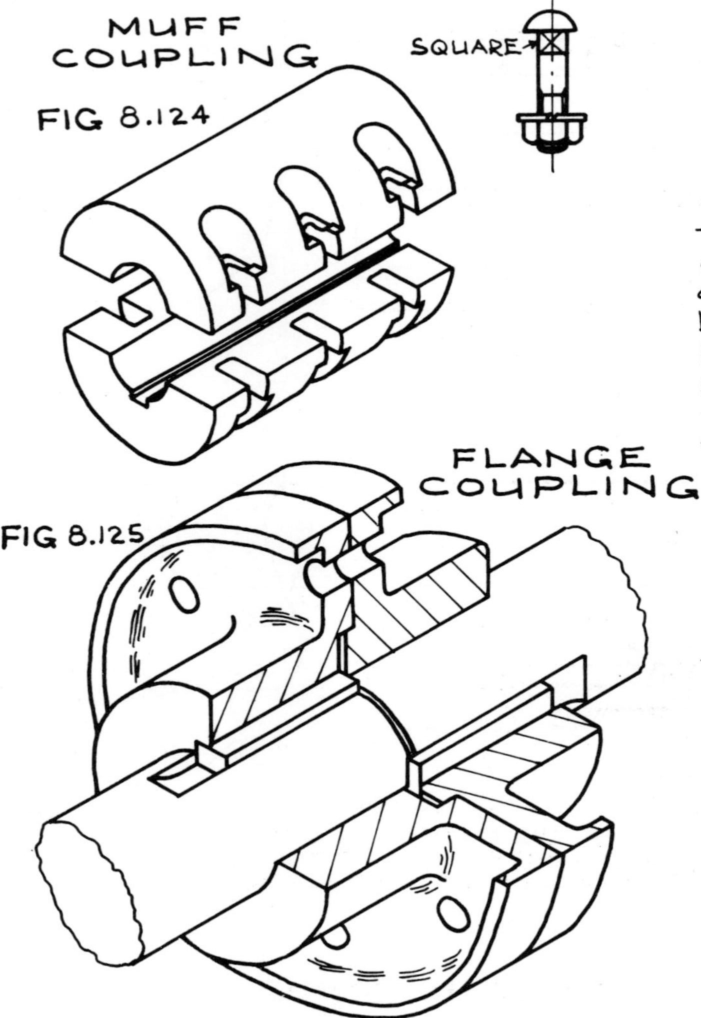

FLANGE COUPLING

FIG 8.125

The FLANGE coupling consists of two cast iron flanges, machined all over and keyed to the shaft ends by means of tapered keys. The keys are driven from the ends of the shafts before assembly. Alignment of the shafts is ensured by the raised part of one flange engaging with the recess in the other. Four, six, or more hex. hd. fitted bolts with nuts and washers are used to hold the flanges tightly together and take the drive. The rim on the flange is a safety measure.

FLEXIBLE COUPLINGS

FIG 8.126

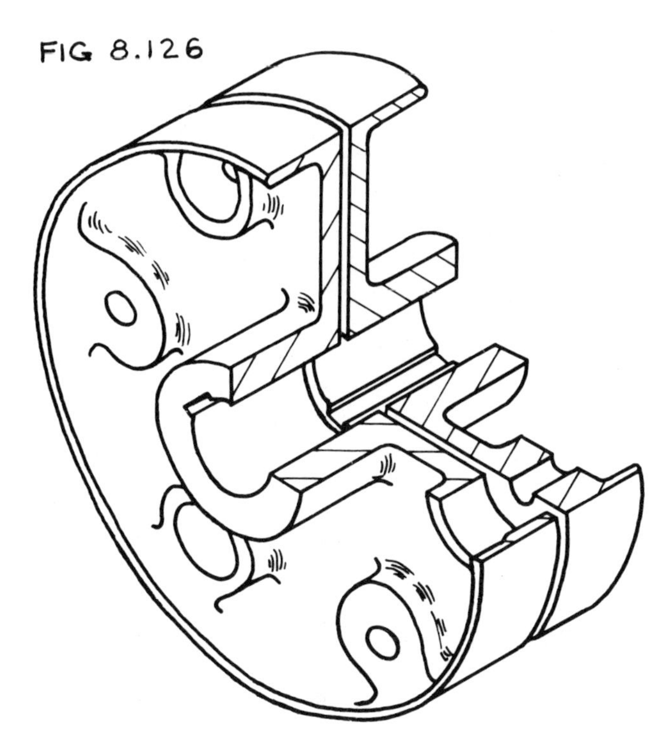

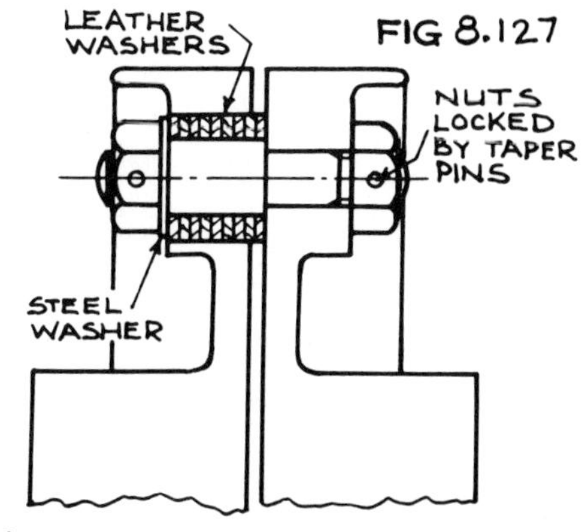

DETAIL OF DRIVING PINS

A PIN TYPE FLEXIBLE COUPLING is shown by Figs. 8.126 & 8.127. The halves are of cast iron, machined, and the driving pins are fixed in alternate holes of each flange.

In a more modern version a rubber bonded to metal pin insert is used in place of pin and washers.

FIG 8.128

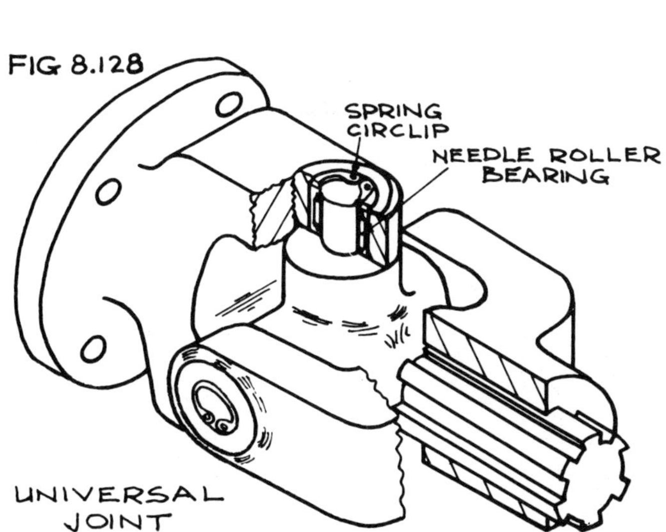

UNIVERSAL JOINT

A universal coupling is used to join two shafts whose axes intersect at an angle of not more than about 20°. An automobile type of coupling is shown in Fig. 8.128. Two such couplings are employed on splined shaft transmission for the orthodox motor car.

The universal coupling has wide application in the machine tool industry.

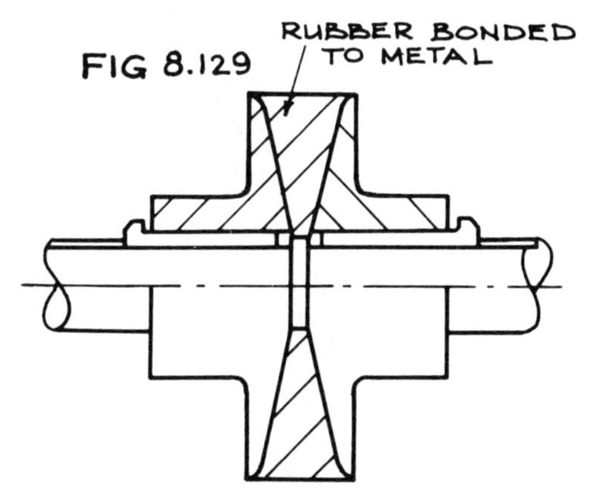

BARREL COUPLING

Fig. 8.129 shows one of the many different types of rubber bonded to metal couplings in wide use today. The rubber is wedge-shaped and this equalises the stresses produced in it during operation.

FIG 8.130

SPIDER TYPE FLEXIBLE COUPLING

THIS TYPE OF COUPLING GIVES ABSOLUTELY SILENT TROUBLE-FREE OPERATION

FIG 8.131

SPACER SPIDER
(TO SMALLER SCALE)

THE ONE-PIECE NITRIDE RUBBER SPIDER ABSORBS VIBRATIONS AND SHOCK LOADS. RAISED DOTS SEPARATE SPIDER FROM METAL COUPLING BODY.

PHOTOGRAPHS ARE BY COURTESY OF :
The Motor Gear & Engineering Co. Ltd. Essex.

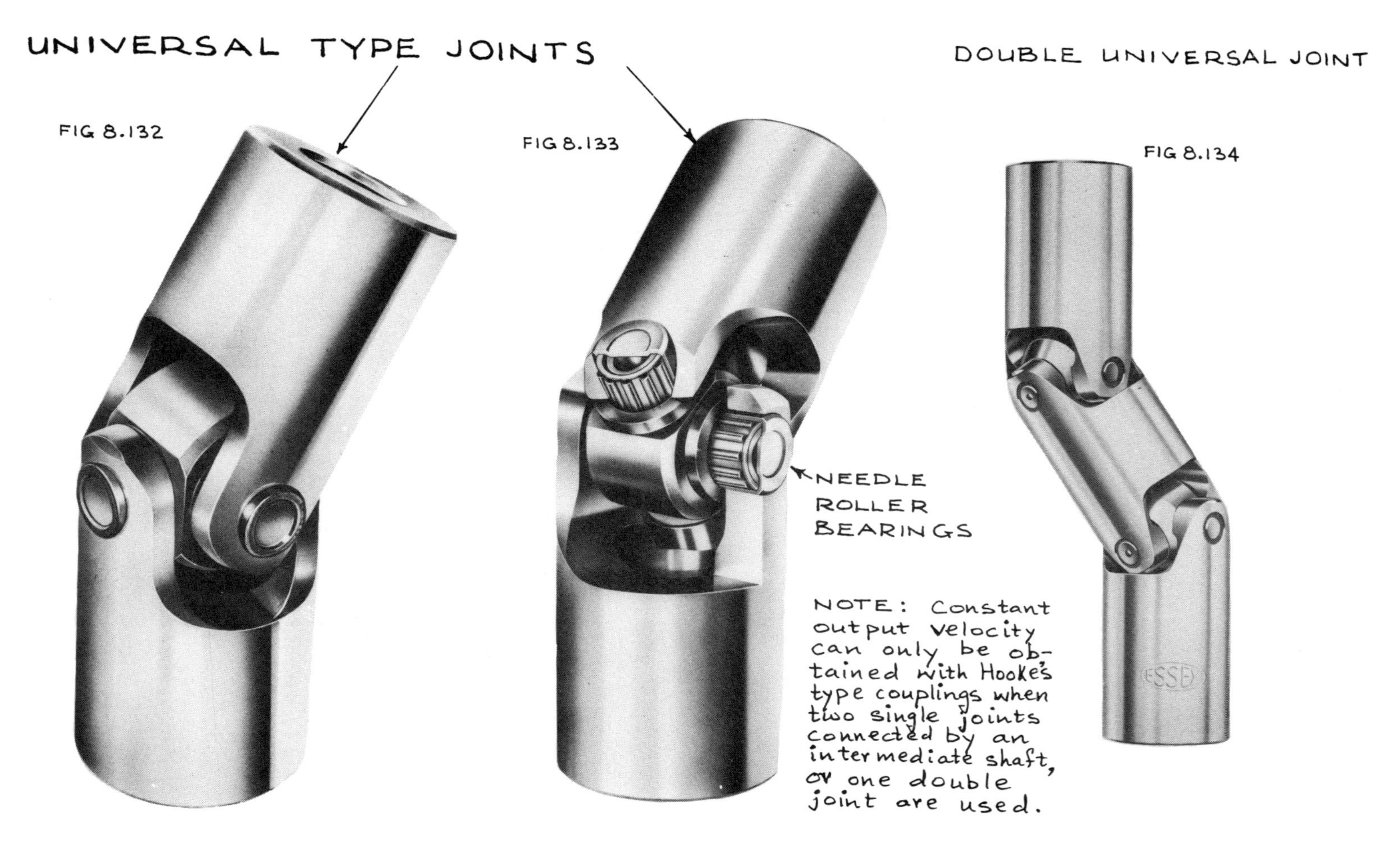
UNIVERSAL TYPE JOINTS
DOUBLE UNIVERSAL JOINT
FIG 8.132
FIG 8.133
FIG 8.134
NEEDLE ROLLER BEARINGS
ESSEX
NOTE: Constant output velocity can only be obtained with Hooke's type couplings when two single joints connected by an intermediate shaft, or one double joint are used.
PHOTOGRAPHS ARE BY COURTESY OF:
The Motor Gear & Engineering Co. Ltd. ESSEX.

TOLERANCED DIMENSIONS

Engineering components which are required to be manufactured to a high degree of accuracy and finish usually necessitate the use of expensive special-purpose machine tools and equipment. The higher the degree of accuracy and finish required, the greater is the cost of production. It is of the utmost importance, therefore, that the engineering designer should bear in mind the very great expense incurred by specifying 'tight' tolerances and he or she should design with limits as wide as possible consistent with the efficient functioning of the component in an assembly.

There are many reasons why it is not possible to consistently machine engineering components to exact size. Variation in finished size may be due to cutting tool wear, wear in machine tool parts, vibration, component material variation, imperfections in the design of the machine tool, etc. The problem for the designer is to decide how much variation from the nominal or basic size can be tolerated without impairing the efficiency of the assembly as a whole.

Sometimes the variation in finished dimensions may be widened by grading of the work pieces according to size by individual gauging. For example, car engine cylinder blocks bored to the larger limit may be fitted with pistons also finished to their large limit size, thus producing an efficient assembly. The process is called 'selective assembly'.

The magnitude of the tolerances will vary for the different dimensions of a component. Less critical dimensions are given a general tolerance which, in some cases, is added as a note to the drawing. Also it is obvious that the tolerance on, say, the diameter of a small watch component must be vastly different from the tolerance on a much larger diameter engine component, each requiring a particular type of clearance fit.

There are, broadly speaking, three types of manufacturing error. These may be classified as errors of size, position, or form. A simple example of drilling a given size of hole in a given position may serve to illustrate all three. First of all the hole may be over-size or under-size, whilst still being cylindrical in shape. The reason for the variation in diameter may be due to the drill or cutting tool being incorrectly sharpened or positioned. Again, the position of the hole may be in error, perhaps caused by an incorrectly sharpened drill wandering. Lastly, the hole may not be cylindrical in form, perhaps again due to wrong sharpening, or vibration, or the thickness of metal being drilled. The student should note that a tolerance on a diameter will permit an ovality equal to the tolerance. If then this form tolerance is unacceptable, a tolerance for circularity will be required. This would be of importance in the operation of pistons or rams where form affects efficiency. Under form tolerance, such qualities as straightness, flatness, parallelism, squareness, or angularity may also be specified. The placing of final tolerances on the component parts of a design calls for a high degree of skill and wide experience so consequently this job is usually left to the chief designer in consultation with the production engineer.

The necessity for interchangeability has led to the development of various systems of limits and fits. Some large firms developed their own system and one widely used system was the *Newall Limit System*, prepared by the Newall Engineering Company Limited. Any limit system attempts to specify suitable tolerances to be applied to a number of types of fit between mating parts, such as running fit, push fit, drive fit, etc. The most widely used system in Britain at present is that introduced by the British Standards Institution, BS 1916, in 1953 and a revised standard, BS 4500, in 1969. The latter standard, BS 4500, is concerned with the International Standards Organization (ISO) Limits and Fits, in metric units. This standard on limits and fits covers a range of basic size diameters up to 3150 mm, which includes 18 grades of tolerances for 27 fundamental deviations from the basic size, for both shafts and holes. The limit of size of any hole and shaft within the range covered may be selected from the prepared tables.

DEFINITIONS OF TERMS EMPLOYED IN A SYSTEM OF LIMITS AND FITS

The *basic size*, or *normal size*, is the size from which the limits are fixed, for both shaft and hole. There is a preference for the *hole basis* system since most holes machined by quantity production methods are produced using a standard diameter cutter such as a drill or reamer. It is therefore more economic to specify a change in shaft diameter to obtain the required type of fit.

The *tolerance* is the difference between the maximum and minimum size of shaft or hole. Tolerances are *graded* according to magnitude by use of numbers prefixed by the letters IT. Lower numbers indicate fine tolerances and higher numbers, up to 16, the wider tolerances.

The *allowance* is the difference between the high limit of size of the shaft and the low limit of size of the hole. The term *minimum clearance* is now more frequently used instead of 'Allowance'.

The three types of fits are:

(i) *Clearance fit*;
(ii) *Transition fit*;
(iii) *Interference fit*.

These types of fits and illustrations of the foregoing basic definitions of terms are shown in figures 8.137 to 8.139.

Note. The student should appreciate that tolerances may be applied to any dimension on a drawing and not just to diameters of holes and shafts.

The recommended method of showing tolerances on individual dimensions on a drawing is shown in figure 8.135.

The maximum limit is placed above the minimum limit.

Alternatively, the nominal size may be stated, with limits of tolereance above and below this size, figure 8.136 (i).

The limits should preferably be equally disposed on either side of normal size, figure 8.136 (ii).

The case where the limits are all to one side of the nominal, one limit being zero, is shown in figure 8.136 (iii). This is an example of *unilateral* tolerancing whereas cases (i) and (ii) are *bilateral*.

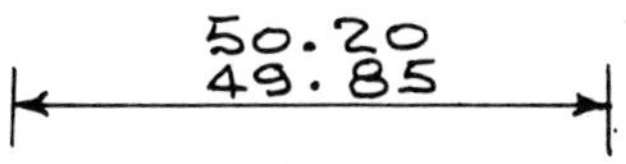

$$\begin{matrix}50.20\\49.85\end{matrix}$$

FIG 8.135

(i) $40^{+0.01}_{-0.03}$

(ii) 35 ± 0.2

(iii) $30^{\;0}_{-0.05}$

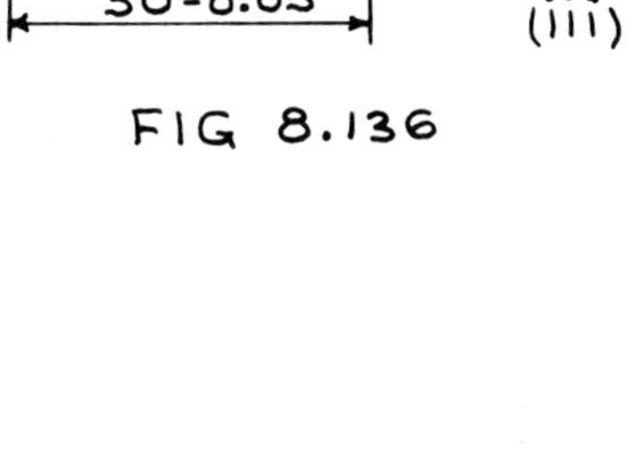

FIG 8.136

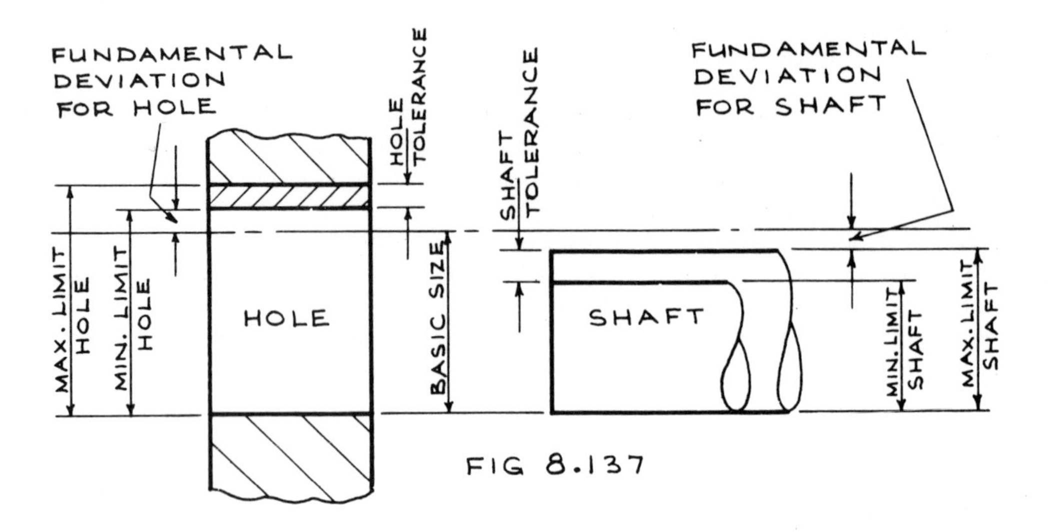

FIG 8.137

DEFINITIONS OF TERMS EMPLOYED IN A SYSTEM OF LIMITS AND FITS

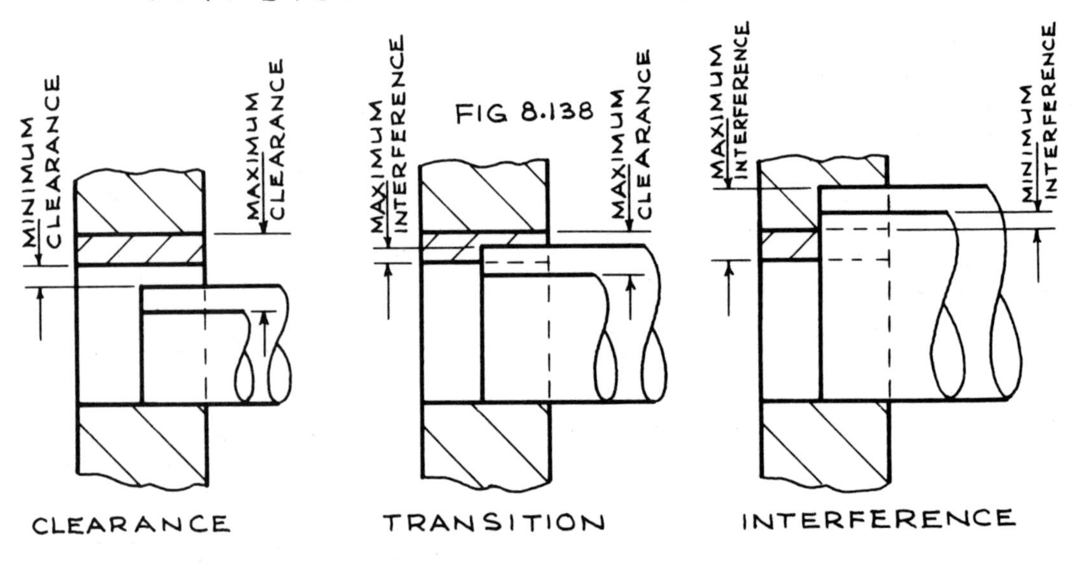

FIG 8.138

TYPES OF FITS

FIG 8.139

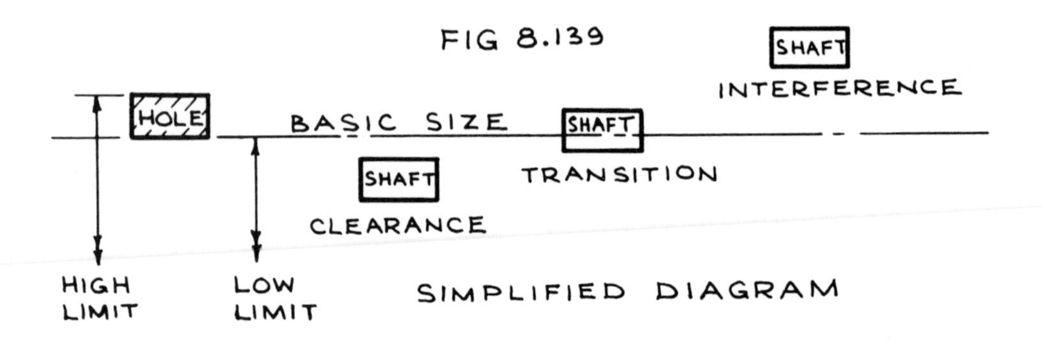

SIMPLIFIED DIAGRAM

Fits which are taken from the ISO limits and fits tabulated in BS 4500 may be designated on a drawing using symbols. For example, 50 H 7 or 50 H 7 $^{(50.025)}_{(50.000)}$ for holes of basic size 50 mm.

CUMULATIVE EFFECTS OF TOLERANCES

Tolerances have a cumulative effect, whether dimensions are added or subtracted, and for this reason datum dimensioning rather than chain dimensioning is preferable where a number of tolerances are involved on the same component.

Chain dimensioning may result in an accumulation of tolerances between hole A and hole B and edge of plate, figure 8.140.

Datum dimensioning avoids such an accumulation of tolerances, figure 8.141.

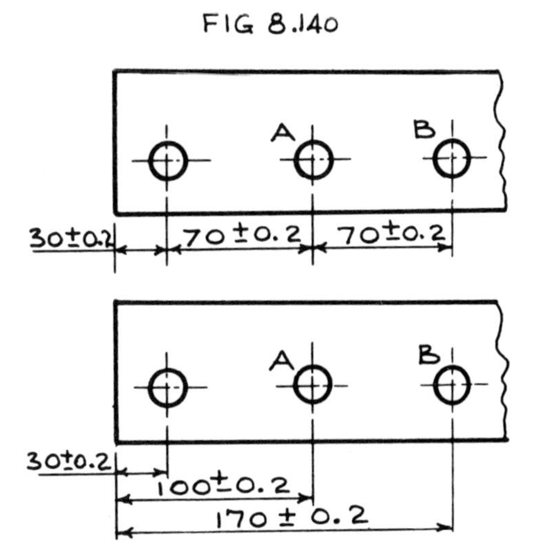

FIG 8.141

TOLERANCE DISPOSITION

The Fundamental Deviation, see figure 8.137, determines the tolerance zone position for either a hole or a shaft relative to the zero line. When the basic size coincides with the minimum limit of the hole tolerance the FD is zero and similarly for a shaft when the basic size coincides with the maximum limit.

In the ISO system, as explained in detail in BS 4500, capital letters designate the FD of the hole and lower case letters that of the shaft. The FD may be positive or negative and increases disproportionately for the extreme positions from the zero, H or h, classification of the hole or shaft respectively. A fit between a shaft and hole, for example, may be specified as 100 mm diameter H7/g6, which is in fact a clearance fit having hole tolerance $^{+35}_{0}$ and shaft tolerance $^{-12}_{-34}$ in thousandths of a millimetre.

When the lower limit of the hole is made equal to the basic size and both limits of the shaft are less than the basic size this is known as the *unilateral hole basis system*. This is the preferred system.

Example 1. A shaft is to be assembled to a hole of basic size 100 mm, with a clearance fit having a minimum clearance of 0.036 mm. If the hole tolerance is 0.054 mm and the shaft tolerance is 0.035 mm determine the largest and smallest shaft dimensions and the maximum clearance. How would you specify this type of fit by reference to BS 4500 and data sheet 4500 A of selected ISO fits for unilateral Hole Basis system?

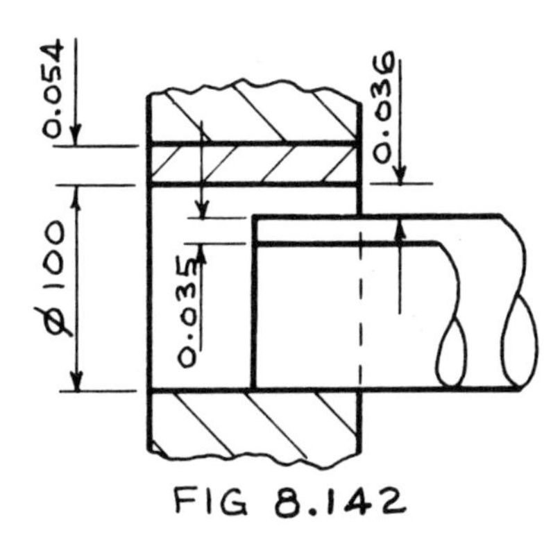

FIG 8.142

Largest shaft diameter
= 100 – 0.036
= 99.964 mm

Smallest shaft diameter
= 100 – 0.036 – 0.035
= 99.929 mm

Maximum clearance is given by largest hole diameter minus smallest shaft diameter.

= 100.054 – 99.929
= 0.0125 mm

This Fit may be specified as 100 dia H8/f7.

Example 2. Two sliding parts of a machine tool fit together by means of a tongue and slot as shown in figure 8.143. Using a basic slot width of 50 mm and a minimum clearance of 0.009 mm determine the sizes for tongue and slot if the maximum clearance is to be 0.050 mm with a tongue tolerance of 0.016 mm.

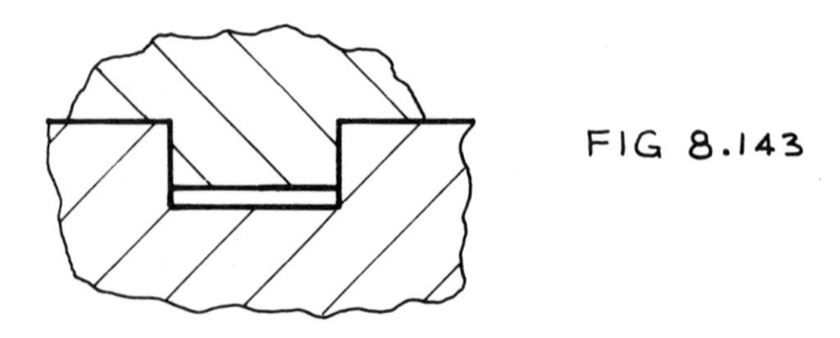

FIG 8.143

Note. Once again use is made of a diagram showing tolerances and clearances in greatly exaggerated form, as in figure 8.144.

Smallest size for tongue = 50 − 0.009 − 0.016 = 49.975 mm
Largest size for tongue = 50 − 0.009 = 49.991 mm
Smallest size for slot = Basic size = 50.000 mm
Largest size for slot = smallest tongue + Max. clearance
= 49.975 + 0.050 = 50.025.

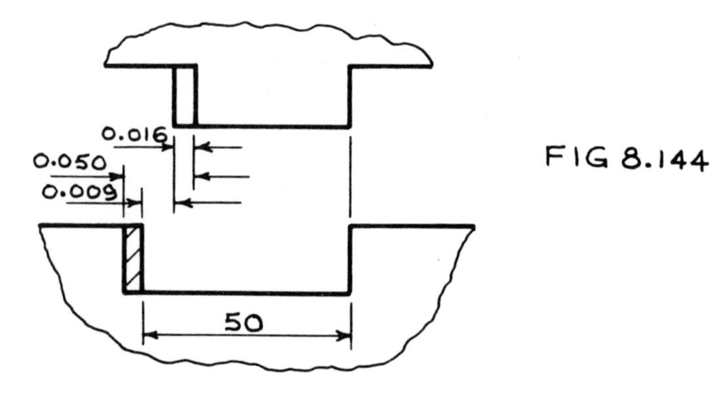

FIG 8.144

The British Standards Institution publications on Limits and Fits, BS 1916 and the later metric version BS 4500, are extremely comprehensive and cover a very wide range of sizes. For example, the number of possible combinations of holes and shafts in one range of hole size only is between four and five thousand! However, for general light and medium engineering purposes a selected range of fits may be usefully referred to. These are published in a single data sheet, BS 4500 A, of the British Standards Institution. This sheet contains 13 unilateral hole based fits using the holes H6, H7, H8 and H11 only.

A commonly used 'running fit' suitable for oil and grease lubricated shaft bearings, such as used in gear-boxes and pumps, may be designated H8/f7 which is a clearance fit. If we suppose the hole nominal size to be 35 mm diameter, then from data sheet 4500 A this would give a hole tolerance of $^{+39}_{\ 0}$ and shaft tolerance of $^{-25}_{-50}$ in thousandths of a millimetre.

A second example may be H7/h6, a 'location fit', which is used for components such as valve guides in cylinder heads where a minimum clearance type of fit is desirable.

Again, a 'heavy press fit' or shrink fit may be required for assembly of cylinder liners to engine blocks and in this case H7/s6 interference type fit may be used for permanent assembly. Shrinkage of the liners in liquid nitrogen is the method usually adopted for this type of assembly.

A transition type fit, H7/k6, which is sometimes referred to as an 'easy keying' or 'push' fit may be used for parts such as hand-wheels, gears, pulleys, clutch members, etc., where slight interference only is necessary.

All of the foregoing examples of fits are to be found on data sheet BS 4500 A but a table containing a secondary selection of fits may be referred to, if required, in the complete British Standards Institution publication.

ASSEMBLY DRAWING EXERCISES

(1) Details of a Milling Machine Steady are given in the detail sheet. Do not copy the views as shown but draw, full size, with the parts assembled:

(a) a sectional elevation, the plane of the section and the direction of the required view being indicated by A-A;

(b) an outside plan projected from view (a);

(c) an outside view as seen when looking in the direction of arrow X.

Insert six important dimensions and add title, scale, etc. (N.I.G.C.E.).

(2) Details of a Swivel Wheel for an Industrial Trolley are given in the detail sheet, also a diagram showing the assembled parts. Draw, full size, the following views showing the parts assembled:

(a) a front elevation as seen looking in the direction of arrow 'X'. Show a portion of the bottom end of a 38 diameter trolley leg, machined to receive part 1 and secured to it by two 10 diameter standard hexagon bolts and nuts;

(b) a sectional end elevation, the plane of section being taken through the vertical centre line of the wheel and as seen looking from the left of view (a);

(c) a plan projected from view (a).(N.I.G.C.E.)

(3) Details of a Fire Isolating Valve are given in the detail sheet. The valve is designed to operate on a fuel oil pipe line and to cut off further supply of fuel oil in the event of an accidental outbreak of fire. This is accomplished by the heat of the fire melting a fusible nut (part 7), which is made from a material of low melting point, allowing the spring-loaded valve (part 5) to snap shut and thus preventing any further flow of fuel oil through the valve.

Draw, full size, using first angle projection, the following views of the assembled valve:

(a) a sectional front elevation as seen looking in the direction of arrow 'X' and taken through the centre of the valve;

(b) a plan;

(c) an end elevation as seen looking from the left of view (a).

The length of the spring on your drawing is to be as shown in the detail (46 mm). The upper surface of the spring rests against the interior face 'S' of the spring compression nut (part 2). The lower surface of the spring rests against the face 'P' of the valve (part 5). The valve will therefore be shown open.

No hidden detail need be shown in any of the views.

Balloon reference the sectional view and show a parts list, quoting a suitable material for each component.

Show the following dimensions on your drawing:

(i) The distance from the horizontal centre line of the valve body to the upper face 'A' of part 2;

(ii) The distance from the horizontal centre line of the valve body to the lower face 'B' of part 7;

(iii) The vertical distance the valve is open.

Add the title, scale and system of projection. (N.I.G.C.E.)

(4) Details of a Control Mechanism are given. In the assembled condition the surface A on the control lever (part 2) faces against the surface B on the sector bracket (part 1) and is held there by the pivot pin (part 3) and the collar (part 4). The upper portion of the control lever has a stud (part 6) screwed into it, which moves along the slot in the sector bracket. The lever can be held in any position by tightening the handle (part 5) on the stud. A forked end (part 7) is joined to the bottom of the control lever by a 4 mm diameter rivet which is not detailed on the drawing.

(a) With the control lever (part 2) in the vertical (mid) position and the forked end and rod (parts 7 and 8) horizontal, using first angle projection draw, full size, the following views of the assembled components:

(i) a front elevation based on the detailed front elevation of the control bracket (i.e. the view on the left);
(ii) a sectional end elevation taken through the vertical centre line of view (i) and located on the right hand side of view (i).

(b) Balloon reference all components, using the sectional end elevation mainly for this purpose.
(c) Add the title, scale and symbol for the projection system.
(d) Measure and show on the front elevation:
(i) the angle of travel of the control lever from one extreme end of the slot to the other;
(ii) the vertical movement of the forked end (part 7) when the control lever is moved from the mid position to either extreme position. (N.I.G.C.E.)

(5) Parts for a Conical Friction Clutch are shown in the detail sheet. Make the following drawings of the assembled clutch and shaft portions, full size and using either first or third angle orthographic projection:

(i) a sectioned longitudinal front elevation, taken on the shafts axis, with clutch in engaged position and forked lever upright;
(ii) a sectioned end elevation taken through the vertical centre line of the forked lever in view (i) and seen looking from the left.

Add a title, scale, projection symbol and the following two dimensions:

(a) the distance between surface 'C', Part 1, and surface 'D', part 2;
(b) the distance between the two end surfaces of the shafts, parts 6 and 8. (N.I.G.C.E.)

(6) Details for a Low Pressure Pneumatic Cylinder for low cost automation work are shown in the sheet. Draw, full size, in third angle projection the following views of the complete assembly;

(a) a longitudinal elevation showing the top half in section on the centre line;
(b) an end elevation projected from the left, showing the left half in section on A-A and the right half in section through the cylinder looking on the piston.

Show the piston at mid-stroke and insert the important dimensions. Omit hidden detail. (N.I.G.C.E.)

(7) Details of an Oil Fog Lubricator are shown in the sheet. Draw, full size, in first or third angle projection the following views of the assembled lubricator:

(a) an elevation, right half in section on B-B, left half outside view and having the axis X-X vertical;
(b) an end elevation projected from the left.

Show the needle valve closed; omit hidden detail and insert important dimensions. (N.I.G.C.E.)

ASSEMBLY DRAWING EXERCISE

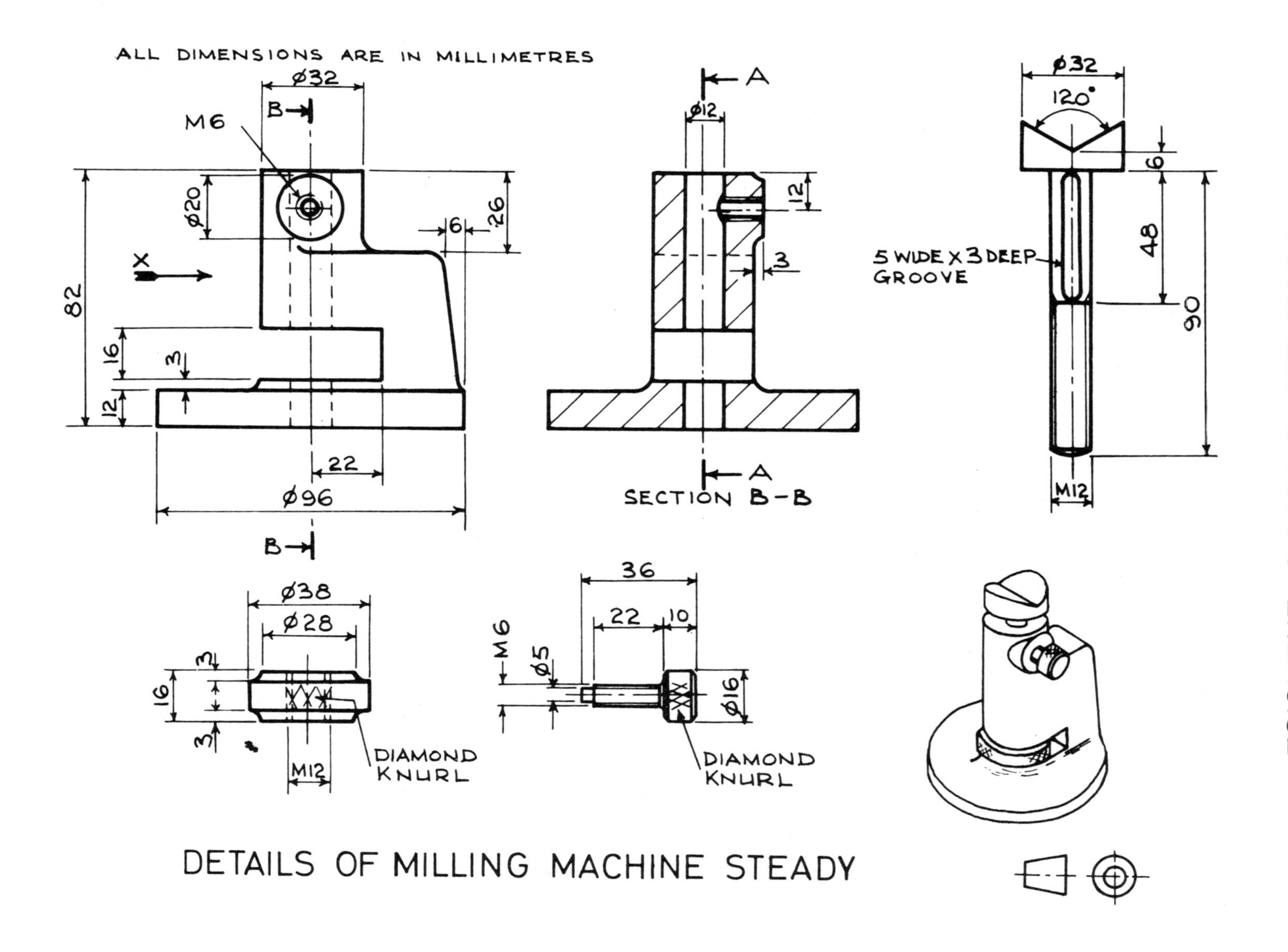

DETAILS OF MILLING MACHINE STEADY

ASSEMBLY DRAWING EXERCISE

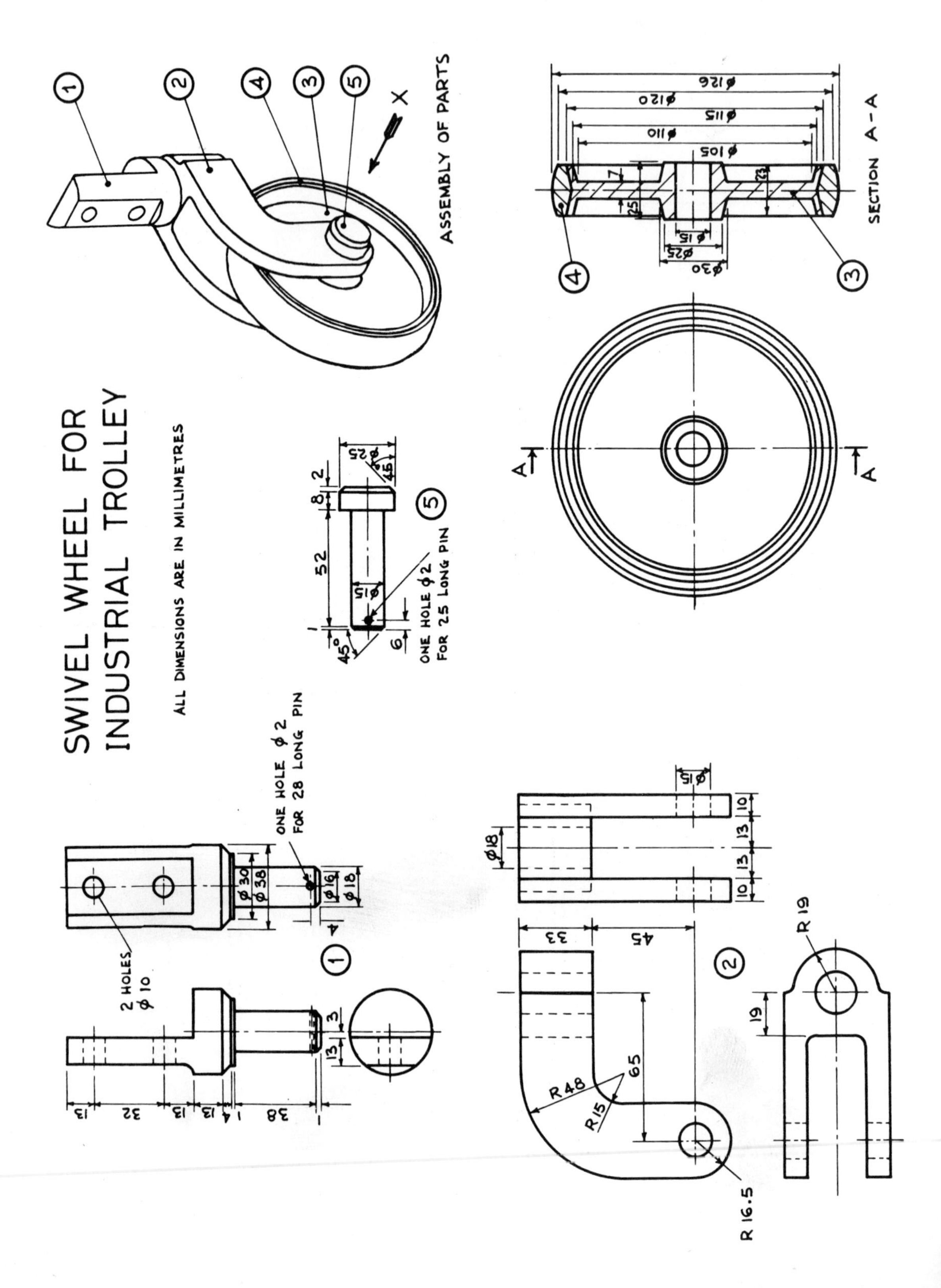

ASSEMBLY DRAWING EXERCISE

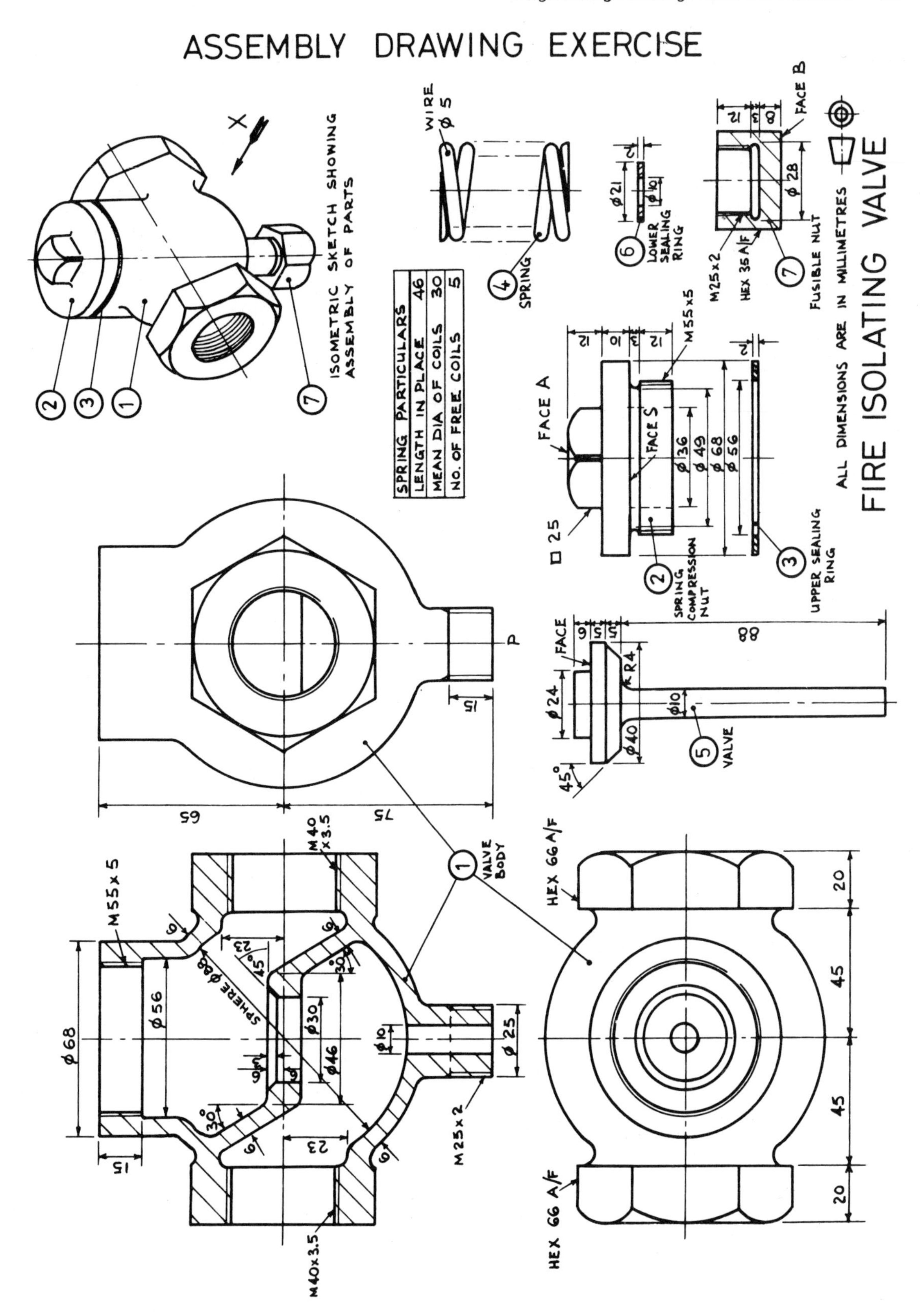

SPRING PARTICULARS	
LENGTH IN PLACE	46
MEAN DIA OF COILS	30
NO. OF FREE COILS	5

ASSEMBLY DRAWING EXERCISE

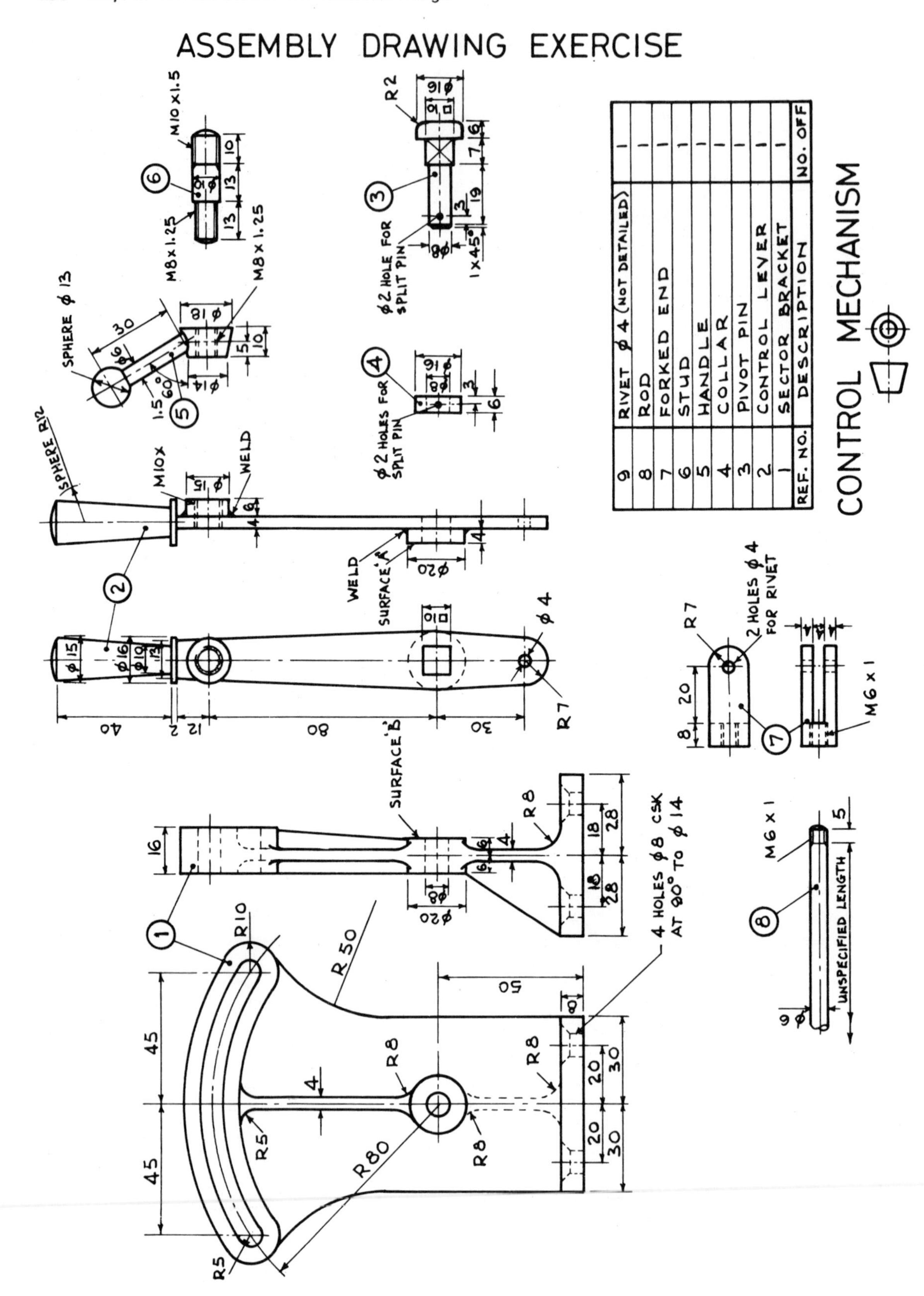

ASSEMBLY DRAWING EXERCISE

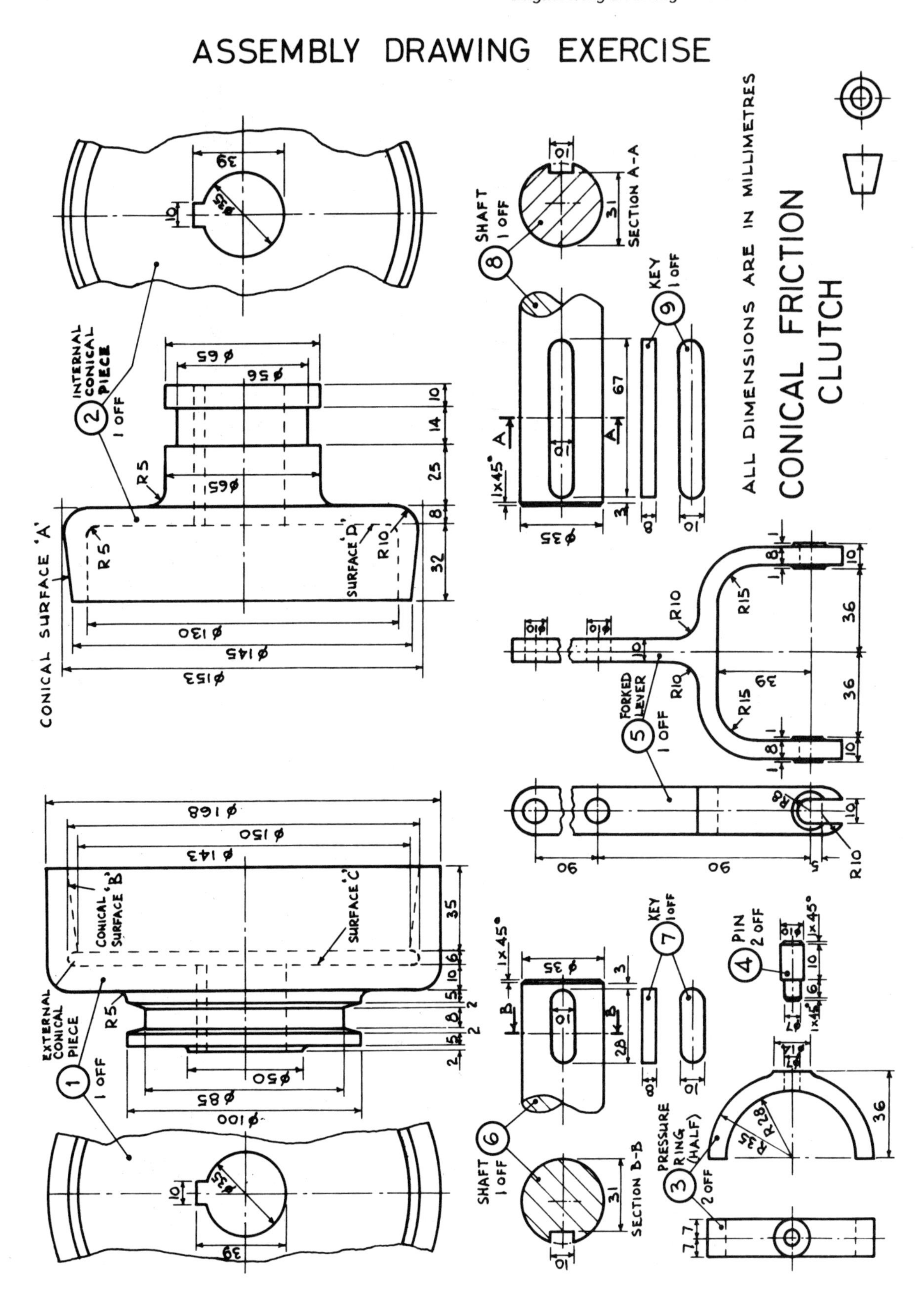

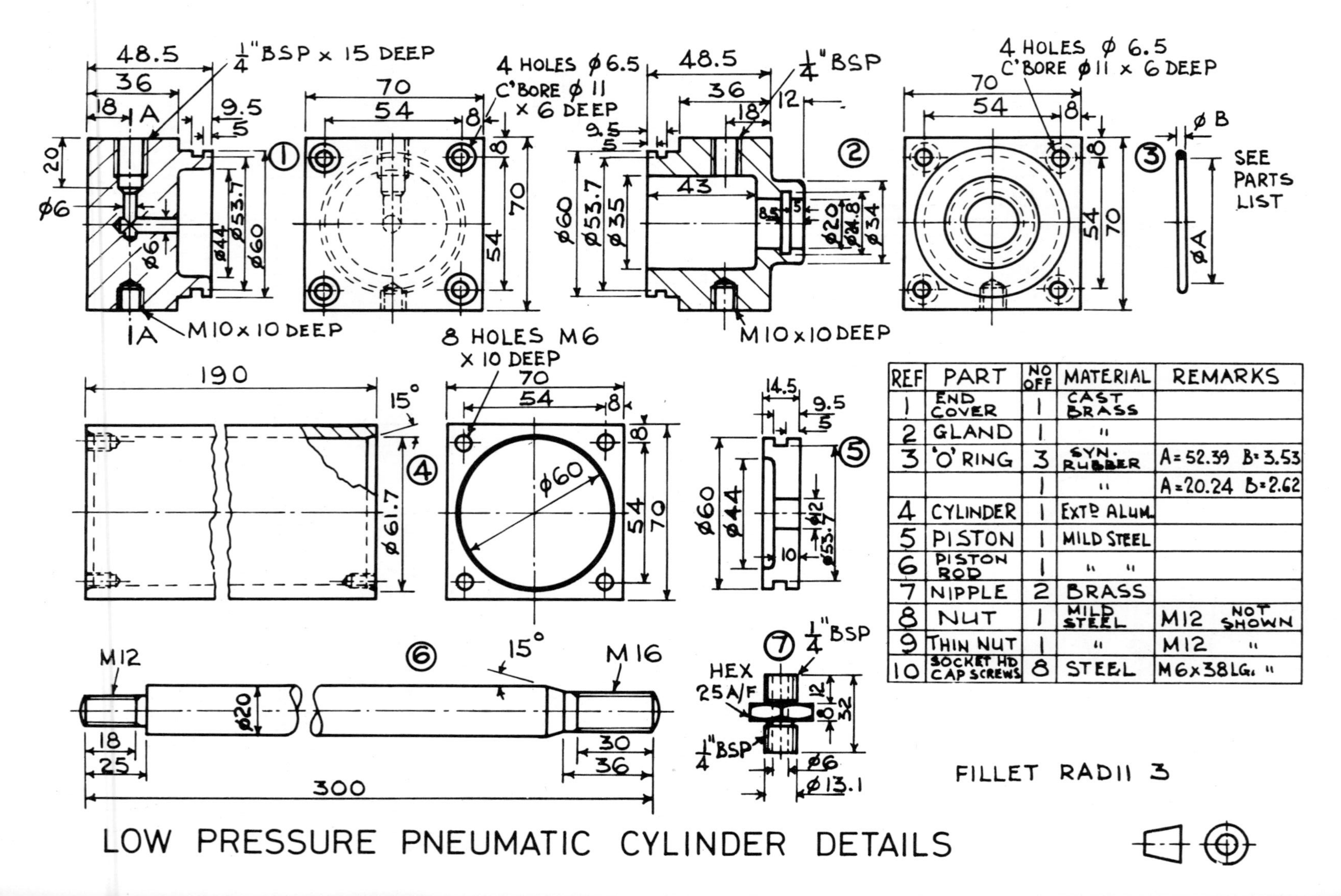

REF	PART	NO OFF	MATERIAL	REMARKS
1	END COVER	1	CAST BRASS	
2	GLAND	1	"	
3	'O' RING	3	SYN. RUBBER	A = 52.39 B = 3.53
		1	"	A = 20.24 B = 2.62
4	CYLINDER	1	EXT^D ALUM.	
5	PISTON	1	MILD STEEL	
6	PISTON ROD	1	" "	
7	NIPPLE	2	BRASS	
8	NUT	1	MILD STEEL	M12 NOT SHOWN
9	THIN NUT	1	"	M12 "
10	SOCKET HD CAP SCREWS	8	STEEL	M6 x 38 LG. "

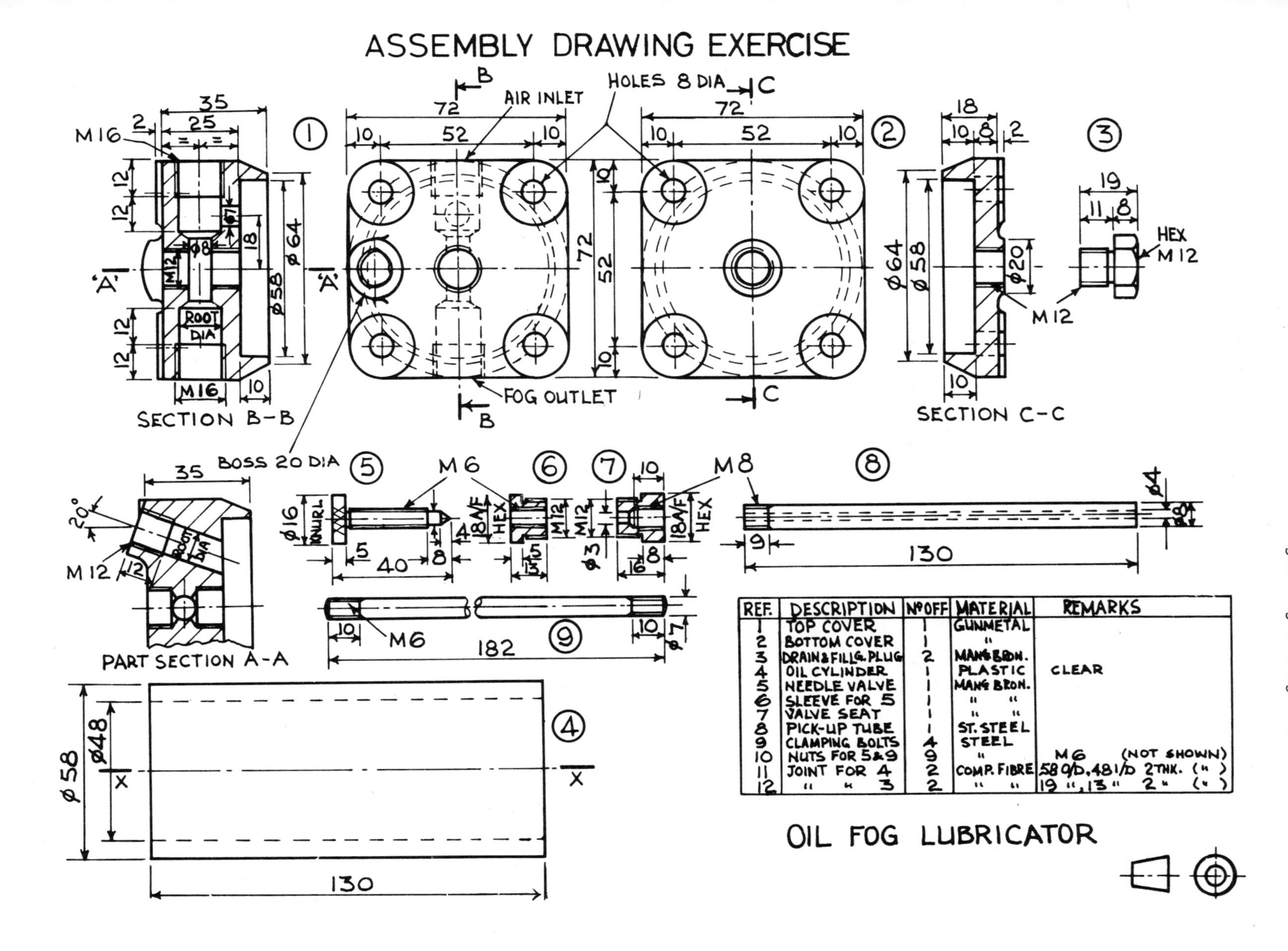

REF.	DESCRIPTION	Nº OFF	MATERIAL	REMARKS
1	TOP COVER	1	GUNMETAL	
2	BOTTOM COVER	1	"	
3	DRAIN & FILLG. PLUG	2	MANG BRON.	
4	OIL CYLINDER	1	PLASTIC	CLEAR
5	NEEDLE VALVE	1	MANG BRON.	
6	SLEEVE FOR 5	1	" "	
7	VALVE SEAT	1	" "	
8	PICK-UP TUBE	1	ST. STEEL	
9	CLAMPING BOLTS	4	STEEL	
10	NUTS FOR 5 & 9	9	"	M6 (NOT SHOWN)
11	JOINT FOR 4	2	COMP. FIBRE	58 O/D, 48 I/D 2 THK. (")
12	" " 3	2	" "	19 ", 13 " 2 " (")

INDEX